"A feast of great thinking and writing about the most profound issues there are, and a treat for any thinking person. . . . Lucidly explains our understanding of the evolution of human moral sentiments and draws out provocative implications for sexual, family, office and societal politics. . . . Mr. Wright writes with a consistent, irreverent wit that does not hide a heartfelt seriousness of purpose." —Steven Pinker,
The New York Times Book Review

"The new field of evolutionary psychology—which seeks to explain human behavior, thought and emotions in terms of Darwinian evolution—finds its most articulate exponent in Robert Wright. In attempting to unravel the evolutionary logic behind friendship, romance, xenophobia, racism, sibling rivalry, and so forth, Wright leavens his presentation with wit and humor, interlacing a biographical profile of Charles Darwin, whose marriage, sex life, personal tragedies and travels in turns are thrust in a neo-Darwinian light. . . . The most sophisticated in-depth exploration to date of the new Darwinian thinking." —*Publishers Weekly*

"Lively, engaging, and well-informed." —*Wall Street Journal*

"An eye-opening, thought-provoking, spine-tingling, mind-boggling, wish-I-had-thought-of-that sort of science book." —Matt Ridley,
Times Literary Supplement (London)

"An engrossing guide, written with wit and an eye to inducting the ignorant into evolutionary psychology." —*The Guardian* (London)

"Lucid and compelling." —*Los Angeles Times*

Robert Wright's

THE MORAL ANIMAL

Robert Wright is a senior editor at *The New Republic* and is coauthor of its TRB column. He has also written for *The Atlantic Monthly*, *The New Yorker*, and *Time*. He previously worked at *The Sciences* magazine, where his writings on science, technology, and philosophy won the National Magazine Award for Essay and Criticism. His first book, *Three Scientists and Their Gods: Looking for Meaning in an Age of Information*, was nominated for a National Book Critics Circle Award. He lives in Washington, D.C., with his wife and two daughters.

THE MORAL ANIMAL

THE MORAL ANIMAL

Evolutionary Psychology and Everyday Life

ROBERT WRIGHT

VINTAGE BOOKS

A Division of Random House, Inc.   New York

FIRST VINTAGE BOOKS EDITION, SEPTEMBER 1995

Copyright © 1994 by Robert Wright

All rights reserved under International and Pan-American
Copyright Conventions. Published in the United States
by Vintage Books, a division of Random House, Inc.,
New York, and simultaneously in Canada by Random House
of Canada Limited, Toronto. Originally published in hardcover
by Pantheon Books, a division of Random House, Inc.,
New York, in 1994.

Illustration credits are on pages 465–466.

The Library of Congress has cataloged the Pantheon edition
as follows:
Wright, Robert, 1957–
The moral animal: the new science of evolutionary psychology
/ Robert Wright.
p. cm.
ISBN 0-679-40773-1
1. Sociobiology. 2. Psychology. 3. Human behavior.
4. Evolution. I. Title.
GN365.3.W75 1994
304.5—dc20
94-7486
Vintage ISBN: 0-679-76399-6

Manufactured in the United States of America

10 9 8

For Lisa

Without thinking what he was doing, he took another drink of brandy. As the liquid touched his tongue he remembered his child, coming in out of the glare: the sullen unhappy knowledgeable face. He said, "Oh God, help her. Damn me, I deserve it, but let her live for ever." This was the love he should have felt for every soul in the world: all the fear and the wish to save concentrated unjustly on the one child. He began to weep; it was as if he had to watch her from the shore drown slowly because he had forgotten how to swim. He thought: This is what I should feel all the time for everyone. . . .

—Graham Greene, *The Power and the Glory*

Contents

THE MORAL ANIMAL

Introduction: **DARWIN AND US**

The Origin of Species contains almost no mention of the human species. The threats the book posed—to the biblical account of our creation, to the comforting belief that we are more than mere animals—were clear enough; Charles Darwin had nothing to gain by amplifying them. Near the end of the final chapter he simply suggested that, through the study of evolution, "light will be thrown on the origin of man and his history." And, in the same paragraph, he ventured that "in the distant future" the study of psychology "will be based on a new foundation."[1]

Distant was right. In 1960, 101 years after the *Origin* appeared, the historian John C. Greene observed, "With respect to the origin of man's distinctively human attributes, Darwin would be disappointed to find matters little advanced beyond his own speculations in *The Descent of Man*. He would be discouraged to hear J. S. Weiner of the Anthropology Laboratory at Oxford University describe this subject as 'one large baffling topic on which our evolutionary insight remains meagre.' . . . In the current emphasis on man's uniqueness as a culture-transmitting animal Darwin might sense a tendency to return to the pre-evolutionary idea of an absolute distinction between man and other animals."[2]

A few years after Greene spoke, a revolution started. Between 1963 and 1974, four biologists—William Hamilton, George Williams, Robert Trivers, and John Maynard Smith—laid down a series of ideas that, taken together, refine and extend the theory of natural selection. These ideas have radically deepened the insight of evolutionary biologists into the social behavior of animals, including us.

At first, the relevance of the new ideas to our species was hazy. Biologists spoke confidently about the mathematics of self-sacrifice among ants, about the hidden logic of courtship among birds; but about human behavior they spoke conjecturally, if at all. Even the two epoch-marking books that synthesized and publicized the new ideas—E. O. Wilson's *Sociobiology* (1975) and Richard Dawkins's *The Selfish Gene* (1976)—said relatively little about humans. Dawkins steered almost entirely clear of the subject, and Wilson confined his discussion of our species to a final, slender, admittedly speculative chapter—28 pages out of 575.

Since the mid-1970s, the human angle has gotten much clearer. A small but growing group of scholars has taken what Wilson called "the new synthesis" and carried it into the social sciences with the aim of overhauling them. These scholars have applied the new, improved Darwinian theory to the human species, and then tested their applications with freshly gathered data. And along with their inevitable failures, they have had great success. Though they still consider themselves an embattled minority (an identity they seem sometimes to secretly enjoy), signs of their rising stature are clear. Venerable journals in anthropology, psychology, and psychiatry are publishing articles by authors who ten years ago were consigned to upstart journals of an expressly Darwinian bent. Slowly but unmistakably, a new worldview is emerging.

Here "worldview" is meant quite literally. The new Darwinian synthesis is, like quantum physics or molecular biology, a body of scientific theory and fact; but, unlike them, it is also a way of seeing everyday life. Once truly grasped (and it is much easier to grasp than either of them) it can entirely alter one's perception of social reality.

The questions addressed by the new view range from the mundane to the spiritual and touch on just about everything that matters: romance, love, sex (Are men and/or women really built for monogamy? What circumstances can make them more so or less so?); friendship and enmity (What is the evolutionary logic behind office politics—or, for that matter, politics in general?); selfishness, self-sacrifice, guilt (Why did natural selection give us that vast guilt repository known as the conscience? Is it truly a guide to "moral" behavior?); social status and social climbing (Is hierarchy inherent in human society?); the differing inclinations of men and women in areas such as friendship and ambition (Are we prisoners of our gender?); racism, xenophobia, war (Why do we so easily exclude large groups of people from the reach of our sympathy?); deception, self-deception, and the unconscious mind (Is intellectual honesty possible?); various psychopathologies (Is getting depressed, neurotic, or paranoid "natural"—and, if so, does that make it any more acceptable?); the love-hate relationship between siblings (Why isn't it pure love?); the tremendous capacity of parents to inflict psychic damage on their children (Whose interests do they have at heart?); and so on.

A QUIET REVOLUTION

The new Darwinian social scientists are fighting a doctrine that has dominated their fields for much of this century: the idea that biology doesn't much matter—that the uniquely malleable human mind, together with the unique force of culture, has severed our behavior from its evolutionary roots; that there is no inherent human nature driving human events, but that, rather, our essential nature is to be driven. As Emile Durkheim, the father of modern sociology, wrote at the turn of the century: human nature is "merely the indeterminate material that the social factor molds and transforms." History shows, said Durkheim, that even such deeply felt emotions as sexual jealousy, a father's love of his child, or the child's love of the father, are "far from being inherent in human nature." The mind, in this view, is basically passive—it is a basin into which, as a person matures, the local culture is gradually poured; if the mind sets any limits at all on the content of culture, they are exceedingly broad. The anthropologist Robert Lowie wrote in 1917 that "the principles of psychology are as incapable of accounting for the phenomena of culture as is gravitation to account for architectural styles."[3] Even psychologists—who might be expected to argue on behalf of the human mind—have often depicted it as little more than a blank slate. Behaviorism, which dominated psychology for a good part of this century, consists largely of the idea that people tend habitually to do what they are rewarded for doing and not do what they are punished for doing; thus is the formless mind given form. In B. F. Skinner's 1948 utopian novel *Walden II*, envy, jealousy, and other antisocial impulses were being eliminated via a strict regime of positive and negative reinforcements.

This view of human nature—as something that barely exists and doesn't much matter—is known among modern Darwinian social scientists as "the standard social science model."[4] Many of them learned it as undergraduates, and some of them spent years under its sway before beginning to question it. After a certain amount of questioning, they began to rebel.

In many ways, what is now happening fits Thomas Kuhn's description of a "paradigm shift" in his well-known book *The Structure of Scientific Revolutions*. A group of mainly young scholars have challenged the settled worldview of their elders, met with bitter resistance, persevered, and begun to flourish. Yet however classic this generational conflict may seem, it features a couple of distinctive ironies.

To begin with, it is, as revolutions go, inconspicuous. The various revolutionaries stubbornly refuse to call themselves by a single, simple name, the sort of thing that would fit easily onto a fluttering banner. They once had such a name—"sociobiology," Wilson's apt and useful term. But Wilson's book drew so much fire, provoked so many charges of malign political intent, so much caricature of sociobiology's substance, that the word became tainted. Most practitioners of the field he defined now prefer to avoid his label.[5] Though bound by allegiance to a compact and coherent set of doctrines, they go by different names: behavioral ecologists, Darwinian anthropologists, evolutionary psychologists, evolutionary psychiatrists. People sometimes ask: What ever happened to sociobiology? The answer is that it went underground, where it has been eating away at the foundations of academic orthodoxy.

The second irony of this revolution is tied to the first. Many features of the new view that the old guard most dislikes and fears are not, in fact, features of the new view. From the beginning, attacks on sociobiology were reflexive—reactions less to Wilson's book than to past books of a Darwinian cast. Evolutionary theory, after all, has a long and largely sordid history of application to human affairs. After being mingled with political philosophy around the turn of the century to form the vague ideology known as "social Darwinism," it played into the hands of racists, fascists, and the most heartless sort of capitalists. It also, around that time, spawned some simplistic ideas about the hereditary basis of behavior—ideas that, conveniently, fed these very political misuses of Darwinism. The resulting aura—of crudeness, both intellectual and ideological—continues to cling to Darwinism in the minds of many academics and laypersons. (Some people think the term Darwinism *means* social Darwinism.) Hence many misconceptions about the new Darwinian paradigm.

INVISIBLE UNITIES

For example: the new Darwinism is often mistaken for an exercise in social division. Around the turn of the century, anthropologists spoke casually of the "lower races," of "savages" who were beyond moral improvement. To the uncritical observer, such attitudes seemed to fit easily enough into a Darwinian framework, as would later supremacist doctrines, including Hitler's. But today's Darwinian anthropologists, in scanning the world's peoples, focus less on surface differences among cultures than on deep unities. Beneath the global crazy quilt of rituals and customs, they see recurring patterns in the structure of family, friendship, politics, courtship, morality. They believe the evolutionary design of human beings explains these patterns: why people in all cultures worry about social status (often more than they realize); why people in all cultures not only gossip, but gossip about the same kinds of things; why in all cultures men and women seem different in a few basic ways; why people everywhere feel guilt, and feel it in broadly predictable circumstances; why people everywhere have a deep sense of justice, so that the axioms "One good turn deserves another" and "An eye for an eye, a tooth for a tooth" shape human life everywhere on this planet.

In a way, it's not surprising that the rediscovery of human nature has taken so long. Being everywhere we look, it tends to elude us. We take for granted such bedrock elements of life as gratitude, shame, remorse, pride, honor, retribution, empathy, love, and so on—just as we take for granted the air we breathe, the tendency of dropped objects to fall, and other standard features of living on this planet.[6] But things didn't have to be this way. We could live on a planet where social life featured none of the above. We could live on a planet where some ethnic groups felt some of the above and others felt others. But we don't. The more closely Darwinian anthropologists look at the world's peoples, the more they are struck by the dense and intricate web of human nature by which all are bound. And the more they see how the web was woven.

Even when the new Darwinians *do* focus on differences—whether among groups of people or among people within groups—they are not generally inclined to explain them in terms of genetic differences. Darwinian anthropologists see the world's undeniably diverse cultures as products of a single human nature responding to widely varying circumstances; evolutionary theory reveals previously invisible links between the circumstances and the cultures (explaining, for example, why some cultures have dowry and others don't). And evolutionary psychologists, contrary to common expectation, subscribe to a cardinal doctrine of twentieth-century psychology and psychiatry: the potency of early social environment in shaping the adult mind. Indeed, a few are preoccupied with this subject, determined to uncover basic laws of psychological development and convinced that they can do so only with Darwinian tools. If we want to know, say, how levels of ambition or of insecurity get adjusted by early experience, we must first ask why natural selection made them adjustable.

This isn't to say that human behavior is infinitely malleable. In tracing the channels of environmental influence, most evolutionary psychologists see some firm banks. The utopian spirit of B. F. Skinner's behaviorism, the sense that a human being can become any sort of animal at all with proper conditioning, is not faring well these days. Still, neither is the idea that the grimmest parts of the human experience are wholly immutable, grounded in "instincts" and "innate drives"; nor the idea that psychological differences among people boil down mainly to genetic differences. They boil down to the *genes*, of course (where else could rules for mental development ultimately reside?), but not necessarily to *differences* in genes. A guiding assumption of many evolutionary psychologists, for reasons we'll come to, is that the most radical differences among people are the ones most likely to be traceable to environment.

In a sense, evolutionary psychologists are trying to discern a second level of human nature, a deeper unity within the species. First the anthropologist notes recurring themes in culture after culture: a thirst for social approval, a capacity for guilt. You might call these, and many other such universals, "the knobs of human nature." Then the psychologist notes that the exact tunings of the knobs seem to differ from person to person. One person's "thirst for approval" knob is set in the comfort zone, down around (relatively) "self-assured," and another person's is up in the excruciating "massively insecure" zone; one person's guilt knob is set low and another person's is painfully high. So the psychologist asks: How do these knobs get set? Genetic differences among individuals surely play a role, but perhaps a larger role is played by genetic commonalities: by a generic, species-wide developmental program that absorbs information from the social environment and adjusts the maturing mind accordingly. Oddly, future progress in grasping the importance of environment will probably come from thinking about genes.

Thus, human nature comes in two forms, both of which have a natural tendency to get ignored. First, there's the kind that's so pervasively apparent as to be taken for granted (guilt, for example). Second, there's the kind whose very function is to generate differences among people as they grow up, and thus naturally conceals itself (a developmental program that *calibrates* guilt). Human nature consists of knobs and of mechanisms for tuning the knobs, and both are invisible in their own way.

There is another source of invisibility, another reason human nature has been slow to come to light: the basic evolutionary logic common to people everywhere is opaque to introspection. Natural selection appears to have hidden our true selves from our conscious selves. As Freud saw, we are oblivious to our deepest motivations—but in ways more chronic and complete (and even, in some cases, more grotesque) than he imagined.

DARWINIAN SELF-HELP

Though this book will touch on many behavioral sciences—anthropology, psychiatry, sociology, political science—evolutionary psychology will be at its center. This young and still inchoate discipline, with its partly fulfilled promise of creating a whole new science of mind, lets us now ask a question that couldn't have been profitably asked in 1859, after the *Origin* appeared, nor in 1959: What does the theory of natural selection have to offer ordinary human beings?

For example: Can a Darwinian understanding of human nature help people reach their goals in life? Indeed, can it help them choose their goals? Can it help distinguish between practical and impractical goals? More profoundly, can it help in deciding which goals are worthy? That is, does knowing how evolution has shaped our basic moral impulses help us decide which impulses we should consider legitimate?

The answers, in my opinion, are: yes, yes, yes, yes, and, finally, yes. The preceding sentence will annoy, if not outrage, many people in the field. (Believe me. I've shown it to some of them.) They have long had to labor under the burden of Darwinism's past moral and political misuses, and they would like to keep the realms of science and value separate. You can't derive basic moral values from natural selection, they say, or indeed from any of nature's workings. If you do, you're committing what philosophers call "the naturalistic fallacy"—the unwarranted inference of "ought" from "is."

I agree: nature isn't a moral authority, and we needn't adopt any "values" that seem implicit in its workings—such as "might makes right." Still, a true understanding of human nature will inevitably affect moral thought deeply and, as I will try to show, legitimately.

This book, with its relevance to questions of everyday life, will have some features of a self-help book. But it will lack many. The next several hundred pages aren't loaded with pithy advice and warm reassurance. A Darwinian viewpoint won't hugely simplify your life, and in some ways will complicate it, by shining harsh light on morally dubious behaviors to which we are prone and whose dubiousness evolution has conveniently hidden from us. The few crisp and upbeat prescriptions I can glean from the new Darwinian paradigm are more than matched by the stubborn and weighty trade-offs, dilemmas, and conundrums it illuminates.

But you can't deny the intensity of the illumination—at least you won't, I hope, be denying it by the end of the book. Although one of my aims is to find practical applications of evolutionary psychology, the prior and central aim is to cover the basic principles of evolutionary psychology—to show how elegantly the theory of natural selection, as understood today, reveals the contours of the human mind. This book is, first, a sales pitch for a new science; only secondarily is it a sales pitch for a new basis of political and moral philosophy.

I've taken pains to keep these two issues separate, to distinguish between the new Darwinism's claims about the human mind and my own claims about the practical emanations of the new Darwinism. Many people who buy the first set of claims, the scientific set, will no doubt reject much of the second set, the philosophical set. But I think few people who buy the first set will deny its relevance to the second set. It is hard, on the one hand, to agree that the new paradigm is by far the most powerful lens through which to look at the human species and then to set the lens aside when examining the human predicament. The human species *is* the human predicament.

DARWIN, SMILES, AND MILL

The Origin of Species wasn't the only seminal book published in England in 1859. There was also the best-selling and genre-christening *Self-Help*, written by the journalist Samuel Smiles. And then there was *On Liberty*, by John Stuart Mill. As it happens, these two books nicely frame the question of what Darwin's book will ultimately come to mean.

Self-Help didn't stress getting in touch with your feelings, extricating yourself from sour relationships, tapping into harmonic cosmic forces, or the various other things that have since given self-help books an air of self-absorption and facile comfort. It preached essential Victorian virtues: civility, integrity, industry, perseverance, and, undergirding them all, iron self-control. A man, Smiles believed, can achieve almost anything "by the exercise of his own free powers of action and self-denial." But he must be ever "armed against the temptation of low indulgences," and must not "defile his body by sensuality, nor his mind by servile thoughts."[7]

On Liberty, by contrast, was a strong polemic against the stifling Victorian insistence on self-restraint and moral conformity. Mill indicted Christianity, with its "horror of sensuality," and complained that " 'thou shalt not' predominates unduly over 'thou shalt.' " He found especially deadening the Calvinist branch, with its belief that, "human nature being radically corrupt, there is no redemption for any one until human nature is killed within him." Mill took a sunnier view of human nature, and suggested that Christianity do the same. "[I]f it be any part of religion to believe that man was made by a good Being, it is more consistent with that faith to believe, that this Being gave all human faculties that they might be cultivated and unfolded, not rooted out and consumed, and that he takes delight in every nearer approach made by his creatures to the ideal conception embodied in them, every increase in any of their capabilities of comprehension, of action, or of enjoyment."[8]

Characteristically, Mill had hit on an important question: Are people inherently bad? Those who believe so have tended, like Samuel Smiles, to be morally conservative—to stress self-denial, abstinence, taming the beast within. Those who believe not have tended, like Mill, to be morally liberal, fairly relaxed about how people choose to behave. Evolutionary psychology, young though it is, has already shed much light on this debate. Its findings are at once comforting and unsettling.

Altruism, compassion, empathy, love, conscience, the sense of justice—all of these things, the things that hold society together, the things that allow our species to think so highly of itself, can now confidently be said to have a firm genetic basis. That's the good news. The bad news is that, although these things are in some ways blessings for humanity as a whole, they didn't evolve for the "good of the species" and aren't reliably employed to that end. Quite the contrary: it is now clearer than ever how (and precisely *why*) the moral sentiments are used with brutal flexibility, switched on and off in keeping with self-interest; and how naturally oblivious we often are to this switching. In the new view, human beings are a species splendid in their array of moral equipment, tragic in their propensity to misuse it, and pathetic in their constitutional ignorance of the misuse. The title of this book is not wholly without irony.

Thus, for all the emphasis in popular treatments of sociobiology on the "biological basis of altruism," and for all its genuine importance, the idea that John Stuart Mill ridiculed—of a corrupt human nature, of "original sin"—doesn't deserve such summary dismissal. And for that reason, I believe, neither does moral conservatism. Indeed, I believe some—*some*—of the conservative norms that prevailed in Victorian England reflect, if obliquely, a surer grasp of human nature than has prevailed in the social sciences for most of this century; and that *some* of the resurgent moral conservatism of the past decade, especially in the realm of sex, rests on an implicit rediscovery of truths about human nature that have long been denied.

If modern Darwinism indeed has some morally conservative emanations, does that mean it has politically conservative emanations? This is a tricky and important question. It's easy enough, and correct, to dismiss social Darwinism as a spasm of malicious confusion. But the question of innate human goodness casts a political shadow that can't be so casually disregarded, for linkage between ideology and views on human nature has a long and distinguished history. Over the past two centuries, as the meanings of political "liberalism" and "conservatism" have changed almost beyond recognition, one distinction between the two has endured: political liberals (such as Mill, in his day) tend to take a rosier view of human nature than conservatives, and to favor a looser moral climate.

Still, it isn't clear that this connection between morals and politics is truly necessary, especially in a modern context. To the extent that the new Darwinian paradigm has reasonably distinct political implications—and as a general rule it just doesn't—they are about as often to the left as to the right. In some ways they are radically to the left.

(Though Karl Marx would find much to dislike in the new paradigm, he would find parts of it very appealing.) What's more, the new paradigm suggests reasons a modern political liberal might subscribe to some morally conservative doctrines as a matter of ideological consistency. At the same time, it suggests that a conservative moral agenda may at times profit from liberal social policies.

DARWINIZING DARWIN

In making the case for a Darwinian outlook, I will use, as Exhibit A, Charles Darwin. His thoughts, emotions, and behavior will illustrate the principles of evolutionary psychology. In 1876, in the first paragraph of his autobiography, Darwin wrote, "I have attempted to write the following account of myself, as if I were a dead man in another world looking back at my own life." (He added, with characteristic grim detachment, "Nor have I found this difficult, for life is nearly over with me.")[9] I like to think that if Darwin were looking back today, with the penetrating hindsight afforded by the new Darwinism, he would see his life somewhat as I'll be depicting it.

Darwin's life will serve as more than illustration. It will be a miniature test of the explanatory power of the modern, refined version of his theory of natural selection. Advocates of evolutionary theory—including him, including me—have long claimed that it is so powerful as to explain the nature of all living things. If we're right, the life of any human being, selected at random, should assume new clarity when looked at from this viewpoint. Well, Darwin hasn't exactly been selected at random, but he'll do as a guinea pig. My claim is that his life—and his social environment, Victorian England—make more sense when looked at from a Darwinian vantage point than from any competing perspective. In this respect, he and his milieu are like all other organic phenomena.

Darwin doesn't *seem* like other organic phenomena. The things that come to mind when we think of natural selection—the ruthless pursuit of genetic self-interest, survival of the fiercest—don't come to mind when we think of Darwin. By all accounts, he was enormously civil and humane (except, perhaps, when circumstance made it hard to be both; he could grow agitated while denouncing slavery, and he might lose his temper if he saw a coachman abusing a horse).[10] His gentleness of manner and his utter lack of pretense, well marked from his youth, were uncorrupted by fame. "[O]f all eminent men that I have ever seen he is beyond comparison the most attractive to me," observed the literary critic Leslie Stephen. "There is something almost pathetic in his simplicity and friendliness."[11] Darwin was, to borrow a phrase from the title of *Self-Help*'s last chapter, a "true gentleman."

Darwin read *Self-Help*. But he needn't have. By then (age fifty-one) he was already a walking embodiment of Smiles's dictum that life is a battle against "moral ignorance, selfishness, and vice." Indeed, one common view is that Darwin was decent to a fault—that, if he needed a self-help book, it was a self-help book of the late-twentieth-century variety, something about how to feel good about yourself, how to look out for number one. The late John Bowlby, one of Darwin's most perceptive biographers, believed that Darwin suffered from "nagging self-contempt" and an "overactive conscience." Bowlby wrote: "While there is so much to admire in the absence of pretension and in the strong moral principles that were an integral part of Darwin's character and that, with much else, endeared him to relatives, friends and colleagues, these qualities were unfortunately developed prematurely and to excessive degree."[12]

Darwin's "excessive" humility and morality, his extreme lack of brutishness, are what make him so valuable as a test case. I will try to show that natural selection, however seemingly alien to his character, can account for it. It is true that Darwin was as gentle, humane, and decent a man as you can reasonably hope to find on this planet. But it is also true that he was fundamentally no different from the rest of us. Even Charles Darwin was an animal.

Part One: **SEX, ROMANCE, AND LOVE**

Chapter 1: **DARWIN COMES OF AGE**

As for an English lady, I have almost forgotten what she is.—
something very angelic and good.
                                        —Letter from the HMS *Beagle* (1835)[1]

Boys growing up in nineteenth-century England weren't generally
advised to seek sexual excitement. And they weren't advised to do
things that might lead them to think about seeking it. The Victorian
physician William Acton, in his book *The Functions and Disorders
of the Reproductive Organs,* warned about exposing a boy to the
"classical works" of literature. "He reads in them of the pleasures,
nothing of the penalties, of sexual indulgences. He is not intuitively
aware that, if the sexual desires are excited, it will require greater
power of will to master them than falls to the lot of most lads; that
if indulged in, the man will and must pay the penalty for the errors
of the boy; that for one that escapes, ten will suffer; that an awful
risk attends abnormal substitutes for sexual intercourse; and that self-
indulgence, long pursued, tends ultimately, if carried far enough, to
early death or self-destruction."[2]

Acton's book was published in 1857, during the mid-Victorian period whose moral tenor it exudes. But sexual repression had long been in the air—before Victoria's ascent to the throne in 1837, even before the date more loosely used to bracket the Victorian era, 1830. Indeed, at the turn of the century, the Evangelical movement that so nourished the new moral austerity was well under way.[3] As G. M. Young noted in *Portrait of an Age,* a boy born in England in 1810— the year after Darwin's birth—"found himself at every turn controlled, and animated, by the imponderable pressure of the Evangelical discipline. . . . " This was not a matter only of sexual restraint, but of restraint generally—an all-out vigilance against indulgence. The boy would learn, as Young put it, that "the world is very evil. An unguarded look, a word, a gesture, a picture, or a novel, might plant a seed of corruption in the most innocent heart. . . . "[4] Another student of Victorianism has described "a life of constant struggle— both to resist temptation and to master the desires of the ego"; by "an elaborate practice of self-discipline, one had to lay the foundation of good habits and acquire the power of self-control."[5]

It was this view that Samuel Smiles, born three years after Darwin, would package in *Self-Help.* As the book's wide success attests, the Evangelical outlook spread well beyond the walls of the Methodist churches that were its wellspring, into the homes of Anglicans, Unitarians, and even agnostics.[6] The Darwin household is a good example. It was Unitarian (and Darwin's father was a freethinker, if a quiet one), yet Darwin absorbed the puritanical strain of his time. It is visible in his burdensome conscience and in the astringent code of conduct he championed. Long after he had given up his faith, he wrote that "the highest stage in moral culture at which we can arrive, is when we recognise that we ought to control our thoughts, and [as Tennyson said] 'not even in inmost thought to think again the sins that made the past so pleasant to us.' Whatever makes any bad action familiar to the mind, renders its performance by so much the easier. As Marcus Aurelius long ago said, 'Such as are thy habitual thoughts, such also will be the character of thy mind; for the soul is dyed by the thoughts.' "[7]

Though Darwin's youth and life were in some ways eccentric, in this one sense they were typical of his era: he lived amid tremendous moral gravity. His world was a place where questions of right and wrong were seen at every turn. What's more, it was a place where these questions seemed answerable—absolutely answerable—though the answers were sometimes painful to bear. It was a world very different from ours, and Darwin's work would do much to make the difference.

AN UNLIKELY HERO

The original career plan for Charles Darwin was to be a doctor. His father, he recalled, felt sure "that I should make a successful physician—meaning by this, one who got many patients." The senior Darwin, himself a successful physician, "maintained that the chief element of success was exciting confidence; but what he saw in me which convinced him that I should create confidence I know not." Nonetheless, Charles at age sixteen dutifully left the cozy family estate in Shrewsbury and, accompanied by his older brother, Erasmus, headed for the University of Edinburgh to study medicine.

Enthusiasm for this calling failed to materialize. At Edinburgh Darwin paid grudging attention to course work, avoided the operating theater (watching surgery, in the days before chloroform, wasn't his cup of tea), and spent much time on extracurricular pursuits: trawling with fishermen to gather oysters, which he then dissected; taking taxidermy lessons to complement his newfound love of hunting; walking and talking with a sponge expert named Robert Grant, who ardently believed in evolution—but didn't, of course, know how it works.

Darwin's father sensed a certain vocational drift and, Charles recalled, "was very properly vehement against my turning an idle sporting man, which then seemed my probable destination."[8] Hence Plan B. Dr. Darwin proposed a career in the clergy.

This may seem strange guidance, coming from a man who didn't believe in God, given to a son who wasn't glaringly devout and who had a more obvious calling in zoology. But Darwin's father was a practical man. And in those days zoology and theology were two sides of one coin. If all living things were God's handiwork, then the study of their ingenious design was the study of God's genius. The most noted proponent of this view was William Paley, author of the 1802 book *Natural Theology; or, evidences of the existence and attributes of the Deity, collected from the appearances of nature.* In it Paley argued that, just as a watch implies a watchmaker, a world full of intricately designed organisms, precisely suited to their tasks, implies a designer.[9] (He was right. The question is whether the designer is a farseeing God or an unconscious process.)

The workaday upshot of natural theology was that a country parson could, without guilt, spend much of his time studying and writing about nature. Hence, perhaps, Darwin's fairly favorable, if not especially spiritual, reaction to the prospect of donning the cloth. "I asked for some time to consider, as from what little I had heard and thought on the subject I had scruples about declaring my belief in all the dogmas of the Church of England; though otherwise I liked the thought of being a country clergyman." He did some reading on divinity and "as I did not then in the least doubt the strict and literal truth of every word in the Bible, I soon persuaded myself that our Creed must be fully accepted." To prepare for the clergy, Darwin went to Cambridge University, where he read his Paley and was "charmed and convinced by the long line of argumentation."[10]

Not for long. Just after finishing at Cambridge, Darwin encountered a strange opportunity: to serve as naturalist aboard the HMS *Beagle.* The rest, of course, is history. Though Darwin didn't conceive of natural selection aboard the *Beagle,* his study of wildlife around the world convinced him that evolution had taken place, and alerted him to some of its most suggestive properties. Two years after the end of the ship's five-year voyage, he saw how evolution works. Darwin's plans to enter the clergy would not survive this insight. As if to provide future biographers with ample symbolism, he had brought along on the voyage his favorite volume of verse, *Paradise Lost.*[11]

As Darwin left England's shores, there was no glaring reason to think people would be writing books about him a century and a half later. His youth, ventured one biographer, in a fairly common judgment, had been "unmarked by the slightest trace of genius."[12] Of course, such claims are always suspect, as the early inauspiciousness of great minds makes for good reading. And this particular claim deserves special doubt, as it rests largely on Darwin's self-appraisals, which didn't tend toward inflation. Darwin reports that he couldn't master foreign languages, and struggled with mathematics, and "was considered by all my masters and by my Father as a very ordinary boy, rather below the common standard in intellect." Maybe, maybe not. Perhaps more stock should be placed in another of his appraisals, about his knack for winning the friendship of men "so much older than me and higher in academical position": "I infer that there must have been something in me a little superior to the common run of youths."[13]

Anyway, the absence of blinding intellectual flash isn't the only thing that has led some biographers to deem Darwin "an unlikely survivor in the immortality stakes."[14] There is also the sense that he just wasn't a formidable man. He was so decent, so sweet, so lacking in untrammeled ambition. And he was something of a country boy, a bit insular and simple. One writer has asked, "Why was it given to Darwin, less ambitious, less imaginative, and less learned than many of his colleagues, to discover the theory sought after by others so assiduously? How did it come about that one so limited intellectually and insensitive culturally should have devised a theory so massive in structure and sweeping in significance?"[15]

One way to answer that question is by contesting its assessment of Darwin (an exercise we'll get to), but an easier way is to contest its assessment of his theory. The idea of natural selection, while indeed "sweeping in significance," is not really "massive in structure." It is a small and simple theory, and it didn't take a huge intellect to conceive it. Thomas Henry Huxley, Darwin's good friend, staunch defender, and fluent popularizer, supposedly chastised himself upon comprehending the theory, exclaiming, "How extremely stupid not to have thought of that!"[16]

All the theory of natural selection says is the following. If within a species there is variation among individuals in their hereditary traits, and some traits are more conducive to survival and reproduction than others, then those traits will (obviously) become more widespread within the population. The result (obviously) is that the species' aggregate pool of hereditary traits changes. And there you have it.

Of course, the change may seem negligible within any given generation. If long necks help animals reach precious leaves, and shorter-necked animals therefore die before reproducing, the species' average neck size barely grows. Still, if variation in neck size arises freshly with new generations (through sexual recombination or genetic mutation, we now know), so that natural selection continues to have a range of neck sizes to "choose" from, then average neck size will keep creeping upward. Eventually, a species that started out with horselike necks will have giraffe-like necks. It will, in other words, be a new species.

Darwin once summed up natural selection in ten words: "[M]ultiply, vary, let the strongest live and the weakest die."[17] Here "strongest," as he well knew, means not just brawniest, but best adapted to the environment, whether through camouflage, cleverness, or anything else that aids survival and reproduction.* The word *fittest* (a coinage Darwin didn't make but did accept) is typically used in place of *strongest,* signifying this broader conception—an organism's "fitness" to the task of transmitting its genes to the next generation, within its particular environment. "Fitness" is the thing that natural selection, in continually redesigning species, perpetually "seeks" to maximize. Fitness is what made us what we are today.

If this seems easy to believe, you probably aren't getting the picture. Your entire body—much more complexly harmonious than any product of human design—was created by hundreds of thousands of incremental advances, and *each increment was an accident;* each tiny step between your ancestral bacterium and you *just happened* to help some intermediate ancestor more profusely get its genes into the next generation. Creationists sometimes say that the odds of a person being produced through random genetic change are about equal to those of a monkey typing the works of Shakespeare. Well, yes. Not the complete works, maybe, but certainly some long, recognizable stretches.

Still, things this unlikely can, through the logic of natural selec-

---

*Actually, Darwin divided the "survival" and "reproductive" aspects of the process. Traits leading to successful mating he attributed to "sexual selection," as distinct from natural selection. But these days, natural selection is often defined broadly, to encompass both aspects: the preservation of traits that are in any way conducive to getting an organism's genes into the next generation.

tion, be rendered plausible. Suppose a single ape gets some lucky break—gene XL, say, which imbues parents with an ounce of extra love for their offspring, love that translates into slightly more assiduous nurturing. In the life of any one ape, that gene probably won't be crucial. But suppose that, *on average*, the offspring of apes with the XL gene are 1 percent more likely to survive to maturity than the offspring of apes without it. So long as this thin advantage holds, the fraction of apes with gene XL will tend to grow, and the fraction without it will tend to shrink, generation by generation by generation. The eventual culmination of this trend is a population in which all animals have the XL gene. The gene, at that point, will have reached "fixation"; a slightly higher degree of parental love will be "species typical" now than before.

Okay, so one lucky break thus flourishes. But how likely is it that the luck will persist—that the *next* random genetic change will *further* increase the amount of parental love? How likely is the "XL" mutation to be followed by an "XXL" mutation? Not at all likely in the case of any one ape. But within the population there are now scads of apes with the XL gene. If any one of them, or any one of their offspring, or grand-offspring, happens to luck out and get the XXL gene, the gene will have a good chance of spreading, if slowly, through the population. Of course, in the meantime, lots more apes will probably get various less auspicious genes, and some of those genes may extinguish the lineage in which they appear. Well, that's life.

Thus does natural selection beat the odds—by not really beating them. The thing that is massively more probable than the charmed lineages that populate the world today—an uncharmed lineage, which reaches a dead end through an unlucky break—happened a massively larger number of times. The dustbin of genetic history overflows with failed experiments, long strings of code that were as vibrant as Shakespearean verse *until* that fateful burst of gibberish. Their disposal is the price paid for design by trial and error. But so long as that price can be paid—so long as natural selection has enough generations to work on, and can cast aside scores of failed experiments for every one it preserves—its creations can be awesome. Natural selection is an inanimate process, devoid of consciousness, yet is a tireless refiner, an ingenious craftsman.*

Every organ inside you is testament to its art—your heart, your lungs, your stomach. All these are "adaptations"—fine products of inadvertent design, mechanisms that are here because they have in the past contributed to your ancestors' fitness. And all are species-typical. Though one person's lungs may differ from another's, sometimes for genetic reasons, almost all the genes involved in lung construction are the same in you as in your next-door neighbor, as in an Eskimo, as in a pygmy. The evolutionary psychologists John Tooby and Leda Cosmides have noted that every page of *Gray's Anatomy* applies to all peoples in the world. Why, they have gone on to ask, should the anatomy of the mind be any different? The working thesis of evolutionary psychology is that the various "mental organs" constituting the human mind—such as an organ inclining people to love their offspring—are species-typical.[18] Evolutionary psychologists are pursuing what is known in the trade as "the psychic unity of humankind."

CLIMATE CONTROL

Between us and the australopithecine, which walked upright but had an ape-sized brain, stand a few million years: 100,000, maybe 200,000 generations. That may not sound like much. But it has taken only around 5,000 generations to turn a wolf into a chihuahua—and, at the same time, along a separate line, into a Saint Bernard. Of course, dogs evolved by artificial, not natural, selection. But as Darwin stressed, the two are essentially the same; in both cases traits are weeded out of a population by criteria that persist for many generations. And in both cases, if the "selective pressure" is strong enough—if genes are weeded out fast enough—evolution can proceed briskly.

One might wonder how the selective pressure could have been

---

*In this book I will sometimes talk about what natural selection "wants" or "intends," or about what "values" are implicit in its workings. I'll always use quotation marks, since these are just metaphors. But the metaphors are worth using, I believe, because they help us come to moral terms with Darwinism.

very strong during recent human evolution. After all, what usually generates the pressure is a hostile environment—droughts, ice ages, tough predators, scarce prey—and as human evolution has proceeded, the relevance of these things has abated. The invention of tools, of fire, the advent of planning and cooperative hunting—these brought growing control over the environment, growing insulation from the whims of nature. How, then, did ape brains turn into human brains in a few million years?

Much of the answer seems to be that the environment of human evolution has been human (or prehuman) beings.[19] The various members of a Stone Age society were each other's rivals in the contest to fill the next generation with genes. What's more, they were each other's tools in that contest. Spreading their genes depended on dealing with their neighbors: sometimes helping them, sometimes ignoring them, sometimes exploiting them, sometimes liking them, sometimes hating them—and having a sense for which people warrant which sort of treatment, and when they warrant it. The evolution of human beings has consisted largely of adaptation to one another.

Since each adaptation, having fixed itself in the population, thus changes the social environment, adaptation only invites more adaptation. Once all parents have the XXL gene, it gives no parent an edge in the ongoing contest to create the most viable and prolific offspring. The arms race continues. In this case, it's an arms race of love. Often, it's not.

It is fashionable in some circles to downplay the whole idea of adaptation, of coherent evolutionary design. Popularizers of biological thought often emphasize not the role of fitness in evolutionary change but the role of randomness and happenstance. Some climate shift may come out of the blue and extinguish unlucky species of flora and fauna, changing the whole context of evolution for any species lucky enough to survive the calamity. A roll of the cosmic dice and suddenly all bets are off. Certainly that happens, and this is indeed one sense in which "randomness" greatly affects evolution. There are other senses as well. For example, new traits on which natural selection passes judgment seem to be randomly generated.[20]

But none of the "randomness" in natural selection should be allowed to obscure its central feature: that the overriding criterion of organic design is fitness. Yes, the dice do get rerolled, and the context of evolution changes. A feature that is adaptive today may not be adaptive tomorrow. So natural selection often finds itself amending outmoded features. This ongoing adjustment to circumstance can give organic life a certain jerry-built quality. (It's the reason people have back trouble; if you were designing a walking organism from scratch rather than incrementally adapting a former tree-dweller, you would never have built such bad backs.) Nonetheless, changes in circumstance are typically gradual enough for evolution to keep pace (even if it has to break into a trot now and then, when selective pressure becomes severe), and it often does so ingeniously.

And all along the way, its definition of good design remains the same. The thousands and thousands of genes that influence human behavior—genes that build the brain and govern neurotransmitters and other hormones, thus defining our "mental organs"—are here for a reason. And the reason is that they goaded our ancestors into getting their genes into the next generation. If the theory of natural selection is correct, then essentially everything about the human mind should be intelligible in these terms. The basic ways we feel about each other, the basic kinds of things we think about each other and say to each other, are with us today by virtue of their past contribution to genetic fitness.

DARWIN'S SEX LIFE

No human behavior affects the transmission of genes more obviously than sex. So no parts of human psychology are clearer candidates for evolutionary explanation than the states of mind that lead to sex: raw lust, dreamy infatuation, sturdy (or at least sturdy-feeling) love, and so on—the basic forces amid which people all over the world, including Charles Darwin, have come of age.

When Darwin left England he was twenty-two and, presumably, flooded with the hormones that young men are, by design, flooded with. He had been sweet on a couple of local girls, especially the pretty, popular, and highly coquettish Fanny Owen. He once let her shoot a hunting gun, and she looked so charming, gamely pretending its kick hadn't hurt her shoulder, that he would recall the incident decades later with evident fondness.[21] From Cambridge he conducted a tenuous flirtation with her by mail, but it isn't clear that he ever so much as kissed her.

While Darwin was at Cambridge, prostitutes were available, not to mention the occasional lower-class girl who might settle for less explicit payment. But university proctors prowled the streets near campus, ready to arrest women who could plausibly be accused of "streetwalking." Darwin had been warned by his brother never to be seen with girls. His closest known connection with illicit sex is when he sent money to a friend who had dropped out of school after siring a bastard.[22] Darwin may well have left the shores of England a virgin.[23] And the next five years, spent mainly on a ninety-foot ship with six dozen males, wouldn't provide many opportunities to change that status, at least not through conventional channels.

For that matter, sex wouldn't be abundantly available on his return either. This was, after all, Victorian England. Prostitutes could be had in London (where Darwin would take up residence), but sex with a "respectable" woman, a woman of Darwin's class, was more elusive—close to impossible in the absence of extreme measures, such as marriage.

The great gulf between these two forms of sex is one of the most famous elements of Victorian sexual morality—the "Madonna-whore" dichotomy. There were two kinds of women: the kind a bachelor would later marry and the kind he might now enjoy, the kind worthy of love and the kind that warranted only lust. A second moral attitude commonly traced to the Victorian age is the sexual double standard. Though this attribution is misleading, since Victorian moralists so strongly discouraged sexual license in men *and* women, it's true that a Victorian man's sexual extravagance raised fewer eyebrows than a woman's. It's also true that this distinction was closely linked to the Madonna-whore dichotomy. The great punishment awaiting a sexually adventurous Victorian woman was permanent consignment to the latter half of the dichotomy, which would greatly restrict her range of available husbands.

There is a tendency these days to reject and scoff at these aspects of Victorian morality. Rejecting them is fine, but to scoff at them is to overestimate our own moral advancement. The fact is that many men still speak openly about "sluts" and their proper use: great for recreation, but not for marriage. And even men (such as well-educated liberal ones) who wouldn't dream of talking like that may in fact act like that. Women sometimes complain about seemingly enlightened men who lavish respectful attention on them but then, after sex on the first or second date, never call again, as if early sex had turned the woman into a pariah. Similarly, while the double standard has waned in this century, it is still strong enough for women to complain about. Understanding the Victorian sexual climate can take us some distance toward understanding today's sexual climate.

The intellectual grounding of Victorian sexual morality was explicit: women and men are inherently different, most importantly in the libido department. Even Victorians who railed against male philandering stressed the difference. Dr. Acton wrote: "I should say that the majority of women (happily for them) are not very much troubled with sexual feeling of any kind. What men are habitually, women are only exceptionally. It is too true, I admit, as the divorce courts show, that there are some few women who have sexual desires so strong that they surpass those of men." This "nymphomania" is "a form of insanity." Still, "there can be no doubt that sexual feeling in the female is in the majority of cases in abeyance . . . and even if roused (which in many instances it never can be) is very moderate compared with that of the male." One problem, said Dr. Acton, was that many young men are misled by the sight of "loose or, at least, low and vulgar women." They thus enter marriage with exaggerated notions of its sexual content. They don't understand that "the best mothers, wives, and managers of households, know little or nothing of sexual indulgences. Love of home, children, and domestic duties, are the only passions they feel."[24]

Some women who consider themselves excellent wives and mothers may hold a different opinion. And they may have strong supporting evidence. Still, the idea that there are *some* differences between the typical male and female sexual appetite, and that the male appetite is less finicky, draws much support from the new Darwinian paradigm. For that matter, it draws support from lots of other places. The recently popular premise that men and women are basically identical in nature seems to have fewer and fewer defenders. It is no longer, for example, a cardinal doctrine of feminism. A whole school of feminists—the "difference feminists," or "essentialists"—now accept that men and women are deeply different. What exactly "deeply" means is something they're often vague about, and many would rather not utter the word *genes* in this context. Until they do, they will likely remain in a state of disorientation, aware that the early feminist doctrine of innate sexual symmetry was incorrect (and that it may have in some ways harmed women) yet afraid to honestly explore the alternative.

If the new Darwinian view of sexuality did nothing more than endorse the coalescing conventional wisdom that men are a pretty libidinous group, it would be of meager value. But in fact it sheds light not just on animal impulses, like lust, but on the subtler contours of consciousness. "Sexual psychology," to an evolutionary psychologist, includes everything from an adolescent's fluctuating self-esteem to the aesthetic judgments men and women make about each other to the moral judgments they make about each other, and, for that matter, the moral judgments they make about members of their own sex. Two good examples are the Madonna-whore dichotomy and the sexual double standard. Both now appear to have roots in human nature—in mental mechanisms that people use to evaluate each other.

This calls for a couple of disclaimers. First, to say something is a product of natural selection is not to say that it is unchangeable; just about any manifestation of human nature can be changed, given an apt alteration of the environment—though the required alteration will in some cases be prohibitively drastic. Second, to say that something is "natural" is not to say that it is good. There is no reason to adopt natural selection's "values" as our own. But presumably if we want to pursue values that are at odds with natural selection's, we need to know what we're up against. If we want to change some disconcertingly stubborn parts of our moral code, it would help to know where they come from. And where they *ultimately* come from is human nature, however complexly that nature is refracted by the many layers of circumstance and cultural inheritance through which it passes. No, there is no "double-standard gene." But yes, to understand the double standard we must understand our genes and how they affect our thoughts. We must understand the process that selected those genes and the strange criteria it used.

We'll spend the next few chapters exploring that process as it has shaped sexual psychology. Then, thus fortified, we'll return to Victorian morality, and to Darwin's own mind, and the mind of the woman he married. All of which will enable us to see our own situation—courtship and marriage at the end of the twentieth century—with new clarity.

Chapter 2: **MALE AND FEMALE**

In the most distinct classes of the animal kingdom, with mammals, birds, reptiles, fishes, insects, and even crustaceans, the differences between the sexes follow almost exactly the same rules; the males are almost always the wooers. . . .
—*The Descent of Man* (1871)[1]

Darwin was wrong about sex.

He wasn't wrong about the males being the wooers. His reading of the basic characters of the two sexes holds up well today. "The female, . . . with the rarest exception, is less eager than the male. . . . [S]he is coy, and may often be seen endeavouring for a long time to escape from the male. Every one who has attended to the habits of animals will be able to call to mind instances of this kind. . . . The exertion of some choice on the part of the female seems almost as general a law as the eagerness of the male."[2]

Nor was Darwin wrong about the *consequences* of this asymmetrical interest. He saw that female reticence left males competing with one another for scarce reproductive opportunities, and thus explained why males so often have built-in weapons—the horns of stags, the hornlike mandibles of stag beetles, the fierce canines of chimpanzees.[3] Males not hereditarily equipped for combat with other males have been excluded from sex, and their traits have thus been discarded by natural selection.

Darwin also saw that the choosiness of females gives great moment to their choices. If they prefer to mate with particular kinds of males, those kinds will proliferate. Hence the ornamentation of so many male animals—a lizard's inflatable throat sack, brightly colored during the mating season; the immense, cumbersome tail of the peacock; and, again, the stag's horns, which seem more elaborate than the needs of combat alone would dictate.[4] These decorations have evolved not because they aid in daily survival—if anything, they complicate it—but because they can so charm a female as to outweigh the everyday burdens they bring. (How it came to be in the genetic interest of females to be charmed by such things is another story, and a point of subtle disagreement among biologists.)[5]

Both of these variants of natural selection—combat among males and discernment by females—Darwin called "sexual selection." He took great pride in the idea, and justifiably so. Sexual selection is a nonobvious extension of his general theory that accounts for seeming exceptions to it (like garish colors that virtually say "Kill me" to predators), and that has not just endured over time but grown in scope.

What Darwin was wrong about was the evolutionary *cause* of female coyness and male eagerness. He saw that this imbalance of interest creates competition among males for precious reproductive slots, and he saw the consequences of this competition; but he didn't see what had created the imbalance. His attempt late in life to explain the phenomenon was unsuccessful.[6] And, in fairness to him, whole generations of biologists would do no better.

Now that there is a consensus on the solution, the long failure to find it seems puzzling. It's a very simple solution. In this sense, sex is typical of many behaviors illuminated by natural selection; though the illumination has gotten truly powerful only within the last three decades, it could in principle have done so a century earlier, so plainly does it follow from Darwin's view of life. There is some subtle logic involved, so Darwin can be forgiven for not having seen the full scope of his theory. Still, if he were around today to hear evolutionary biologists talk about sex, he might well sink into one of his self-effacing funks, exclaiming at his obtuseness in not getting the picture sooner.

PLAYING GOD

The first step toward understanding the basic imbalance of the sexes is to assume hypothetically the role natural selection plays in designing a species. Take the human species, for example. Suppose you're in charge of installing, in the minds of human (or prehuman) beings, rules of behavior that will guide them through life, the object of the game being to maximize each person's genetic legacy. To oversimplify a bit: you're supposed to make each person behave in such a way that he or she is likely to have lots of offspring—offspring, moreover, who themselves have lots of offspring.

Obviously, this isn't the way natural selection actually works. It doesn't consciously design organisms. It doesn't consciously do anything. It blindly preserves hereditary traits that happen to enhance survival and reproduction.* Still, natural selection works *as if* it were consciously designing organisms, so pretending you're in charge of organism design is a legitimate way to figure out which tendencies evolution is likely to have ingrained in people and other animals. In fact, this is what evolutionary biologists spend a good deal of time doing: looking at a trait—mental or otherwise—and figuring out what, if any, engineering challenge it is a solution to.

When playing the Administrator of Evolution, and trying to maximize genetic legacy, you quickly discover that this goal implies different tendencies for men and women. Men can reproduce hundreds of times a year, assuming they can persuade enough women to cooperate, and assuming there aren't any laws against polygamy—which there assuredly weren't in the environment where much of our evolution took place. Women, on the other hand, can't reproduce more often than once a year. The asymmetry lies partly in the high price of eggs; in all species they're bigger and rarer than minuscule,

---

*Actually, one premise of the new Darwinian paradigm is that natural selection's guiding light is a bit more complex than "survival and reproduction." But that nuance won't matter until chapter seven.

mass-produced sperm. (That, in fact, is biology's official definition of a female: the one with the larger sex cells.) But the asymmetry is exaggerated by the details of mammalian reproduction; the egg's lengthy conversion into an organism happens inside the female, and she can't handle many projects at once.

So, while there are various reasons why it could make Darwinian sense for a woman to mate with more than one man (maybe the first man was infertile, for example), there comes a time when having more sex just isn't worth the trouble. Better to get some rest or grab a bite to eat. For a man, unless he's really on the brink of collapse or starvation, that time never comes. Each new partner offers a very real chance to get more genes into the next generation—a much more valuable prospect, in the Darwinian calculus, than a nap or a meal. As the evolutionary psychologists Martin Daly and Margo Wilson have succinctly put it: for males "there is always the possibility of doing better."[7]

There's a sense in which a female can do better, too, but it has to do with quality, not quantity. Giving birth to a child involves a huge commitment of time, not to mention energy, and nature has put a low ceiling on how many such enterprises she can undertake. So each child, from her (genetic) point of view, is an extremely precious gene machine. Its ability to survive and then, in turn, produce its own young gene machines is of mammoth importance. It makes Darwinian sense, then, for a woman to be selective about the man who is going to help her build each gene machine. She should size up an aspiring partner before letting him in on the investment, asking herself what he'll bring to the project. This question then entails a number of subquestions that, in the human species especially, are more numerous and subtle than you might guess.

Before we go into these questions, a couple of points must be made. One is that the woman needn't literally ask them, or even be aware of them. Much of the relevant history of our species took place before our ancestors were smart enough to ask much of anything. And even in the more recent past, after the arrival of language and self-awareness, there has been no reason for every evolved behavioral tendency to fall under conscious control. In fact, sometimes it is emphatically *not* in our genetic interest to be aware of exactly what we are doing or why. (Hence Freud, who was definitely onto something, though some evolutionary psychologists would say he didn't know exactly what.) In the case of sexual attraction, at any rate, everyday experience suggests that natural selection has wielded its influence largely via the emotional spigots that turn on and off such feelings as tentative attraction, fierce passion, and swoon-inducing infatuation. A woman doesn't typically size up a man and think: "He seems like a worthy contributor to my genetic legacy." She just sizes him up and feels attracted to him—or doesn't. All the "thinking" has been done—unconsciously, metaphorically—by natural selection. Genes leading to attractions that wound up being good for her ancestors' genetic legacies have flourished, and those leading to less productive attractions have not.

Understanding the often unconscious nature of genetic control is the first step toward understanding that—in many realms, not just sex—we're all puppets, and our best hope for even partial liberation is to try to decipher the logic of the puppeteer. The full scope of the logic will take some time to explain, but I don't think I'm spoiling the end of the movie by noting here that the puppeteer seems to have exactly zero regard for the happiness of the puppets.

The second point to grasp before pondering how natural selection has "decided" to shape the sexual preferences of women (and of men) is that it isn't foresightful. Evolution is guided by the environment in which it takes place, and environments change. Natural selection had no way of anticipating, for example, that someday people would use contraception, and that their passions would thus lead them into time-consuming and energy-sapping sex that was sure to be fruitless; or that X-rated videotapes would come along and lead indiscriminately lustful men to spend leisure time watching them rather than pursuing real, live women who might get their genes to the next generation.

This isn't to say that there's anything wrong with "unproductive" sexual recreation. Just because natural selection created us doesn't mean we have to slavishly follow its peculiar agenda. (If anything, we might be tempted to spite it for all the ridiculous baggage it's saddled us with.) The point is just that it isn't correct to say that people's minds are designed to maximize their fitness, their genetic legacy. What the theory of natural selection says, rather, is that people's minds were designed to maximize fitness *in the environment in which those minds evolved*. This environment is known as the EEA—the environment of evolutionary adaptation.[8] Or, more memorably: the "ancestral environment." Throughout this book, the ancestral environment will lurk in the background. At times, in pondering whether some mental trait is an evolutionary adaptation, I will ask whether it seems to be in the "genetic interest" of its bearer. For example: Would indiscriminate lust be in the genetic interest of men? But this is just a kind of shorthand. The question, properly put, is always whether a trait would be in the "genetic interest" of someone in the EEA, not in modern America or Victorian England or anywhere else. Only traits that would have propelled the genes responsible for them through the generations in our ancestral social environment should, in theory, be part of human nature today.[9]

What was the ancestral environment like? The closest thing to a twentieth-century example is a hunter-gatherer society, such as the !Kung San of the Kalahari Desert in Africa, the Inuit (Eskimos) of the Arctic region, or the Ache of Paraguay. Inconveniently, hunter-gatherer societies are quite different from one another, rendering simple generalization about the crucible of human evolution difficult. This diversity is a reminder that the idea of a single EEA is actually a fiction, a composite drawing; our ancestral social environment no doubt changed much in the course of human evolution.[10] Still, there are recurring themes among contemporary hunter-gatherer societies, and they suggest that some features probably stayed fairly constant during much of the evolution of the human mind. For example: people grew up near close kin in small villages where everyone knew everyone else and strangers didn't show up very often. People got married—monogamously or polygamously—and a female typically was married by the time she was old enough to be fertile.

This much, at any rate, is a safe bet: whatever the ancestral environment was like, it wasn't much like the environment we're in now. We aren't designed to stand on crowded subway platforms, or to live in suburbs next door to people we never talk to, or to get hired or fired, or to watch the evening news. This disjunction between the contexts of our design and of our lives is probably responsible for much psychopathology, as well as much suffering of a less dramatic sort. (And, like the importance of unconscious motivation, it is an observation for which Freud gets some credit; it is central to his *Civilization and Its Discontents*.)

To figure out what women are inclined to seek in a man, and vice versa, we'll need to think more carefully about our ancestral social environment(s). And, as we'll see, thinking about the ancestral environment also helps explain why females in our species are less sexually reserved than females in many other species. But for purposes of making the single, largest point of this chapter—that, whatever the typical level of reserve for females in our species, it is higher than the level for males—the particular environment doesn't much matter. For this point depends only on the premise that an individual female can, over a lifetime, have many fewer offspring than an individual male. And that has been the case, basically, forever: since before our ancestors were human, before they were primates, before they were mammals—way, way back through the evolution of our brain, down to its reptilian core. Female snakes may not be very smart, but they're smart enough to know, unconsciously, at least, that there are some males it's not a good idea to mate with.

Darwin's failure, then, was a failure to see what a deeply precious commodity females are. He saw that their coyness had made them precious, but he didn't see that they were *inherently* precious—precious by virtue of their biological role in reproduction, and the resulting slow rate of female reproduction. Natural selection had seen this—or, at least, had "seen" it—and female coyness is the result of this implicit comprehension.

ENLIGHTENMENT DAWNS

The first large and clear step toward human comprehension of this logic was made in 1948 by the British geneticist A. J. Bateman. Bateman took fruit flies and ran them through a dating game. He would place five males and five females in a chamber, let them follow their hearts, and then, by examining the traits of the next generation, figure out which offspring belonged to which parents. He found a clear pattern. Whereas almost all females had about the same number of offspring, regardless of whether they mated with one, two, or three males, male legacies differed according to a simple rule: the more females you mate with, the more offspring you have. Bateman saw the import: natural selection encourages "an undiscriminating eagerness in the males and a discriminating passivity in the females."[11]

Bateman's insight long lay essentially unappreciated. It took nearly three decades, and several evolutionary biologists, to give it the things it lacked: full and rigorous elaboration on the one hand, and publicity on the other.

The first part—the rigor—came from two biologists who are good examples of how erroneous some stereotypes about Darwinism are. In the 1970s, opposition to sociobiology often took the form of charges that its practitioners were closet reactionaries, racists, fascists, etcetera. It is hard to imagine two people less vulnerable to such charges than George Williams and Robert Trivers, and it is hard to name anyone who did more than they to lay the foundations of the new paradigm.

Williams, a professor emeritus at the State University of New York, has worked hard to dispel vestiges of social Darwinism, with its underlying assumption that natural selection is a process somehow worthy of obedience or emulation. Many biologists share his view, and stress that we can't derive our moral values from its "values." But Williams goes further. Natural selection, he says, is an "evil" process, so great is the pain and death it thrives on, so deep is the selfishness it engenders.

Trivers, who was an untenured professor at Harvard when the new paradigm was taking shape and is now at Rutgers University, is much less inclined than Williams toward moral philosophy. But he evinces an emphatic failure to buy the right-wing values associated with social Darwinism. He speaks proudly of his friendship with the late Black Panther leader Huey Newton (with whom he once co-authored an article on human psychology). He rails against the bias of the judicial system. He sees conservative conspiracies where some people don't.

In 1966 Williams published his landmark work, *Adaptation and Natural Selection: A Critique of Some Current Evolutionary Thought*.

Slowly this book has acquired a nearly holy stature in its field. It is a basic text for biologists who think about social behavior, including human social behavior, in light of the new Darwinism.[12] Williams's book dispelled confusions that had long plagued the study of social behavior, and it laid down foundational insights that would support whole edifices of work on the subjects of friendship and sex. In both cases Trivers would be instrumental in building the edifices.

Williams amplified and extended the logic behind Bateman's 1948 paper. He cast the issue of male versus female genetic interests in terms of the "sacrifice" required for reproduction. For a male mammal, the necessary sacrifice is close to zero. His "essential role may end with copulation, which involves a negligible expenditure of energy and materials on his part, and only a momentary lapse of attention from matters of direct concern to his safety and well-being." With little to lose and much to gain, males can profit, in the currency of natural selection, by harboring "an aggressive and immediate willingness to mate with as many females as may be available." For the female, on the other hand, "copulation may mean a commitment to a prolonged burden, in both the mechanical and physiological sense, and its many attendant stresses and dangers." Thus, it is in her genetic interest to "assume the burdens of reproduction" only when circumstances seem propitious. And "one of the most important circumstances is the inseminating male"; since "unusually fit fathers tend to have unusually fit offspring," it is "to the female's advantage to be able to pick the most fit male available. . . . "[13]

Hence courtship: "the advertisement, by a male, of how fit he is." And just as "it is to his advantage to pretend to be highly fit whether he is or not," it is to the female's advantage to spot false advertising. So natural selection creates "a skilled salesmanship among the males and an equally well-developed sales resistance and discrimination among the females."[14] In other words: males should, in theory, tend to be show-offs.

A few years later, Trivers used the ideas of Bateman and Williams to create a full-blown theory that ever since has been shedding light on the psychology of men and women. Trivers began by replacing Williams's concept of "sacrifice" with "investment." The difference may seem slight, but nuances can start intellectual avalanches, and this one did. The term *investment*, linked to economics, comes with a ready-made analytical framework.

In a now-famous paper published in 1972, Trivers formally defined "parental investment" as "any investment by the parent in an individual offspring that increases the offspring's chance of surviving (and hence [the offspring's] reproductive success) at the cost of the parent's ability to invest in other offspring."[15] Parental investment includes the time and energy consumed in producing an egg or a sperm, achieving fertilization, gestating or incubating the egg, and rearing the offspring. Plainly, females will generally make the higher investment up until birth, and, less plainly but in fact typically, this disparity continues after birth.

By quantifying the imbalance of investment between mother and father in a given species, Trivers suggested, we could better understand many things—for example, the extent of male eagerness and female coyness, the intensity of sexual selection, and many subtle aspects of courtship and parenthood, fidelity and infidelity. Trivers saw that in our species the imbalance of investment is not as stark as in many others. And he correctly suspected that (as we'll see in the next chapter) the result is much psychological complexity.

At last, with Trivers's paper "Parental Investment and Sexual Selection," the flower had bloomed; a simple extension of Darwin's theory—so simple that Darwin would have grasped it in a minute—had been glimpsed in 1948, clearly articulated in 1966, and was now, in 1972, given full form.[16] Still, the concept of parental investment lacked one thing: publicity. It was E. O. Wilson's book *Sociobiology* (1975) and Richard Dawkins's *The Selfish Gene* (1976) that gave Trivers's work a large and diverse audience, getting scores of psychologists and anthropologists to think about human sexuality in modern Darwinian terms. The resulting insights are likely to keep accumulating for a long time.

TESTING THE THEORY

Theories are a dime a dozen. Even strikingly elegant theories, which, like the theory of parental investment, seem able to explain much with little, often turn out to be worthless. There is justice in the complaint (from creationists, among others) that some theories about the evolution of animal traits are "just so stories"—plausible, but nothing more. Still, it is possible to separate the merely plausible from the compelling. In some sciences, testing theories is so straightforward that it is only a slight exaggeration (though it is always, in a certain strict sense, an exaggeration) to talk of theories being "proved." In others, corroboration is roundabout—an ongoing, gradual process by which confidence approaches the threshold of consensus, or fails to. Studying the evolutionary roots of human nature, or of anything else, is a science of the second sort. About each theory we ask a series of questions, and the answers nourish belief or doubt or ambivalence.

One question about the theory of parental investment is whether human behavior in fact complies with it in even the most basic ways. Are women more choosy about sex partners than men? (This is not to be confused with the very different question, to which we'll return, of which sex, if either, is choosier about *marriage* partners.) Certainly there is plenty of folk wisdom suggesting as much. More concretely, there's the fact that prostitution—sex with someone you don't know and don't care to know—is a service sought overwhelmingly by males, now as in Victorian England. Similarly, virtually all pornography that relies sheerly on visual stimulation—pictures or films of anonymous people, spiritless flesh—is consumed by males. And various studies have shown men to be, on average, much more open to casual, anonymous sex than women. In one experiment, three fourths of the men approached by an unknown woman on a college campus agreed to have sex with her, whereas none of the women approached by an unknown man were willing.[17]

It used to be common for doubters to complain that this sort of evidence, drawn from Western society, reflects only its warped values. This tack has been problematic since 1979, when Donald Symons published *The Evolution of Human Sexuality,* the first comprehensive anthropological survey of human sexual behavior from the new Darwinian perspective. Drawing on cultures East and West, industrial and preliterate, Symons demonstrated the great breadth of the patterns implied by the theory of parental investment: women tend to be relatively selective about sex partners; men tend to be less so, and tend to find sex with a wide variety of partners an extraordinarily appealing concept.

One culture Symons discussed is about as far from Western influence as possible: the indigenous culture of the Trobriand Islands in Melanesia. The prehistoric migration that populated these islands broke off from the migrations that peopled Europe at least tens of thousands of years ago, and possibly more than 100,000 years ago. The Trobrianders' ancestral culture was separated from Europe's ancestral culture even earlier than was that of Native Americans.[18] And indeed, when visited by the great anthropologist Bronislaw Malinowski in 1915, the islands proved startlingly remote from the currents of Western thought. The natives, it seemed, hadn't even gotten the connection between sex and reproduction. When one seafaring Trobriander returned from a voyage of several years to find his wife with two children, Malinowski was tactful enough not to suggest that she had been unfaithful. And "when I discussed the matter with others, suggesting that one at least of these children could not be his, my interlocutors did not understand what I meant."[19]

Some anthropologists have doubted that the Trobrianders could have been so ignorant. And although Malinowski's account of this issue seems to have the ring of authority, there is no way of knowing whether he got the story straight. But it is important to understand that he could, in principle, be right. The evolution of human sexual psychology seems to have preceded the discovery by humans of what sex is for. Lust and other such feelings are natural selection's way of getting us to act as if we wanted lots of offspring and knew how to get them, whether or not we actually do.[20] Had natural selection *not* worked this way—had it instead harnessed human intelligence so that our pursuit of fitness was entirely conscious and calculated—then life would be very different. Husbands and wives would, for example, spend no time having extramarital affairs with contraception; they would either scrap the contraception or scrap the sex.

Another un-Western thing about Trobriand culture was the lack of Victorian anxiety about premarital sex. By early adolescence, both girls and boys were encouraged to mate with a series of partners to

their liking. (This freedom is found in some other preindustrial societies, though the experimentation typically ends, and marriage begins, before a girl reaches fertility.) But Malinowski left no doubt
about which sex was choosier. "[T]here is nothing roundabout in a
Trobriand wooing. . . . Simply and directly a meeting is asked for
with the avowed intention of sexual gratification. If the invitation is
accepted, the satisfaction of the boy's desire eliminates the romantic
frame of mind, the craving for the unattainable and mysterious. If
he is rejected, there is not much room for personal tragedy, for he
is accustomed from childhood to having his sexual impulses thwarted
by some girl, and he knows that another intrigue cures this type of
ill surely and swiftly. . . . " And: "In the course of every love affair
the man has constantly to give small presents to the woman. To the
natives the need of one-sided payment is self-evident. This custom
implies that sexual intercourse, even where there is mutual attachment, is a service rendered by the female to the male."[21]

There were certainly cultural forces reinforcing coyness among
Trobriand women. Though a young woman was encouraged to have
an active sex life, her advances would be frowned on if too overt and
common because of the "small sense of personal worth that such
urgent solicitation implies."[22] But is there any reason to believe this
norm was anything other than a culturally mediated reflection of
deeper genetic logic? Can anyone find a single culture in which
women with unrestrained sexual appetites *aren't* viewed as more aberrant than comparably libidinous men? If not, isn't it an astonishing
coincidence that all peoples have independently arrived at roughly
the same cultural destination, with no genetic encouragement? Or is
it the case that this universal cultural element was present half a million
or more years ago, before the species began splitting up? That seems
a long time for an essentially arbitrary value to endure, without being
extinguished in a single culture.

This exercise holds a couple of important lessons. First: one good
reason to suspect an evolutionary explanation for something—some
mental trait or mechanism of mental development—is that it's universal, found everywhere, even in cultures that are as far apart as two
cultures can be.[23] Second: the general difficulty of explaining such
universality in utterly cultural terms is an example of how the Dar-

winian view, though not *proved* right in the sense that mathematical theorems are proved right, can still be the view that, by the rules of science, wins; its chain of explanation is shorter than the alternative chain and has fewer dubious links; it is a simpler and more potent theory. If we accept even the three meager assertions made so far— (1) that the theory of natural selection straightforwardly implies the "fitness" of women who are choosy about sexual partners and of men who often aren't; (2) that this choosiness and unchoosiness, respectively, is observed worldwide; and (3) that this universality can't be explained with equal simplicity by a competing, purely cultural, theory—if we accept these things, and if we're playing by the rules of science, we have to endorse the Darwinian explanation: male license and (relative) female reserve are to some extent innate.

Still, it is always good to have more evidence. Though absolute "proof" may not be possible in science, varying degrees of confidence are. And while evolutionary explanations rarely attain the 99.99 percent confidence sometimes found in physics or chemistry, it's always nice to raise the level from, say, 70 to 97 percent.

One way to strengthen an evolutionary explanation is to show that its logic is obeyed generally. If women are choosy about sex because they can have fewer kids than men (by virtue of investing more in them), and if females in the animal kingdom generally can have fewer offspring than males, then female animals in general should be choosier than males. Evolutionary theories can generate falsifiable predictions, as good scientific theories are expected to do, even though evolutionary biologists don't have the luxury of rerunning evolution in their labs, with some of its variables controlled, and predicting the outcome.

This particular prediction has been abundantly confirmed. In species after species, females are coy and males are not. Indeed, males are so dim in their sexual discernment that they may pursue things other than females. Among some kinds of frogs, mistaken homosexual courtship is so common that a "release call" is used by males who find themselves in the clutches of another male to notify him that they're both wasting their time.[24] Male snakes, for their part, have been known to spend a while with dead females before moving on to a live prospect.[25] And male turkeys will avidly court a stuffed replica of a female turkey. In fact, a replica of a female turkey's head suspended fifteen inches from the ground will generally do the trick. The male circles the head, does its ritual displays, and then (confident, presumably, that its performance has been impressive) rises into the air and comes down in the proximity of the female's backside, which turns out not to exist. The more virile males will show such interest even when a wooden head is used, and a few can summon lust for a wooden head with no eyes or beak.[26]

Of course, such experiments only confirm in vivid form what Darwin had much earlier said was obvious: males are very eager. This raises a much-cited problem with testing evolutionary explanations: the odd sense in which a theory's "predictions" are confirmed. Darwin didn't sit in his study and say, "My theory implies coy, picky females and mindlessly lustful males," and then take a walk to see if he could find examples. On the contrary, the many examples are what prompted him to wonder which implication of natural selection had created them—a question not correctly answered until midway through the following century, after even more examples had piled up. This tendency for Darwinian "predictions" to come after their evident fulfillment has been a chronic gripe of Darwin's critics. People who doubt the theory of natural selection, or just resist its application to human behavior, complain about the retrofitting of fresh predictions to preexisting results. This is often what they have in mind when they say evolutionary biologists spend their time dreaming up "just so stories" to explain everything they see.

In a sense, dreaming up plausible stories *is* what evolutionary biologists do. But that's not by itself a damning indictment. The power of a theory, such as the theory of parental investment, is gauged by how much data it explains and how simply, regardless of when the data surfaced. After Copernicus showed that assuming the Earth to revolve around the Sun accounted elegantly for the otherwise perplexing patterns that stars trace in the sky, it would have been beside the point to say, "But you cheated. You knew about the patterns all along." Some "just so stories" are plainly better than others, and they win. Besides, how much choice do evolutionary biologists have? There's not much they can do about the fact that the database on animal life began accumulating millennia before Darwin's theory.

But there is one thing they can do. Often a Darwinian theory generates, in addition to the pseudopredictions that the theory was in fact designed to explain, additional predictions—real predictions, untested predictions, which can be used to further evaluate the theory. (Darwin elliptically outlined this method in 1838, at age twenty-nine—more than twenty years before *The Origin of Species* was published. He wrote in his notebook: "The line of argument pursued throughout my theory is to establish a point as a probability by induction, & to apply it as hypothesis to other points. & see whether it will solve them.")[27] The theory of parental investment is a good example. For there are a few oddball species, as Williams noted in 1966, in which the male's investment in the offspring roughly matches, or even exceeds, the female's. If the theory of parental investment is right, these species should defy sex stereotypes.

Consider the spindly creatures known as pipefish. Here the male plays a role like a female kangaroo's: he takes the eggs into a pouch and plugs them into his bloodstream for nutrition. The female can thus start on another round of reproduction while the male is playing nurse. This may not mean that she can have many more offspring than he in the long run—after all, it took her a while to produce the eggs in the first place. Still, the parental investment isn't grossly imbalanced in the usual direction. And, predictably, female pipefish tend to take an active role in courtship, seeking out the males and initiating the mating ritual.[28]

Some birds, such as the phalarope (including the two species known as sea snipes), exhibit a similarly abnormal distribution of parental investment. The males sit on the eggs, leaving the females free to go get some wild oats sown. Again, we see the expected departure from stereotype. It is the phalarope *females* who are larger and more colorful—a sign that sexual selection is working in reverse, as females compete for males. One biologist observed that the females, in classically male fashion, "quarrel and display among themselves" while the males patiently incubate the eggs.[29]

If the truth be told, Williams knew that these species defy stereotype when he wrote in 1966. But subsequent investigation has confirmed his "prediction" more broadly. Extensive parental investment by males has been shown to have the expected consequences in other birds, in the Panamanian poison-arrow frog, in a water bug whose males cart fertilized eggs around on their backs, and in the (ironically named, it turns out) Mormon cricket. So far Williams's prediction has encountered no serious trouble.[30]

APES AND US

There is another major form of evolutionary evidence bearing on differences between men and women: our nearest relatives. The great apes—chimpanzees, pygmy chimps (also called bonobos), gorillas, and orangutans—are not, of course, our ancestors; all have evolved since their path diverged from ours. Still, those forks in the road have come between eight million years ago (for chimps and bonobos) and sixteen million years ago (for orangutans).[31] That's not long, as these things go. (A reference point: The australopithecine, our presumed ancestor, whose skull was ape-sized but who walked upright, appeared between six and four million years ago, shortly after the chimpanzee off-ramp. *Homo erectus*—the species that had brains midway in size between ours and apes', and used them to discover fire—took shape around 1.5 million years ago.)[32]

The great apes' nearness to us on the evolutionary tree legitimizes a kind of detective game. It's possible—though hardly certain—that when a trait is shared by all of them and by us, the reason is common descent. In other words: the trait existed in our common, sixteen-million-year-old proto-ape ancestor, and has been in all our lineages ever since. The logic is roughly the same as tracking down four distant cousins, finding that they all have brown eyes, and inferring that at least one of their two common great-great-grandparents had brown eyes. It's far from being an airtight conclusion, but it has more credence than it had when you had seen only one of the cousins.[33]

Lots of traits are shared by us and the great apes. For many of the traits—five-fingered hands, say—pointing this out isn't worth the trouble; no one doubts the genetic basis of human hands anyway. But in the case of human mental traits whose genetic substratum is still debated—such as the differing sexual appetites of men and women—this inter-ape comparison can be useful. Besides, it's worth taking a minute to get acquainted with our nearest relatives. Who knows how much of our psyche we share by common descent with some or all of them?

Orangutan males are drifters. They wander in solitude, looking for females, who tend to be stationary, each in her own home range. A male may settle down long enough to monopolize one, two, or even more of these ranges, though vast monopolies are discouraged by the attendant need to fend off vast numbers of rivals. Once the mission is accomplished, and the resident female gives birth, the male is likely to disappear. He may return years later, when pregnancy is again possible.[34] In the meantime, he doesn't bother to write.

For a gorilla male, the goal is to become leader of a pack comprising several adult females, their young offspring, and maybe a few young adult males. As dominant male, he will get sole sexual access to the females; the young males generally mind their manners (though a leader may, as he ages and his strength ebbs, share females with them).[35] On the downside, the leader does have to confront any male interlopers, each of which aims to make off with one or more of his females and thus is in an assertive mood.

The life of the male chimpanzee is also combative. He strives to climb a male hierarchy that is long and fluid compared to a gorilla hierarchy. And, again, the dominant male—working tirelessly to protect his rank through assault, intimidation, and cunning—gets first dibs on any females, a prerogative he enforces with special vigor when they're ovulating.[36]

Pygmy chimps, or bonobos (they're actually a distinct species from chimpanzees), may be the most erotic of all primates. Their sex comes in many forms and often serves purposes other than reproduction. Periodic homosexual behavior, such as genital rubbings between females, seems to be a way of saying, "Let's be friends." Still, broadly speaking, the bonobos' sociosexual outline mirrors that of the common chimpanzees: a pronounced male hierarchy that helps determine access to females.[37]

Amid the great variety of social structure in these species, the basic theme of this chapter stands out, at least in minimal form: males seem very eager for sex and work hard to find it; females work less hard. This isn't to say the females don't like sex. They love it, and may initiate it. And, intriguingly, the females of the species most closely related to humans—chimpanzees and bonobos—seem particularly amenable to a wild sex life, including a variety of partners. Still, female apes don't do what male apes do: search high and low, risking life and limb, to find sex, and to find as much of it, with as many different partners, as possible; it has a way of finding them.

FEMALE CHOICE

That female apes are, on balance, more reticent than male apes doesn't necessarily mean that they actively screen their prospective partners. To be sure, the partners get screened; those who dominate other males mate, while those who get dominated may not. This competition is exactly what Darwin had in mind in defining one of the two kinds of sexual selection, and these species (like our own) illustrate how it favors the evolution of big, mean males. But what about the other kind of sexual selection? Does the female participate in the screening, choosing the male that seems the most auspicious contributor to her project?

Female choice is notoriously hard to spot, and signs of its long-term effect are often ambiguous. Are males larger and stronger than females just because tougher males have bested their rivals and gotten to mate? Or, in addition, have females come to prefer tough males, since females with this genetically ingrained preference have had tougher and therefore more prolific sons, whose many daughters then inherited their grandmother's taste?

Notwithstanding such difficulties, it's fairly safe to say that in one sense or another, females are choosy in all the great ape species. A female gorilla, for example, though generally confined to sex with a single, dominant male, normally emigrates in the course of her lifetime. When an alien male approaches her pack, engaging its leader in mutual threats and maybe even a fight, she will, if sufficiently impressed, decide to follow him.[38]

In the case of chimps, the matter is more subtle. The dominant, or alpha, male can have any female he wants, but that's not necessarily because she prefers him; he shuts off alternatives by frightening other males. He can frighten her too, so that any spurning of low-ranking males may reflect only her fear. (Indeed, the spurning has been known to disappear when the alpha isn't looking.)[39] But there is a wholly different kind of chimp mating—a sustained, private consortship that may be a prototype for human courtship. A male and female chimp will leave the community for days or even weeks. And although the female may be forcibly abducted if she resists an invitation, there are times when she successfully resists, and times when she chooses to go peacefully, even though nearby males would gladly aid her in any resistance.[40]

Actually, even going unpeacefully can involve a kind of choice. Female orangutans are a good example. They do often seem to exercise positive choice, favoring some males over others. But sometimes they resist a mating and are forcibly subdued and—insofar as this word can be applied to nonhumans—raped. There is evidence that the rapists, often adolescents, usually fail to impregnate.[41] But suppose that they succeed with some regularity. Then a female, in sheerly Darwinian terms, is better off mating with a good rapist, a big, strong, sexually aggressive male; her male offspring will then be more likely to be big, strong, and sexually aggressive (assuming sexual aggressiveness varies at least in part because of genetic differences)—and therefore prolific. So female resistance should be favored by natural selection as a way to avoid having a son who is an inept rapist (assuming it doesn't bring injury to the female).

This isn't to say that a female primate, her protests notwithstanding, "really wants it," as human males have been known to assume. On the contrary, the more an orangutan "really wants it," the less she'll resist, and the less powerful a screening device her reticence will be. What natural selection "wants" and what any individual wants needn't be the same, and in this case they're somewhat at odds. The point is simply that, even when females demonstrate no clear preference for certain kinds of males, they may be, in practical terms, preferring a certain kind of male. And this de facto discretion may be de jure. It may be an adaptation, favored by natural selection precisely *because* it has this filtering effect.

In the broadest sense, the same logic could apply in any primate species. Once females in general begin putting up the slightest resistance, then a female that puts up a little extra resistance is exhibiting a valuable trait. For whatever it takes to penetrate resistance, the sons of strong resisters are more likely to have it than the sons of weak resisters. (This assumes, again, that the relative possession by different males of "whatever it takes" reflects underlying genetic differences.) Thus, in sheerly Darwinian terms, coyness becomes its own reward. And this is true regardless of whether the male's means of approach is physical or verbal.

ANIMALS AND THE UNCONSCIOUS

A common reaction to the new Darwinian view of sex is that it makes perfect sense as an explanation for animal behavior—which is to say, for the behavior of *nonhuman* animals. People may chuckle appreciatively at a male turkey that tries to mate with a poor rendition of a female's head, but if you then point out that many a human male regularly gets aroused after looking at two-dimensional representations of a nude woman, they don't see the connection. After all, the man surely knows that it's only a photo he's looking at; his behavior may be pathetic, but it isn't comic.

And maybe it isn't. But if he "knows" it's a photo, why is he getting so excited? And why are women so seldom whipped up into an onanistic frenzy by pictures of men?

Resistance to lumping humans and turkeys under one set of Darwinian rules has its points. Yes, our behavior is under more subtle, presumably more "conscious," control than is turkey behavior. Men can decide not to get aroused by something—or, at least, can decide not to look at something they know will arouse them. Sometimes they even stick with those decisions. And although turkeys can make what look like comparable "choices" (a turkey hounded by a shotgun-wielding man may decide that now isn't the time for romance), it is plainly true that the complexity and subtlety of options available to a human are unrivaled in the animal kingdom. So too is the human's considered pursuit of very long-run goals.

It all feels very rational, and in some ways it is. But that doesn't mean it isn't in the service of Darwinian ends. To a layperson, it may seem natural that the evolution of reflective, self-conscious brains would liberate us from the base dictates of our evolutionary past. To an evolutionary biologist, what seems natural is roughly the opposite: that human brains evolved not to insulate us from the mandate to survive and reproduce, but to follow it more effectively, if more pliably; that as we evolve from a species whose males forcibly abduct females into a species whose males whisper sweet nothings, the whispering will be governed by the same logic as the abduction—it is a means of manipulating females to male ends, and its form serves this function. The basic emanations of natural selection are refracted from the older, inner parts of our brain all the way out to its freshest tissue. Indeed, the freshest tissue would never have appeared if it didn't toe natural selection's bottom line.

Of course, a lot *has* happened since our ancestors parted ways with the great apes' ancestors, and one can imagine a change in evolutionary context that would have removed our lineage from the logic that so imbalances the romantic interests of male and female in most species. Don't forget about the seahorses, sea snipes, Panamanian poison-arrow frogs, and Mormon crickets, with their reversed sex roles. And, less dramatically, but a bit closer to home, there are the gibbons, another of our primate cousins, whose ancestors waved good-bye to ours about twenty million years ago. At some point in gibbon evolution, circumstances began to encourage much male parental investment. The males regularly stick around and help provide for the kids. In one gibbon species the males actually carry the infants, something male apes aren't exactly known for. And talk about marital harmony: gibbon couples sing a loud duet in the morning, pointedly advertising their familial stability for the information of would-be home-wreckers.[42]

Well, human males too have been known to carry around infants, and to stay with their families. Is it possible that at some time over the last few million years something happened to us rather like what happened to the gibbons? Have male and female sexual appetites converged at least enough to make monogamous marriage a reasonable goal?

Chapter 3: **MEN AND WOMEN**

Judging from the social habits of man as he now exists, and from most savages being polygamists, the most probable view is that primeval man aboriginally lived in small communities, each with as many wives as he could support and obtain, whom he would have jealously guarded against all other men. Or he may have lived with several wives by himself, like the Gorilla. . . .
—*The Descent of Man* (1871)[1]

One of the more upbeat ideas to have emerged from an evolutionary view of sex is that human beings are a "pair-bonding" species. In its most extreme form, the claim is that men and women are designed for a lifetime of deep, monogamous love. This claim has not emerged from close scrutiny in pristine condition.

The pair-bond hypothesis was popularized by Desmond Morris in his 1967 book *The Naked Ape*. This book, along with a few other 1960s books (Robert Ardrey's *The Territorial Imperative*, for example), represent a would-be watershed in the history of evolutionary thought. That they found large readerships signaled a new openness to Darwinism, an encouraging dissipation of the fallout from its past political misuses. But there was no way, in the end, that these books could start a Darwinian renaissance within academia. The problem was simple: they didn't make sense.

One example surfaced early in Morris's pair-bonding argument. He was trying to explain why human females are generally faithful to their mates. This is indeed a good question (if you believe they are, that is). For high fidelity would place women in a distinct minority within the animal kingdom. Though female animals are generally less licentious than males, the females of many species are far from prudes, and this is particularly true of our nearest ape relatives. Female chimpanzees and bonobos are, at times, veritable sex machines. In explaining how women came to be so virtuous, Morris referred to the sexual division of labor in an early hunter-gatherer economy. "To begin with," he wrote, "the males had to be sure that their females were going to be faithful to them when they left them alone to go hunting. So the females had to develop a pairing tendency."[2]

Stop right there. It was in the reproductive interests of the *males* for the *females* to develop a tendency toward fidelity? So natural selection obliged the males by making the necessary changes in the females? Morris never got around to explaining how, exactly, natural selection would perform this generous feat.

Maybe it's unfair to single Morris out for blame. He was a victim of his times. The trouble was an atmosphere of loose, hyper-teleological thinking. One gets the impression, reading Morris's book, and Ardrey's books, of a natural selection that peers into the future, decides what needs to be done to make things generally better for the species, and takes the necessary steps. But natural selection doesn't work that way. It doesn't peer ahead, and it doesn't try to make things generally better. Every single, tiny, blindly taken step either happens to make sense in immediate terms of genetic self-interest or it doesn't. And if it doesn't, you won't be reading about it a million years later. This was an essential message of George Williams's 1966 book, a message that had barely begun to take hold when Morris's book appeared.

One key to good evolutionary analysis, Williams stressed, is to focus on the fate of the gene in question. If a woman's "fidelity gene" (or her "infidelity gene") shapes her behavior in a way that helps get copies of *itself* into future generations in large numbers, then that gene will by definition flourish. Whether the gene, in the process, gets mixed in with her husband's genes or with the mailman's genes is by itself irrelevant. As far as natural selection is concerned, one vehicle is as good as the next. (Of course, when we talk about "a gene" for anything—fidelity, infidelity, altruism, cruelty—we are usefully oversimplifying; complex traits result from the interaction of numerous genes, each of which, typically, was selected for its incremental addition to fitness.)

A new wave of evolutionists has used this stricter view of natural selection to think with greater care about the question that rightly interested Morris: Are human males and females born to form enduring bonds with one another? The answer is hardly an unqualified yes for either sex. Still, it is closer to a yes for both sexes than it is in the case of, say, chimpanzees. In every human culture on the anthropological record, marriage—whether monogamous or polygamous, permanent or temporary—is the norm, and the family is the atom of social organization. Fathers everywhere feel love for their children, and that's a lot more than you can say for chimp fathers and bonobo fathers, who don't seem to have much of a clue as to which youngsters are theirs. This love leads fathers to help feed and defend their children, and teach them useful things.[3]

At some point, in other words, extensive *male parental investment* entered our evolutionary lineage. We are, as they say in the zoology literature, high in MPI. We're not so high that male parental investment typically rivals female parental investment, but we're a lot higher than the average primate. We indeed have something important in common with the gibbons.

High MPI has in some ways made the everyday goals of male and female humans dovetail, and, as any two parents know, it can give them a periodic source of common and profound joy. But high MPI has also created whole new ways for male and female aims to diverge, during both courtship and marriage. In Robert Trivers's 1972 paper on parental investment, he remarked, "One can, in effect, treat the sexes as if they were different species, the opposite sex being a resource relevant to producing maximum surviving offspring."[4] Trivers was making a specific analytical point, not a sweeping rhetorical one. But to a distressing extent—and an extent that was unclear before his paper—this metaphor does capture the overall situation; even with high MPI, and in some ways because of it, a basic underlying dynamic between men and women is mutual exploitation. They seem, at times, designed to make each other miserable.

WHY WE'RE HIGH IN MPI

There is no shortage of clues as to why men are inclined to help rear their young. In our recent evolutionary past lie several factors that can make parental investment worthwhile from the point of view of the male's genes.[5] In other words, because of these factors, genes inclining a male to love his offspring—to worry about them, defend them, provide for them, educate them—could flourish at the expense of genes that counseled continued remoteness.

One factor is the vulnerability of offspring. Following the generic male sexual strategy—roaming around, seducing and abandoning everything in sight—won't do a male's genes much good if the resulting offspring get eaten. That seems to be one reason so many bird species are monogamous, or at least relatively monogamous. Eggs left alone while the mother went out and hunted worms wouldn't last long. When our ancestors moved from the forests out onto the savanna, they had to cope with fleet predators. And this was hardly the only new danger to the young. As the species got smarter and its posture more upright, female anatomy faced a paradox: walking upright implied a narrow pelvis, and thus a narrow birth canal, but the heads of babies were larger than ever. This is presumably why human infants are born prematurely in comparison to other primates. From early on, baby chimps can cling to their mother while she walks around, her hands unencumbered. Human babies, though, seriously compromise a mother's food gathering. For many months, they're mounds of helpless flesh: tiger bait.

Meanwhile, as the genetic payoff of male investment was growing, the cost of investment was dropping. Hunting seems to have figured heavily in our evolution. With men securing handy, dense packages of protein, feeding a family was practical. It is probably no coincidence that monogamy is more common among carnivorous mammals than among vegetarians.

On top of all of this, as the human brain got bigger, it probably depended more on early cultural programming. Children with two parents may have had an educational edge over children with only one.

Characteristically, natural selection appears to have taken this cost-benefit calculus and transmuted it into feeling—in particular, the sensation of love. And not just love for the *child;* the first step toward becoming a solid parental unit is for the man and woman to develop a strong mutual attraction. The genetic payoff of having two parents devoted to a child's welfare is the reason men and women can fall into swoons over one another, including swoons of great duration.

Until recently, this claim was heresy. "Romantic love" was thought to be an invention of Western culture; there were reports of cultures in which choice of mate had nothing to do with affection, and sex carried no emotional weight. But lately anthropologists mindful of the Darwinian logic behind attachment have taken a second look, and such reports are falling into doubt.[6] Love between man and woman appears to have an innate basis. In this sense, the "pair-bonding" hypothesis stands supported, though not for all the reasons Desmond Morris imagined.

At the same time, the term *pair bonding*—and for that matter, the term *love*—conveys a sense of permanence and symmetry that, as any casual observer of our species can see, is not always warranted. To fully appreciate how large is the gap between idealized love and the version of love natural to people, we need to do what Trivers did in his 1972 paper: focus not on the emotion itself, but on the abstract evolutionary logic it embodies. What are the respective genetic interests of males and females in a species with internal fertilization, an extended period of gestation, prolonged infant dependence on mother's milk, and fairly high male parental investment? Seeing these interests clearly is the only way to appreciate how evolution not only invented romantic love, but, from the beginning, corrupted it.

WHAT DO WOMEN WANT?

For a species low in male parental investment, the basic dynamic of courtship, as we've seen, is pretty simple: the male really wants sex; the female isn't so sure.[7] She may want time to (unconsciously) assess

59

the quality of his genes, whether by inspecting him or by letting him battle with other males for her favor. She may also pause to weigh the chances that he carries disease. And she may try to extract a precopulation gift, taking advantage of the high demand for her eggs. This "nuptial offering"—which technically constitutes a tiny male parental investment, since it nourishes her and her eggs—is seen in a variety of species, ranging from primates to black-tipped hanging flies. (The female hanging fly insists on having a dead insect to eat during sex. If she finishes it before the male is finished, she may head off in search of another meal, leaving him high and dry. If she isn't so quick, the male may repossess the leftovers for subsequent dates.)[8] These various female concerns can usually be addressed fairly quickly; there's no reason for courtship to drag on for weeks.

But now throw high MPI into the equation—male investment not just at the time of sex, but extending up to and well beyond birth. Suddenly the female is concerned not only with the male's genetic investment, or with a free meal, but with what he'll bring to the offspring after it materializes. In 1989 the evolutionary psychologist David Buss published a pioneering study of mate preferences in thirty-seven cultures around the world. He found that in every culture, females placed more emphasis than males on a potential mate's financial prospects.[9]

That doesn't mean women have a specific, evolved preference for *wealthy* men. Most hunter-gatherer societies have very little in the way of accumulated resources and private property. Whether this accurately reflects the ancestral environment is controversial; hunter-gatherers have, over the last few millennia, been shoved off of rich land into marginal habitats and thus may not, in this respect, be representative of our ancestors. But if indeed all men in the ancestral environment were about equally affluent (that is, not very), women may be innately attuned not so much to a man's wealth as to his social status; among hunter-gatherers, status often translates into power—influence over the divvying up of resources, such as meat after a big kill. In modern societies, in any event, wealth, status, and power often go hand in hand, and seem to make an attractive package in the eyes of the average woman.

Ambition and industry also seem to strike many women as auspicious—and Buss found that this pattern, too, is broadly international.[10] Of course, ambition and industriousness are things a female might look for even in a low-MPI species, as indices of genetic quality. Not so, however, for her assessment of the male's *willingness* to invest. A female in a high-MPI species may seek signs of generosity, trustworthiness, and, especially, an enduring commitment to her in particular. It is a truism that flowers and other tokens of affection are more prized by women than by men.

Why should women be so suspicious of men? After all, aren't males in a high-MPI species designed to settle down, buy a house, and mow the lawn every weekend? Here arises the first problem with terms like *love* and *pair bonding*. Males in high-MPI species are, paradoxically, capable of greater treachery than males in low-MPI species. For the "optimal male course," as Trivers noted, is a "mixed strategy."[11] Even if long-term investment is their main aim, seduction and abandonment can make genetic sense, provided it doesn't take too much, in time and other resources, from the offspring in which the male *does* invest. The bastard youngsters may thrive even without paternal investment; they may, for that matter, attract investment from some poor sap who is under the impression that they're his. So males in a high-MPI species should, in theory, be ever alert for opportunistic sex.

Of course, so should males in a low-MPI species. But this doesn't amount to exploitation, since the female has no chance of getting much more from another male. In a high-MPI species, she does, and a failure to get it from any male can be quite costly.

The result of these conflicting aims—the female aversion to exploitation, the male affinity for exploiting—is an evolutionary arms race. Natural selection may favor males that are good at deceiving females about their future devotion and favor females that are good at spotting deception; and the better one side gets, the better the other side gets. It's a vicious spiral of treachery and wariness—even if, in a sufficiently subtle species, it may assume the form of soft kisses, murmured endearments, and ingenuous demurrals.

At least it's a vicious spiral *in theory*. Moving beyond all this theoretical speculation and into the realm of concrete evidence— actually glimpsing the seamy underside of kisses and endearments—

is tricky. Evolutionary psychologists have made only meager progress. True, one study found that males, markedly more than females, report depicting themselves as more kind, sincere, and trustworthy than they actually are.[12] But that sort of false advertising may be only half the story, and the other half is much harder to get at. As Trivers didn't note in his 1972 paper, but did note four years later, one effective way to deceive someone is to believe what you're saying. In this context, that means being blinded by love—to feel deep affection for a woman who, after a few months of sex, may grow markedly less adorable.[13] This, indeed, is the great moral escape hatch for men who persist in a pattern of elaborate seduction and crisp, if anguished, abandonment. "I loved her at the time," they can movingly recall, if pressed on the matter.

This isn't to say that a man's affections are chronically delusional, that every swoon is tactical self-deception. Sometimes men *do* make good on their vows of eternal devotion. Besides, in one sense, an out-and-out lie is impossible. There's no way of knowing in mid-swoon, either at the conscious or unconscious level, what the future holds. Maybe some more genetically auspicious mate will show up three years from now; then again, maybe the man will suffer some grave misfortune that renders him unmarketable, turning his spouse into his only reproductive hope. But, in the face of uncertainty as to how much commitment lies ahead, natural selection would likely err on the side of exaggeration, so long as it makes sex more likely and doesn't bring counterbalancing costs.

There probably would have been *some* such costs in the intimate social environment of our evolution. Leaving town, or at least village, wasn't a simple matter back then, so blatantly false promises might quickly catch up with a man—in the form of lowered credibility or even shortened life span; the anthropological archives contain stories about men who take vengeance on behalf of a betrayed sister or daughter.[14]

Also, the supply of potentially betrayable women wasn't nearly what it is in the modern world. As Donald Symons has noted, in the average hunter-gatherer society, every man who can snare a wife does, and virtually every woman is married by the time she's fertile. There probably was no thriving singles scene in the ancestral environment, except one involving adolescent girls during the fruitless phase between first menstruation and fertility. Symons believes that the lifestyle of the modern philandering bachelor—seducing and abandoning available women year after year after year, without making any of them targets for ongoing investment—is not a distinct, evolved sexual strategy. It is just what happens when you take the male mind, with its preference for varied sex partners, and put it in a big city replete with contraceptive technology.

Still, even if the ancestral environment wasn't full of single women sitting alone after one-night stands muttering "Men are scum," there were reasons to guard against males who exaggerate commitment. Divorce can happen in hunter-gatherer societies; men do up and leave after fathering a child or two, and may even move to another village. And polygamy is often an option. A man may vow that his bride will stay at the center of his life, and then, once married, spend half his time trying to woo another wife—or, worse still, succeed, and divert resources away from his first wife's children. Given such prospects, a woman's genes would be well served by her early and careful scrutiny of a man's likely devotion. In any event, the gauging of a man's commitment does seem to be part of human female psychology; and male psychology does seem inclined to sometimes encourage a false reading.

That male commitment is in limited supply—that each man has only so much time and energy to invest in offspring—is one reason females in our species defy stereotypes prevalent elsewhere in the animal kingdom. Females in *low*-MPI species—that is, in most sexual species—have no great rivalry with one another. Even if dozens of them have their hearts set on a single, genetically optimal male, he can, and gladly will, fulfill their dreams; copulation doesn't take long. But in a high-MPI species such as ours, where a female's ideal is to *monopolize* her dream mate—steer his social and material resources toward her offspring—competition with other females is inevitable. In other words: high male parental investment makes sexual selection work in two directions at once. Not only have males evolved to compete for scarce female eggs; females have evolved to compete for scarce male investment.

Sexual selection, to be sure, seems to have been more intense among men than among women. And it has favored different sorts of traits in the two. After all, the things women do to gain investment from men are different from the things men do to gain sexual access to women. (Women aren't—to take the most obvious example—designed for physical combat with each other, as men are.) The point is simply that, whatever each sex must do to get what it wants from the other, both sexes should be inclined to do it with zest. Females in a high-MPI species will hardly be passive and guileless. And they will sometimes be the natural enemies of one another.

WHAT DO MEN WANT?

It would be misleading to say that males in a high-male-parental-investment species are selective about mates, but in theory they are at least *selectively* selective. They will, on the one hand, have sex with just about anything that moves, given an easy chance, like males in a low-MPI species. On the other hand, when it comes to finding a female for a long-term joint venture, discretion makes sense; males can undertake only so many ventures over a lifetime, so the genes that the partner brings to the project—genes for robustness, brains, whatever—are worth scrutinizing.

The distinction was nicely drawn by a study in which both men and women were asked about the minimal level of intelligence they would accept in a person they were "dating." The average response, for both male and female, was: average intelligence. They were also asked how smart a person would have to be before they would consent to sexual relations. The women said: Oh, in that case, markedly *above* average. The men said: Oh, in that case, markedly *below* average.[15]

Otherwise, the responses of male and female moved in lockstep. A partner they were "steadily dating" would have to be much smarter than average, and a marriageable partner would have to be smarter still. This finding, published in 1990, confirmed a prediction Trivers had made in his 1972 paper on parental investment. In a high-MPI species, he wrote, "a male would be selected to differentiate between a female he will only impregnate and a female with whom he will also raise young. Toward the former he should be more eager for sex and less discriminating in choice of sex partner than the female toward him, but toward the latter he should be about as discriminating as she toward him."[16]

As Trivers knew, the nature of the discrimination, if not its intensity, should still differ between male and female. Though both seek general genetic quality, tastes may in other ways diverge. Just as women have special reason to focus on a man's ability to provide resources, men have special reason to focus on the ability to produce babies. That means, among other things, caring greatly about the age of a potential mate, since fertility declines until menopause, when it falls off abruptly. The last thing evolutionary psychologists would expect to find is that a plainly postmenopausal woman is sexually attractive to the average man. They don't find it. (According to Bronislaw Malinowski, Trobriand Islanders considered sex with an old woman "indecorous, ludicrous, and unaesthetic.")[17] Even before menopause, age matters, especially in a long-term mate; the younger a woman, the more children she can bear. In every one of Buss's thirty-seven cultures, males preferred younger mates (and females preferred older mates).

The importance of youth in a female mate may help explain the extreme male concern with physical attractiveness in a spouse (a concern that Buss also documented in all thirty-seven cultures). The generic "beautiful woman"—yes, she has actually been assembled, in a study that collated the seemingly diverse tastes of different men—has large eyes and a small nose. Since her eyes will look smaller and her nose larger as she ages, these components of "beauty" are also marks of youth, and thus of fertility.[18] Women can afford to be more open-minded about looks; an oldish man, unlike an oldish woman, is probably fertile.

Another reason for the relative flexibility of females on the question of facial attractiveness may be that a woman has other things to (consciously or unconsciously) worry about. Such as: Will he provide for the kids? When people see a beautiful woman with an ugly man, they typically assume he has lots of money or status. Researchers have actually gone to the trouble of showing that people make this inference, and that the inference is often correct.[19]

When it comes to assessing character—to figuring out if you can *trust* a mate—a male's discernment may again differ from a female's, because the kind of treachery that threatens his genes is different from the kind that threatens hers. Whereas the woman's natural fear is the withdrawal of his investment, his natural fear is that the investment is misplaced. Not long for this world are the genes of a man who spends his time rearing children who aren't his. Trivers noted in 1972 that, in a species with high male parental investment and internal fertilization, "adaptations should evolve to help guarantee that the female's offspring are also his own."[20]

All of this may sound highly theoretical—and of course it is. But this theory, unlike the theory about male love sometimes being finely crafted self-delusion, is readily tested. Years after Trivers suggested that anticuckoldry technology might be built into men, Martin Daly and Margo Wilson found some. They realized that if indeed a man's great Darwinian peril is cuckoldry, and a woman's is desertion, then male and female jealousy should differ.[21] Male jealousy should focus on *sexual* infidelity, and males should be quite unforgiving of it; a female, though she'll hardly applaud a partner's extracurricular activities, since they consume time and divert resources, should be more concerned with *emotional* infidelity—the sort of magnetic commitment to another woman that could eventually lead to a much larger diversion of resources.

These predictions have been confirmed—by eons of folk wisdom and, over the past few decades, by considerable data. What drives men craziest is the thought of their mate in bed with another man; they don't dwell as much as women do on any attendant emotional attachment, or the possible loss of the mate's time and attention. Wives, for their part, do find the sheerly sexual infidelity of husbands traumatic, and do respond harshly to it, but the long-run effect is often a self-improvement campaign: lose weight, wear makeup, "win him back." Husbands tend to respond to infidelity with rage; and even after it subsides, they often have trouble contemplating a continued relationship with the infidel.[22]

Looking back, Daly and Wilson saw that this basic pattern had been recorded (though not stressed) by psychologists before the theory of parental investment came along to explain it. But evolutionary psychologists have now confirmed the pattern in new and excruciating detail. David Buss placed electrodes on men and women and had them envision their mates doing various disturbing things. When men imagined sexual infidelity, their heart rates took leaps of a magnitude typically induced by three successive cups of coffee. They sweated. Their brows wrinkled. When they imagined instead a budding emotional attachment, they calmed down, though not quite to their normal level. For women, things were reversed: envisioning *emotional* infidelity—redirected love, not supplementary sex—brought the deeper physiological distress.[23]

The logic behind male jealousy isn't what it used to be. These days some adulterous women use contraception and thus don't, in fact, dupe their husbands into spending two decades shepherding another man's genes. But the weakening of the logic doesn't seem to have weakened the jealousy. For the average husband, the fact that his wife inserted a diaphragm before copulating with her tennis instructor will not be a major source of consolation.

The classic example of an adaptation that has outlived its logic is the sweet tooth. Our fondness for sweetness was designed for an environment in which fruit existed but candy didn't. Now that a sweet tooth can bring obesity, people try to control their cravings, and sometimes they succeed. But their methods are usually roundabout, and few people find them easy; the basic sense that sweetness feels good is almost unalterable (except by, say, repeatedly pairing a sweet taste with a painful shock). Similarly, the basic impulse toward jealousy is very hard to erase. Still, people can muster some control over the impulse, and, moreover, can muster much control over some forms of its expression, such as violence, given a sufficiently powerful reason. Prison, for example.

WHAT ELSE DO WOMEN WANT?

Before further exploring the grave imprint that cuckoldry has left on the male psyche, we might ask why it would exist. Why would a woman cheat on a man, if that won't increase the number of her progeny—and if, moreover, she thus risks incurring the wrath, and losing the investment, of her mate? What reward could justify such a gamble? There are more possible answers to this question than you might imagine.

First, there is what biologists call "resource extraction." If female humans, like female hanging flies, can get gifts in exchange for sex, then the more sex partners, the more gifts. Our closest primate relatives act out this logic. Female bonobos are often willing to provide sex in exchange for a hunk of meat. Among common chimpanzees, the food-for-sex swap is less explicit but is evident; male chimps are more likely to give meat to a female when she exhibits the red vaginal swelling that signifies ovulation.[24]

Human females, of course, *don't* advertise their ovulation. One theory about this "cryptic ovulation" sees it as an adaptation designed to expand the period during which they can extract resources. Men may lavish gifts on them well before or past ovulation and receive sex in return, blissfully oblivious to the fruitlessness of their conquest. Nisa, a woman in a !Kung San hunter-gatherer village, spoke candidly with an anthropologist about the material rewards of multiple sex partners. "One man can give you very little. One man gives you only one kind of food to eat. But when you have lovers, one brings you something and another brings you something else. One comes at night with meat, another with money, another with beads. Your husband also does things and gives them to you."[25]

Another reason women might copulate with more than one man—and another advantage of concealed ovulation—is to leave several men under the impression that they *might* be the father of particular offspring. Across primate species, there is a rough correlation between a male's kindness to youngsters and the chances that he is their father. The dominant male gorilla, with his celestial sexual stature, can rest pretty much assured that the youngsters in his troop are his; and, although not demonstrative by comparison with a human father, he is indulgent of them and reliably protective. At the other end of the spectrum, male langur monkeys kill infants sired by others as a kind of sexual icebreaker, a prelude to pairing up with the (former) mother.[26] What better way to return her to ovulation—by putting an emphatic end to her breast-feeding—and to focus her energies on the offspring to come?

Anyone tempted to launch into a sweeping indictment of langur morality should first note that infanticide on grounds of infidelity has been acceptable in various human societies. In two societies men have been known to demand, upon marrying women with a past, that their babies be killed.[27] And among the Ache hunter-gatherers of Paraguay, men sometimes collectively decide to kill a newly fatherless child. Even leaving murder aside, life can be hard on children without a devoted father. Ache children raised by stepfathers after their biological fathers die are half as likely to live to age fifteen as children whose parents stay alive and together.[28] For a woman in the ancestral environment, then, the benefits of multiple sex partners could have ranged from their not killing her youngster to their defending or otherwise aiding her youngster.

This logic doesn't depend on the sex partners' consciously mulling it over. Male gorillas and langurs, like the Trobriand Islanders as depicted by Malinowski, are not conscious of biological paternity. Still, the behavior of males in all three cases reflects an implicit recognition. Genes making males unconsciously sensitive to cues that certain youngsters may or may not be carrying their genes have flourished. A gene that says, or at least whispers, "Be nice to children if you've had a fair amount of sex with their mothers" will do better than a gene that says, "Steal food from children even if you were having regular sex with their mothers months before birth."

This "seeds of confusion" theory of female promiscuity has been championed by the anthropologist Sarah Blaffer Hrdy. Hrdy has described herself as a feminist sociobiologist, and she may take a more than scientific interest in arguing that female primates tend to be "highly competitive . . . sexually assertive individuals."[29] Then again, male Darwinians may get a certain thrill from saying males are built for lifelong sex-a-thons. Scientific theories spring from many sources. The only question in the end is whether they work.

Both of these theories of female promiscuity—"resource extraction" and "seeds of confusion"—could in principle apply to a mateless woman as well as a married one. Indeed, both would make sense for a species with little or no male parental investment, and thus may help explain the extreme promiscuity of female chimpanzees and bonobos. But there is a third theory that grows uniquely out of the dynamics of male parental investment, and thus has special application to wives: the "best of both worlds" theory.

In a high-MPI species, the female seeks two things: good genes and high ongoing investment. She may not find them in the same package. One solution would be to trick a devoted but not especially brawny or brainy mate into raising the offspring of another male. Again, cryptic ovulation would come in handy, as a treachery facilitator. It's fairly easy for a man to keep rivals from impregnating his mate if her brief phase of fertility is plainly visible; but if she appears equally fertile all month, surveillance becomes a problem. This is exactly the confusion a female would want to create if her goal is to draw investment from one man and genes from another.[30] Of course, the female may not consciously "want" this "goal." And she may not be consciously aware of when she's ovulating. But at some level she may be keeping track.

Theories involving so much subconscious subterfuge may sound too clever by half, especially to people not steeped in the cynical logic of natural selection. But there is some evidence that women are more sexually active around ovulation.[31] And two studies have found that women going to a singles bar wear more jewelry and makeup when near ovulation.[32] These adornments, it seems, have the advertising value of a chimpanzee's pink genital swelling, attracting a number of men for the woman to choose from. And these decked-out women did indeed tend to have more physical contact with men in the course of the evening.

Another study, by the British biologists R. Robin Baker and Mark Bellis, found that women who cheat on their mates are more likely to do so around ovulation. This suggests that often the secret lover's genes, not just his resources, are indeed what they're after.[33]

Whatever the reason(s) women cheat on their mates (or, as biologists value-neutrally put it, have "extra-pair copulations"), there's no denying that they do. Blood tests show that in some urban areas more than one fourth of the children may be sired by someone other than the father of record. And even in a !Kung San village, which, like the ancestral environment, is so intimate as to make covert liaison tricky, one in fifty children was found to have misassigned paternity.[34] Female infidelity appears to have a long history.

Indeed, if female infidelity *weren't* a long-standing part of life in this species, why would distinctively maniacal male jealousy have evolved? At the same time, that men so often invest heavily in the children of their mates suggests that cuckoldry hasn't been rampant;

if it had, genes encouraging this investment would long ago have run into a dead end.[35] The minds of men are an evolutionary record of the past behavior of women. And vice versa.

If a "psychological" record seems too vague, consider a more plainly physiological bit of data: human testicles—or, more exactly, the ratio of average testes weight to average male body weight. Chimpanzees and other species with high relative testes weights have "multimale breeding systems," in which females are quite promiscuous.[36] Species with low relative testes weights are either monogamous (gibbons, for example) or polygynous (gorillas), with one male monopolizing several families. (*Polygamous* is the more general term, denoting a male *or* a female that has more than one mate.) The explanation is simple. When females commonly breed with many different males, male genes can profit by producing lots of semen for their transportation. Which male gets his DNA into a given egg may be a question of sheer volume, as competing hordes of sperm do subterranean battle. A species' testicles are thus a record of its females' sexual adventure over the ages. In our species, relative testes weight falls between that of the chimpanzee and the gorilla, suggesting that women, while not nearly as wild as chimpanzee females, are, by nature, somewhat adventurous.

Of course, adventurous doesn't mean unfaithful. Maybe women in the ancestral environment had their wild, unattached periods— during which fairly weighty testicles paid off for men—as well as their devoted, monogamous periods. Then again, maybe not. Consider a truer record of female infidelity: variable sperm density. You might think that the number of sperm cells in a husband's ejaculate would depend only on how long it's been since he last had sex. Wrong. According to work by Baker and Bellis, the quantity of sperm depends heavily on the amount of time a man's mate has been out of his sight lately.[37] The more chances a woman has had to collect sperm from other males, the more profusely her mate sends in his own troops. Again: that natural selection designed such a clever weapon is evidence of something for the weapon to combat.

It is also evidence that natural selection is fully capable of designing equally clever psychological weapons, ranging from furious jealousy to the seemingly paradoxical tendency of some men to be sexually aroused by the thought of their mate in bed with another man. Or, more generally: the tendency of men to view women as possessions. In a 1992 paper called "The Man Who Mistook His Wife for a Chattel," Wilson and Daly wrote that "men lay claim to particular women as songbirds lay claim to territories, as lions lay claim to a kill, or as people of both sexes lay claim to valuables. . . . [R]eferring to man's view of woman as 'proprietary' is more than a metaphor: Some of the same mental algorithms are apparently activated in the marital and mercantile spheres."[38]

The theoretical upshot of all this is another evolutionary arms race. As men grow more attuned to the threat of cuckoldry, women should get better at convincing a man that their adoration borders on awe, their fidelity on the saintly. And they may partly convince themselves too, just for good measure. Indeed, given the calamitous fallout from infidelity uncovered—likely desertion by the offended male, and possible violence—female self-deception may be finely honed. It could be adaptive for a married woman to not *feel* chronically concerned with sex, even if her unconscious mind is keeping track of prospects and will notify her when ardor is warranted.

THE MADONNA-WHORE DICHOTOMY

Anticuckoldry technology could come in handy not just when a man has a mate, but earlier, in choosing her. If available females differ in their promiscuity, and if the more promiscuous ones tend to make less faithful wives, natural selection might incline men to discriminate accordingly. Promiscuous women would be welcome as short-term sex partners—indeed, preferable, in some ways, since they can be had with less effort. But they would make poor wife material, a dubious conduit for male parental investment.

What emotional mechanisms—what complex of attractions and aversions—would natural selection use to get males to uncomprehendingly follow this logic? As Donald Symons has noted, one candidate is the famed Madonna-whore dichotomy, the tendency of men to think in terms of "two kinds of women"—the kind they respect and the kind they just sleep with.[39]

One can imagine courtship as, among other things, a process of placing a woman in one category or the other. The test would run roughly as follows. If you find a woman who appears genetically suitable for investment, start spending lots of time with her. If she seems quite taken by you, and yet remains sexually aloof, stick with her. If, on the other hand, she seems eager for sex right away, then by all means oblige her. But if the sex does come that easily, you might want to shift from investment mode into exploitation mode. Her eagerness could mean she'll always be an easy seduction—not a desirable quality in a wife.

Of course, in the case of any particular woman, sexual eagerness may *not* mean she'll always be an easy seduction; maybe she just finds this one man irresistible. But if there is any general correlation between the speed with which a woman succumbs to a man and her likelihood of later cheating on him, then that speed is a statistically valid cue to a matter of great genetic consequence. Faced with the complexity and frequent unpredictability of human behavior, natural selection plays the odds.

Just to add a trifle more ruthlessness to this strategy: the male may actually *encourage* the early sex for which he will ultimately punish the woman. What better way to check for the sort of self-restraint that is so precious in a woman whose children you may invest in? And, if self-restraint proves lacking, what faster way to get the wild oats sown before moving on to worthier terrain?

In its extreme, pathological form—the Madonna-whore *complex*—this dichotomization of women leaves a man unable to have sex with his wife, so holy does she seem. Obviously, this degree of worship isn't likely to have been favored by natural selection. But the more common, more moderate version of the Madonna-whore distinction has the earmarks of an efficient adaptation. It leads men to shower worshipful devotion on the sexually reserved women they want to invest in—exactly the sort of devotion these women will demand before allowing sex. And it lets men guiltlessly exploit the women they don't want to invest in, by consigning them to a category that merits contempt. This general category—the category of reduced, sometimes almost subhuman, moral status—is, as we'll see, a favorite tool of natural selection's; it is put to especially effective use during wars.

In polite company, men sometimes deny that they think differently of a woman who has slept with them casually. And wisely so. To admit that they do would sound morally reactionary. (Even to admit as much to themselves might make it hard to earnestly assure such a woman that they'll still respect her in the morning—sometimes a vital part of foreplay.)

As many modern wives can attest, sleeping with a man early in courtship doesn't doom the prospect of long-term commitment. A man's (largely unconscious) assessment of a woman's likely fidelity presumably involves many things—her reputation, how she looks at other men, how honest she seems generally. And anyway, even in theory the male mind shouldn't be designed to make virginity a prerequisite for investment. The chances of finding a virgin wife vary from man to man and from culture to culture—and to judge by some hunter-gatherer societies, they would have been quite low in the ancestral environment. Presumably males are designed to do the best they can under the circumstances. Though in prudish Victorian England some men may have insisted on virgin wives, the term *Madonna-whore dichotomy* is actually a misnomer for what is surely a more flexible mental tendency.[40]

Still, the flexibility is bounded. There is some level of female promiscuity above which male parental investment plainly makes no genetic sense. If a woman seems to have an unbreakable habit of sleeping with a different man each week, the fact that all women in that culture do the same thing doesn't make her any more logical a spouse. In such a society, men should in theory give up entirely on concentrated parental investment and focus solely on trying to mate with as many women as possible. That is, they should act like chimpanzees.

VICTORIAN SAMOANS

The Madonna-whore dichotomy has long been dismissed as an aberration, another pathological product of Western culture. In particular, the Victorians, with their extraordinary emphasis on virginity and their professed disdain for illicit sex, are held responsible for nourishing, even inventing, the pathology. If only men in Darwin's day had been more relaxed about sex, like the men in non-Western, sexually liberated societies. How different things would be now!

The trouble is, those idyllic, non-Western societies seem to have existed only in the minds of a few misguided, if influential, academics. The classic example is Margaret Mead, one of several prominent anthropologists who early this century reacted to the political misuses of Darwinism by stressing the malleability of the human species and asserting the near absence of human nature. Mead's best-known book, *Coming of Age in Samoa,* created a sensation upon its appearance in 1928. She seemed to have found a culture nearly devoid of many Western evils: status hierarchies, intense competition, and all kinds of needless anxieties about sex. Here in Samoa, Mead wrote, girls postpone marriage "through as many years of casual love-making as possible." Romantic love "as it occurs in our civilisation," bound up with ideas of "exclusiveness, jealousy and undeviating fidelity," simply "does not occur in Samoa."[41] What a wonderful place!

It is hard to exaggerate the influence of Mead's findings on twentieth-century thought. Claims about human nature are always precarious, vulnerable to the discovery of even a single culture in which its elements are fundamentally lacking. For much of this century, such claims have been ritually met with a single question: "What about Samoa?"

In 1983 the anthropologist Derek Freeman published a book called *Margaret Mead and Samoa: The Making and Unmaking of an Anthropological Myth.* Freeman had spent nearly six years in Samoa (Mead had spent nine months, and hadn't spoken the language when she arrived), and was well versed in accounts of its earlier history, before Western contact had much changed it. His book left Mead's reputation as a great anthropologist in serious disarray. He depicted her as a naïf, a twenty-three-year-old idealist who went to Samoa steeped in fashionable cultural determinism, chose not to live among the natives, and then, dependent for her data on scheduled interviews, was duped by Samoan girls who made a game of misleading her. Freeman assaulted Mead's data broadly—the supposed dearth of status competition, the simple bliss of Samoan adolescence—but for present purposes what matters is the sex: the purportedly minor significance of jealousy and male possessiveness, the seeming indifference of men to the Madonna-whore dichotomy.

Actually, on close examination, Mead's point-by-point findings turn out to be less radical than her glossy, well-publicized generalizations. She conceded that Samoan males took a certain pride in the conquest of a virgin. She also noted that each tribe had a ceremonial virgin—a girl of good breeding, often a chief's daughter, who was carefully guarded until, upon marriage, she was manually deflowered, with the blood from her hymen proving her purity. But this girl, Mead insisted, was an aberration, "excepted" from the "free and easy experimentation" that was the norm. Parents of lower rank "complacently ignore" their daughters' sexual experimentation.[42] Mead granted, almost under her breath, that a virginity test was "theoretically" performed "at weddings of people of all ranks," but she dismissed the ceremony as easily and often evaded.

Freeman raised the volume of Mead's more hushed observations and pointed out some things she had failed entirely to note. The value of virgins was so great in the eyes of marriageable men, he wrote, that an adolescent female of any social rank was monitored by her brothers, who would "upbraid, and sometimes beat" her if they found her with "a boy suspected of having designs on her virginity." As for the suspected boy, he was "liable to be assaulted with great ferocity." Young men who fared poorly in the mating game sometimes secured a mate by sneaking in at night, forcibly deflowering a woman, and then threatening to disclose her corruption unless she agreed to marriage (perhaps in the form of elopement, the surest way to avoid a virginity test). A woman found on her wedding day not to be a virgin was publicly denounced with a term meaning, roughly, "whore." In Samoan lore, one deflowered woman is described as a "wanton woman, like an empty shell exposed by the ebbing tide!" A song performed at defloration ceremonies went like this: "All others have failed to achieve entry, all others have failed to achieve entry. . . . He is first by being foremost, being first he is foremost; O to be foremost!"[43] These are not the hallmarks of a sexually liberated culture.

It now appears that some of the supposed Western aberrations that Mead found lacking in Samoa had if anything been *suppressed* by Western influence. Missionaries, Freeman noted, had made virginity testing less public—performed in a house, behind a screen. In "former days," as Mead herself wrote, if the tribe's ceremonial virgin was found at her wedding to be less than virginal, "her female relatives fell upon and beat her with stones, disfiguring and sometimes fatally injuring the girl who had shamed their house."[44]

So too with the Samoan jealousy that, Mead stressed, was so muted by Western standards: Westerners may have done the muting. Mead noted that a husband who caught his wife in adultery might be appeased by a harmless ritual that, as she depicted it, would end in an air of bonhomie. The male offender would bring men of his family, sit outside the victimized husband's house in supplication, offering finery in recompense, until forgiveness was forthcoming and everyone buried the hatchet over dinner. Of course, "in olden days," Mead observed, the offended man might "take a club and together with his relatives go out and kill those who sit without."[45]

That violence became less frequent under Christian influence is, of course, a testament to human malleability. But if we are ever to fathom the complex parameters of that malleability, we must be clear about which is the core disposition and which is the modifying influence. Time and again, Mead, along with her whole cohort of mid-twentieth-century cultural determinists, got things backwards.

Darwinism helps set the record straight. A new generation of Darwinian anthropologists is combing old ethnographies and conducting new field studies, finding things past anthropologists didn't stress, or even notice. Many candidates for "human nature" are emerging. And one of the more viable is the Madonna-whore dichotomy. In exotic cultures from Samoa to Mangaia to the land of the Ache in South America, a reputation for extreme promiscuity is something men actively avoid in a long-term mate.[46] And an analysis of folklore reveals the "good girl/bad girl" polarity to be a chronically recurring image—in the Far East, in Islamic states, in Europe, even in pre-Columbian America.[47]

Meanwhile, in the psychology laboratory, David Buss has found evidence that men do dichotomize between short-term and long-term mates. Cues suggesting promiscuity (a low-cut dress, perhaps, or aggressive body language) make a woman more attractive as a short-term mate and less attractive as a long-term mate. Cues suggesting a lack of sexual experience work the other way around.[48]

For now, the hypothesis that the Madonna-whore dichotomy has at least some inherent basis rests on strong theoretical expectation and considerable, though hardly exhaustive, anthropological and psychological evidence. There is also, of course, the testimony of experienced mothers from many eras who have warned their daughters what will happen if a man gets the impression that they're "that kind of girl": he won't "respect" them anymore.

FAST AND SLOW WOMEN

The Madonna-whore distinction is a dichotomy imposed on a continuum. In real life, women aren't either "fast" or "slow"; they are promiscuous to various degrees, ranging from not at all to quite. So the question of why some women are of one type and others of the other has no meaning. But there is meaning in the question of why women are nearer one end of the spectrum than the other—why women differ in their general degree of sexual reserve. And for that matter, what about men? Why do some men seem capable of unswerving monogamy, and others so inclined to depart from that ideal to various degrees? Is this difference—between Madonnas and whores, between dads and cads—in the genes? The answer is a definite yes. But the only reason the answer is definite is that the phrase "in the genes" is so ambiguous as to be essentially meaningless.

Let's start with the popular conception of "in the genes." Are some women, from the moment their father's sperm meets their mother's egg, all but destined to be Madonnas, while others are almost certain to be whores? Are some men equally bound to be cads, and others dads?

For both men and women, the answer is: unlikely, but not impossible. As a rule, two extremely different alternative traits will not both be preserved by natural selection. One or the other is usually at least slightly more conducive to genetic proliferation. However marginal its edge, it should win out, given enough time.[49] That's why almost all of the genes in you are also in the average inhabitant of any land in the world. But there is something called "frequency-dependent" selection, in which the value of a trait declines as it becomes more common, so that natural selection places a ceiling on its predominance, thus leaving room for the alternative.

Consider the bluegill sunfish.[50] The average bluegill male grows up, builds a bunch of nests, waits for females to lay eggs, then fertilizes the eggs and guards them. He is an upstanding member of the community. But he may have as many as 150 nests to tend, a fact that leaves him vulnerable to a second, less responsible kind of male, a drifter. The drifter sneaks around, surreptitiously fertilizes eggs, and then darts off, leaving them to be tended by their duped custodian. At a certain stage in life, drifters even don the color and behavior of females to mask their covert operations.

You can see how the balance between drifters and their victims is maintained. The drifters must do fairly well in reproductive terms; otherwise they wouldn't be around. But as this success makes their fraction of the population grow, the success itself diminishes, because the relative supply of upstanding, exploitable males—the drifters' meal ticket—shrinks. This is a situation in which success is its own punishment. The more drifters there are, the fewer offspring per drifter there are.

In theory, the drifter fraction of the population should grow until the average drifter is having as many offspring as the average upstanding bluegill. At that point, any shift in the fraction—growth or shrinkage—will change the value of the two strategies in a way that tends to reverse the shift. This equilibrium is known as an "evolutionarily stable" state, a term coined by the British biologist John Maynard Smith, who, during the 1970s, fully developed the idea of frequency-dependent selection.[51] Bluegill drifters, presumably, long ago reached their evolutionarily stable fraction of the population, which seems to be about one fifth.

The dynamics of sexual treachery are different for humans than for bluegill sunfish, in part because of the mammalian penchant for internal fertilization. But Richard Dawkins has shown, with an abstract analysis applicable to our species, that Maynard Smith's logic can, in principle, fit us too. In other words: one can imagine a situation in which neither coy nor fast women, and neither cads nor dads, have a monopoly on the ideal strategy. Rather, the success of each strategy varies with the prevalence of the three other strategies, and the population tends toward equilibrium. For example, with one set of assumptions, Dawkins found that five-sixths of the females would be coy, and five-eighths of the males would be faithful.[52]

Now, having comprehended this fact, you are advised to forget it. Don't just forget the fractions themselves, which, obviously, grow out of arbitrary assumptions within a highly artificial model. Forget the whole idea that each individual would be firmly bound to one strategy or the other.

As Maynard Smith and Dawkins have noted, evolution equilibrates to an equally stable state if you assume that the magic proportions are found *within* individuals—that is, if each female is coy on five-sixths of her mating opportunities, and each male is coy on five-eighths of his. And that's true even if the fractions are *randomly* realized—if each person just rolls dice on each encounter to decide what to do. Imagine how much more effective it is for the person to ponder each situation (consciously or unconsciously) and make an informed guess as to which strategy is more propitious under the circumstances.

Or imagine a different kind of flexibility: a developmental program that, during childhood, assesses the local social environment and then, by adulthood, inclines the person toward the strategy more likely to pay off. To put this in bluegill terms: imagine a male that during its early years checks out the local environment, calculates the prevalence of exploitable, upstanding males, and *then* decides—or, at least, "decides"—whether to become a drifter. This plasticity should eventually dominate the population, pushing the two more rigid strategies into oblivion.

The moral of the story is that limberness, given the opportunity, usually wins out over stiffness. In fact, limberness seems to have won a partial victory even in the bluegill sunfish, which isn't exactly known for its highly developed cerebral cortex. Though some genes incline a male bluegill to one strategy, and others to the other, the inclination isn't complete; the male absorbs local data before "deciding" which strategy to adopt.[53] Obviously, when you move from fish to us, the likely extent of flexibility grows. We have huge brains whose whole reason for being is deft adjustment to variable conditions. Given the many things about a person's social environment that can alter the value of being a Madonna versus a whore, a cad versus a dad— including the way other people react to the person's particular assets and liabilities—natural selection would be uncharacteristically obtuse not to favor genes that build brains sensitive to these things.

So too in many other realms. The value of being a given "type" of person—cooperative, say, or stingy—has depended, during evolution, on things that vary from time to time, place to place, person to person. Genes that irrevocably committed our ancestors to one personality type should in theory have lost out to genes that let the personality solidify gracefully.

This is not a matter of consensus. There are in the literature a few articles with titles like "The Evolution of the 'Con Artist.' "[54] And, to return to the realm of Madonnas and whores, there is a theory that some women are innately inclined to pursue a "sexy son" strategy: they mate promiscuously with sexually attractive men (handsome, brainy, brawny, and so on), risking the high male parental investment they might extract if more Madonnaish but gaining the likelihood that any sons will be, like their fathers, attractive and therefore prolific. Such theories are interesting, but they all face the same obstacle: with con artists as with promiscuous women, however effective the strategy, it is even more effective when flexible—when it can be abandoned amid signs of likely failure.[55] And the human brain is a pretty flexible thing.

To stress this flexibility isn't to say that all people are born psychologically identical, that all differences in personality emerge from environment. There plainly are important genetic differences for such traits as nervousness and extroversion. The "heritability" of these traits is around .4; that is, about 40 percent of individual differences in these traits (within the particular populations geneticists have studied) can be explained by genetic differences. (By comparison, the heritability of height is around .9; about 10 percent of the difference in height among individuals is due to nutritional and other environmental differences.) The question is *why* the undoubtedly important genetic variation in personality exists. Do different degrees of genetic disposition toward extroversion represent different personality "types," the products of a very elaborate process of frequency-dependent selection? (Though frequency dependence is classically analyzed in terms of two or three distinct strategies, it could also yield a more finely graded array.) Or are the differing genetic dispositions just "noise"—some incidental by-product of evolution, not specifically favored by natural selection? No one knows, and evolutionary psychologists differ in their suspicions.[56] What they agree on is that a big part of the story of personality differences is the evolution of malleability, of "developmental plasticity."

This emphasis on psychological development doesn't leave us back where social scientists were twenty-five years ago, attributing everything they saw to often unspecified "environmental forces." A primary—perhaps *the* primary—promise of evolutionary psychology is to help specify the forces, to generate good theories of personality development. In other words: evolutionary psychology can help us see not only the "knobs" of human nature, but also how the knobs are tuned. It not only shows us that (and why) men in all cultures are quite attracted to sexual variety, but can suggest what circumstances make some men more obsessed with it than others; it not only shows us that (and why) women in all cultures are more sexually reserved, but promises to help us figure out how some women come to defy this stereotype.

A good example lies in Robert Trivers's 1972 paper on parental investment. Trivers noted two patterns that social scientists had already uncovered: (1) the more attractive an adolescent girl, the more likely she is to "marry up"—marry a man of higher socioeconomic status; and (2) the more sexually active an adolescent girl, the less likely she is to marry up.

To begin with, these two patterns make Darwinian sense independently. A wealthy, high-status male often has a broad range of aspiring wives to choose from. So he tends to choose a good-looking woman who is also relatively Madonnaish. Trivers took the analysis further. Is it possible, he asked, "that females adjust their reproductive strategies in adolescence to their own assets"?[57] In other words, maybe adolescent girls who get early social feedback affirming their beauty make the most of it, becoming sexually reserved and thus encouraging long-term investment by high-status males who are looking for pretty Madonnas. Less attractive women, with less chance to hit the jackpot via sexual reserve, become more promiscuous, extracting small chunks of resources from a series of males. Though this promiscuity may somewhat lower their values as wives, it wouldn't, in the ancestral environment, have doomed their chances of finding a husband. In the average hunter-gatherer society, almost any fertile woman can find a husband, even if he's far from ideal, or she has to share him with another woman.

DARWINISM AND PUBLIC POLICY

The Trivers scenario doesn't imply a *conscious* decision by attractive women to guard their jewels (though that may play a role, and, what's more, parents may be genetically inclined to encourage a daughter's sexual reserve with special force when she is pretty). By the same token, we aren't necessarily talking about unattractive women who "realize" they can't be choosy and start having sex on less than ideal Darwinian terms. The mechanism at work might well be subconscious, a gradual molding of sexual strategy—read: "moral values"— by adolescent experience.

Theories like this one matter. There has been much talk about the problem of unwed motherhood among teenagers, especially poor teenagers. But no one really knows how sexual habits get shaped, or how firmly fixed they then are. There is much talk about boosting "self-esteem," but little understanding of what self-esteem is, what it's for, or what it does.

Evolutionary psychology can't yet confidently provide the missing basis for these discussions. But the problem isn't a shortage of plausible theories; it's a shortage of studies to test the theories. The Trivers theory has sat in limbo for two decades. In 1992, one psychologist did find what the theory predicts—a correlation between a woman's self-perception and her sexual habits: the less attractive she thinks she is, the more sex partners she has had. But another scholar didn't find the predicted correlation—and, more to the point, *neither* study was conducted specifically to test Trivers's theory, of which both scholars were unaware.[58] For now, this is the state of evolutionary psychology: so much fertile terrain, so few farmers.

Eventually, the main drift of Trivers's theory, if not the theory itself, will likely be vindicated. That is: women's sexual strategies probably depend on the likely (genetic) profitability of each strategy, given prevailing circumstances. But those circumstances go beyond what Trivers stressed—a particular woman's desirability. Another factor is the general availability of male parental investment. This factor surely fluctuated in the ancestral environment. For example, a village that had just invaded a neighboring village might have a suddenly elevated ratio of women to men—not just because of male casualties, but because victorious warriors commonly kill or vanquish enemy men and keep their women.[59] Overnight, a young woman's prospects for receiving a man's undivided investment could thus plummet. Famine, or sudden abundance, might also alter investment patterns. Given these currents of change, any genes that helped women navigate them would, in theory, have flourished.

There is tentative evidence that they did. According to a study by the anthropologist Elizabeth Cashdan, women who perceive men in general as pursuing no-obligation sex are more likely to wear provocative clothes and have sex often than women who see men as generally willing to invest in offspring.[60] Though some of these women may be conscious of the connection between local conditions and their lifestyle, that isn't necessary. Women surrounded by men who are unwilling or unable to serve as devoted fathers may simply feel a deepened attraction to sex without commitment—feel, in other words, a relaxation of "moral" constraint. And perhaps if market conditions suddenly improve—if the male to female ratio rises, or if men for some other reason shift toward a high-investment strategy—women's sexual attractions, and moral sensibilities, shift accordingly.

All of this is necessarily speculative at this early stage in evolutionary psychology's growth. But already we can see the sort of light that will increasingly be shed. For example, "self-esteem" almost

Emma and Charles Darwin around the time of their wedding. Evolutionary psychology sheds light on various aspects of their relationship, including Emma's relative coolness toward impending marriage. She wrote to Charles, "I shall always look upon the event of the 29th as a most happy one on my part though perhaps not so great or so good as you do."

Above: George Williams in the early 1960s. Williams helped formulate the theory that explains why in most species males are sexually assertive and females more reserved. He also saw a way to test the theory; if it is correct, then this pattern should be weak, or even reversed, in species whose males invest heavily in offspring. *Below:* A male phalarope incubates eggs. In keeping with Williams's theory, female phalaropes compete more intensely for mates than males do.

Above: The biologists Robert Trivers and John Maynard Smith in India in 1980. *Below:* A !Kung San father and daughter. In all human cultures, males invest heavily in offspring—less heavily than phalarope males, perhaps, but more heavily than males in the primate species most closely related to us. This male parental investment, Trivers has observed, helps explain much about human sexual psychology, including the kinds of duplicity to which men and women are prone.

!Kung San women collecting food. Hunter-gatherer societies—such as the !Kung of southern Africa, the Ache of South America, or the Ainu of Japan—are the closest thing anthropologists have had to a real-life model of the social context of human evolution. These societies reconcile motherhood and work fairly simply, whether, as here, by bringing children to work or by using kin and friends for babysitting. The anthropologist Marjorie Shostak, who took this picture, wrote, "The isolated mother burdened with bored small children is not a scene that has parallels in !Kung daily life." Her remark illustrates a general point that may account for much current suffering: the human mind wasn't designed for the modern world.

certainly isn't the same, either in its sources or in its effects, for boys and girls. For teenage girls, feedback reflecting great beauty may, as Trivers suggested, bring high self-esteem, which in turn encourages sexual restraint. For boys, extremely high self-esteem could well have the opposite effect: it may lead them to seek with particular intensity the short-term sexual conquests that are, in fact, more open to a good-looking, high-status male. In many high schools, a handsome, star athlete is referred to, only half-jokingly, as a "stud." And, for those who insist on scientific verification of the obvious: good-looking men do have more sex partners than the average man.[61] (Women report putting more emphasis on a sex partner's looks when they don't expect the relationship to last; they are apparently willing, unconsciously, to trade off parental investment for good genes.)[62]

Once a high-self-esteem male is married, he may not be notable for his devotion. Presumably his various assets still make philandering a viable lifestyle, even if it's now covert. (And you never know when an outside escapade will take on a life of its own, and lead to desertion.) Men with more moderate self-esteem may make more committed, if otherwise less desirable, husbands. With fewer chances at extramarital dalliance, and perhaps more insecurity about their own mate's fidelity, they may focus their energy and attention toward family. Meanwhile, men with *extremely* low self-esteem, given continued frustration with women, may eventually resort to rape. There is ongoing debate within evolutionary psychology over whether rape is an adaptation, a designed strategy that any boy might grow up to adopt, given sufficiently discouraging feedback from his social environment. Certainly rape surfaces in a wide variety of cultures, and often under the expected circumstances: when men have had trouble finding attractive women by legitimate means. One (non-Darwinian) study found the typical rapist to possess "deep-seated doubts about his adequacy and competency as a person. He lacks a sense of confidence in himself as a man in both sexual and nonsexual areas."[63]

A second sort of light shed by the new Darwinian paradigm may illuminate links between poverty and sexual morality. Women living in an environment where few men have the ability and/or desire to support a family might naturally grow amenable to sex without commitment. (Often in history—including Victorian England—women in the "lower classes" have had a reputation for loose morals.)[64] It is too soon to assert this confidently, or to infer that inner-city sexual mores would change markedly if income levels did. But it is noteworthy, at least, that evolutionary psychology, with its emphasis on the role of environment, may wind up highlighting the social costs of poverty, and thus at times lend strength to liberal policy prescriptions, defying old stereotypes of Darwinism as right-wing.

Of course, one could argue that various policy implications ensue from any given theory. And one can dream up wholly different kinds of Darwinian theories about how sexual strategies get shaped.[65] The one thing one *can't* do, I submit, is argue that evolutionary psychology is irrelevant to the whole discussion. The idea that natural selection, acutely attentive to the most subtle elements of design in the lowliest animals, should build huge, exquisitely pliable brains and not make them highly sensitive to environmental cues regarding sex, status, and various other things known to figure centrally in our reproductive prospects—that idea is literally incredible. If we want to know when and how a person's character begins to assume distinct shape, if we want to know how resistant to change the character will subsequently be, we have to look to Darwin. We don't yet know the answers, but we know where they'll come from, and that knowledge helps us phrase the questions more sharply.

THE FAMILY THAT STAYS TOGETHER

Much of the attention paid to the "short-term" sexual strategies of women—whether unattached women willing to settle for a one-night stand, or attached women sneaking out on a mate—is fairly recent. Sociobiological discussion during the 1970s, at least in its popular form, tended to depict men as wild, libidinous creatures who roamed the landscape looking for women to dupe and exploit; women were often depicted as dupes and exploitees. The shift in focus is due largely to the growing number of female Darwinian social scientists, who have patiently explained to their male colleagues how a woman's psyche looks from the inside.

Even after this restoration of balance, there remains one important sense in which men and women will tend to be, respectively, exploiter and exploited. As a marriage progresses, the temptation to desert should—in the *average* case—shift toward the man. The reason isn't, as people sometimes assume, that the Darwinian *costs* of marital breakup are greater for the woman. True, if she has a young child and her marriage dissolves, that child may suffer—whether because she can't find a husband willing to commit to a woman with another man's child, or because she finds one who neglects or mistreats the child. But, in Darwinian terms, this cost is borne equally by the deserting husband; the child who will thus suffer is his child too, after all.

The big difference between men and women comes, rather, on the *benefits* side of the desertion ledger. What can each partner gain from a breakup in the way of future reproductive payoff? The husband can, in principle, find an eighteen-year-old woman with twenty-five years of reproduction ahead. The wife—even aside from the trouble she'll have finding a husband if she already has a child—cannot possibly find a mate who will give her twenty-five years worth of reproductive potential. This difference in outside opportunity is negligible at first, when both husband and wife are young. But as they age, it grows.

Circumstance can subdue or heighten it. A poor, low-status husband may not have a chance to desert and may, indeed, provide his wife with reason to desert, especially if she has no children and can thus find another mate readily. A husband who rises in status and wealth, on the other hand, will thus strengthen his incentive to desert while weakening his wife's. But all other things being equal, it is the husband's restlessness that will tend to grow as the years pass.

All this talk of "desertion" may be misleading. Though divorce is available in many hunter-gatherer cultures, so is polygyny; in the ancestral environment, gaining a second wife didn't necessarily mean leaving the first. And so long as it didn't, there was no good Darwinian reason to desert. Staying near offspring, giving protection and guidance, would have made more genetic sense. Thus, males may be designed less for opportune desertion than for opportune polygyny. But in the modern environment, with institutionalized monogamy, a polygynous impulse will find other outlets, such as divorce.

As a mother's children grow self-sufficient, the urgency of hanging onto male parental investment drops. There are plenty of middle-aged women who, especially if they're financially secure, can take or leave their husbands. Still, there is no Darwinian force *driving* them to leave their husbands, nothing about leaving him that will sharply advance their genetic interests. The thing most likely to drive a post-menopausal woman out of a marriage is her husband's malicious marital discontent. Many a woman seeks divorce, but that doesn't mean *her* genes are ultimately the problem.

Among all the data on contemporary marriage, two items stand out as especially telling. First is the 1992 study which found that the husband's dissatisfaction with a marriage is the single strongest predictor of divorce.[66] Second is that men are much more likely than women to remarry after a divorce.[67] The second fact—and the biological force behind the second fact—is probably a good part of the reason for the first.

Objections to this sort of analysis are predictable: "But people leave marriages for *emotional* reasons. They don't add up the number of their children and pull out their calculators. Men are driven away by dull, nagging wives, or by the profound soul-searching of a mid-life crisis. Women are driven out by abusive or indifferent husbands, or lured away by a sensitive, caring man."

All true. But, again: emotions are just evolution's executioners. Beneath all the thoughts and feelings and temperamental differences that marriage counselors spend their time sensitively assessing are the stratagems of the genes—cold, hard equations composed of simple variables: social status, age of spouse, number of children, their ages, outside opportunities, and so on. Is the wife *really* duller and more nagging than she was twenty years ago? Possibly, but it's also possible that the husband's tolerance for nagging has dropped now that she's forty-five and has no reproductive future. And the promotion he just got, which has already drawn some admiring glances from a young woman at work, hasn't helped. Similarly, we might ask the young childless wife who finds her husband intolerably insensitive why the insensitivity wasn't so oppressive a year ago, before he lost his job and she met the kindly, affluent bachelor who seems to be flirting with her. Of course, maybe her husband's abuses are quite real, in which case they signal his disaffection, and perhaps his impending departure—and merit just the sort of preemptive strike the wife is now mustering.

Once you start seeing everyday feelings and thoughts as genetic weapons, marital spats take on new meaning. Even the ones that aren't momentous enough to bring divorce are seen as incremental renegotiations of contract. The husband who on his honeymoon said he didn't want an "old-fashioned wife" now sarcastically suggests that it wouldn't be too taxing for her to cook dinner once in a while. The threat is as clear as it is implicit: I'm willing and able to break the contract if you're not willing to renegotiate.

PAIR BONDS REVISITED

All told, things are not looking good for Desmond Morris's version of the pair-bond hypothesis. We do not seem to be too much like our famously, just about unswervingly, monogamous primate relatives, the gibbons, to which we have been optimistically compared. This should come as no great surprise. Gibbons aren't very social. Each family lives on a large home range—sometimes more than a hundred acres—that buffers it from extramarital dalliances. And gibbons chase off any intruders that might want to steal or borrow a mate.[68] We, by contrast, have evolved in large social groups that are rife with genetically profitable alternatives to fidelity.

We do have, to be sure, the earmarks of high male parental investment. For hundreds of thousands of years, and maybe longer, natural selection has been inclining males to love their children, thus giving them a feeling females had been enjoying for the previous several hundred million years of mammalian evolution. Natural selection has also, during that time, been inclining men and women to love each other (or, at least, to "love" each other, with the meaning of that word varying greatly, and seldom approaching the constancy of devotion it reaches between parent and offspring). Still, love or no love, gibbons we aren't.

So what *are* we? Just how far from being naturally monogamous is our species? Biologists often answer this question anatomically. We've already seen anatomical evidence—testes weight and the fluctuations in sperm density—suggesting that human females are not devoutly monogamous by nature. There is also anatomical evidence bearing on the question of precisely how far from monogamous males naturally are. As Darwin noted, in highly polygynous species the contrast in body size between male and female—the "sexual dimorphism"—is great. Some males monopolize several females, while other males get shut out of the genetic sweepstakes altogether, so there is immense evolutionary value in being a big male, capable of intimidating other males. Male gorillas, who mate with lots of females if they win lots of fights and no females if they win none, are gargantuan—twice as heavy as females. Among the monogamous gibbons, small males breed about as prolifically as bigger ones, and sexual dimorphism is almost imperceptible. The upshot is that sexual dimorphism is a good index of the intensity of sexual selection among males, which in turn reflects how polygynous a species is. When placed on the spectrum of sexual dimorphism, humans get a "mildly polygynous" rating.[69] We're much less dimorphic than gorillas, a bit less than chimps, and markedly more than gibbons.

One problem with this logic is that competition among human, and even prehuman, males has been largely mental. Men don't have the long canine teeth that male chimps use to fight for alpha rank and thus supreme mating rights. But men do employ various stratagems to raise their social status, and thus their attractiveness. So some, and maybe much, of the polygyny in our evolutionary past would be reflected not in gross physiology but in distinctively male mental traits. If anything, the less than dramatic difference in size between men and women paints an overly flattering picture of men's monogamous tendencies.[70]

How have societies over the years coped with the basic sexual asymmetry in human nature? Asymmetrically. A huge majority—980 of the 1,154 past or present societies for which anthropologists have data—have permitted a man to have more than one wife.[71] And that number includes most of the world's hunter-gatherer societies, societies that are the closest thing we have to a living example of the context of human evolution.

The more zealous champions of the pair-bond thesis have been known to minimize this fact. Desmond Morris, hell-bent on proving the natural monogamy of our species, insisted in *The Naked Ape* that the only societies worth paying much attention to are modern industrial societies, which, coincidentally, fall into the 15 percent of societies that have been avowedly monogamous. "[A]ny society that has failed to advance has in some sense failed, 'gone wrong,' " he wrote. "Something has happened to it to hold it back, something that is working against the natural tendencies of the species. . . . " So "the small, backward, and unsuccessful societies can largely be ignored." In sum, said Morris (who was writing back when Western divorce rates were about half what they are now): "whatever obscure, backward tribal units are doing today, the mainstream of our species expresses its pair-bonding character in its most extreme form, namely long-term monogamous matings."[72]

Well, that's one way to get rid of unsightly, inconvenient data: declare them aberrant, even though they vastly outnumber the "mainstream" data.

Actually, there *is* a sense in which polygynous marriage has not been the historical norm. For 43 percent of the 980 polygynous cultures, polygyny is classified as "occasional." And even where it is "common," multiple wives are generally reserved for a relatively few men who can afford them or qualify for them via formal rank. For eons and eons, most marriages have been monogamous, even though most societies haven't been.

Still, the anthropological record suggests that polygyny is natural in the sense that men given the opportunity to have more than one wife are strongly inclined to seize it. The record also suggests something else: that polygyny has its virtues as a way of handling the basic imbalance between what men and women want. In our culture, when a man whose wife has given him a few children grows restless and "falls in love" with a younger woman, we say: Okay, you can marry her, but we insist that you desert your first wife and that a stigma be placed on your kids and that, if you don't make much money, your kids and former wife suffer miserably. Some other cultures have tended to say: Okay, you can marry her, but only if you can really afford a second family; and you can't desert your first family, and there won't be any stigma placed on your kids.

Maybe some of today's nominally monogamous societies, those in which half of all marriages actually fail, should just go whole hog. Maybe we should fully erase the already fading stigma of divorce.

Maybe we should simply make sure that men who stray from their families remain legally responsible for them, and keep supporting them in the manner to which they had become accustomed. Maybe we should, in short, permit polygyny. A lot of presently divorced women, and their children, might be better off.

The only way to intelligently address this option is to first ask a simple question (one that turns out to have a counterintuitive answer): How did a strict cultural insistence on monogamy, which seems to go against the grain of human nature, and several millennia ago was almost unheard of, ever come to be?

Chapter 4: **THE MARRIAGE MARKET**

It is hardly possible to read Mr. M'Lennan's work and not admit that almost all civilised nations still retain some traces of such rude habits as the forcible capture of wives. What ancient nation, as the same author asks, can be named that was originally monogamous?

—*The Descent of Man* (1871)[1]

Something about the world doesn't seem to make sense. On the one hand, it is run mostly by men. On the other hand, in most parts of it, polygamy is illegal. If men really are the sort of animals described in the two previous chapters, why did they let this happen?

Sometimes this paradox gets explained away as a compromise between the male and female natures. In an old-fashioned, Victorian-style marriage, men get routine subservience in exchange for keeping their wanderlust more or less under control. The wives cook, clean, take orders, and put up with all the unpleasant aspects of a regular male presence. In return, the husbands graciously agree to stick around.

This theory, however appealing, is beside the point. Granted, within any monogamous marriage there is compromise. And within any two-man prison cell there is compromise. But that doesn't mean prisons were invented by a compromise among criminals. Compromise between men and women is the way monogamy endures (when it does), but it's no explanation of how monogamy got here.

The first step toward answering the "Why monogamy" question is to understand that, for some monogamous societies on the anthropological record—including many hunter-gatherer cultures—the question isn't all that perplexing. These societies have hovered right around the subsistence level. In such a society, where little is stowed away for a rainy day, a man who stretches his resources between two families may end up with few or no surviving children. And even if he were willing to gamble on a second family, he'd have trouble attracting a second wife. Why should she settle for half of a poor man if she can have all of one? Out of love? But how often will love malfunction so badly? Its very purpose, remember, is to attract her to men who will be good for her progeny. Besides, why should her family—and in preindustrial societies families often shape a bride's "choice" forcefully and pragmatically—tolerate such foolishness?

Roughly the same logic holds if a society is somewhat above the subsistence level but all men are about equally above it. A woman who chooses half a husband over a whole one is still settling for much less in the way of material well-being.

The general principle is that economic equality among men—especially, but not only, if near the subsistence level—tends to short-circuit polygyny. This tendency by itself dispels a good part of the monogamy mystery, for more than half of the known monogamous societies have been classified as "nonstratified" by anthropologists.[2] What really demand explanation are the six dozen societies in the history of the world, including the modern industrial nations, that have been monogamous yet economically stratified. These are true freaks of nature.

The paradox of monogamy amid uneven affluence has been stressed especially by Richard Alexander, one of the first biologists to broadly apply the new paradigm to human behavior. When monogamy is found in subsistence-level cultures, Alexander calls it "ecologically imposed." When it appears in more affluent, more stratified cultures, he calls it "socially imposed."[3] The question is why society imposed it.

The term *socially imposed* may offend some people's romantic ideals. It seems to imply that, in the absence of bigamy laws, women would flock toward money, gleefully signing on as second or third wife so long as there was enough of it to go around. Nor is the term *flock* used lightly here. There is a tendency for polygyny to occur in bird species whose males control territories of sharply differing quality or quantity. Some female birds are quite happy to share a male so long as he has a lot more real estate than any male they could have monopolized.[4] Most human females would like to think they are guided by a more ethereal sort of love, and that they have somewhat more pride than a long-billed marsh wren.

And of course they do. Even in polygynous cultures, women are often less than eager to share a man. But, typically, they would rather do that than live in poverty with the undivided attention of a ne'er-do-well. It is easy for well-educated, upper-class women to scoff at the idea that any self-respecting woman would willingly suffer the degradation of polygyny, or to deny that women place great emphasis on a husband's income. But upper-class women seldom even *meet* a man with a low income, much less face the prospect of marrying one. Their milieu is so economically homogeneous that they don't have to worry about finding a minimally adequate provider; they can refocus their search, and spend their time pondering a prospective mate's taste in music and literature. (And these tastes are themselves cues to a man's socioeconomic status. This is a reminder that the Darwinian evaluation of a mate needn't be *consciously* Darwinian.)

In favor of Alexander's belief that there's something artificial about highly stratified yet monogamous societies is the fact that polygyny tends to lurk stubbornly beneath their surface. Though being a mistress is even today considered at least mildly scandalous, a number of women seem to prefer that role to the alternative: a greater commitment from a man of lesser means—or, perhaps, a commitment from no man.

Since Alexander began stressing the two kinds of monogamous societies, his distinction has drawn a second, more subtle, kind of support. The anthropologists Steven J. C. Gaulin and James S. Boster have shown that dowry—a transfer of assets from the bride's to the groom's family—is found almost exclusively in societies with socially imposed monogamy. Thirty-seven percent of these stratified nonpolygynous societies have had dowry, whereas 2 percent of all nonstratified nonpolygynous societies have. (For polygynous societies the figure is around 1 percent.)[5] Or, to put it another way: although only 7 percent of societies on record have had socially imposed monogamy, they account for 77 percent of societies with a dowry tradition. This suggests that dowry is the product of a market disequilibrium, a blockage of marital commerce; monogamy, by limiting each man to a single wife, makes wealthy men artificially precious commodities, and dowry is the price paid for them. Presumably, if polygyny were legalized, the market would right itself more straightforwardly: males with the most money (and perhaps with the most charm and the ruggedest physiques and whatever else might partly outweigh considerations of wealth) would, rather than fetch large dowries, have multiple wives.

WINNERS AND LOSERS

If we adopt this way of looking at things—if we abandon a Western ethnocentric perspective and hypothetically accept the Darwinian view that men (consciously or unconsciously) want as many sex-providing and child-making machines as they can comfortably afford, and women (consciously or unconsciously) want to maximize the resources available to their children—then we may have the key to explaining why monogamy is with us today: whereas a polygynous society is often depicted as something men would love and women would hate, there is really no natural consensus on the matter within either sex. Obviously, women who are married to a poor man and would rather have half of a rich one aren't well served by the institution of monogamy. And, obviously, the poor husband they would gladly desert wouldn't be well served by polygyny.

Nor are these superficially ironic preferences confined to people near the bottom of the income scale. Indeed, in sheerly Darwinian terms, *most* men are probably better off in a monogamous system and *most* women worse off. This is an important point, and warrants a brief illustrative detour.

Consider a crude and offensive but analytically useful model of the marital marketplace. One thousand men and one thousand women are ranked in terms of their desirability as mates. Okay, okay: there isn't, in real life, full agreement on such things. But there are clear patterns. Few women would prefer an unemployed and rudderless man to an ambitious and successful one, all other things being even roughly equal; and few men would choose an obese, unattractive, and dull woman over a shapely, beautiful, sharp one. For the sake of intellectual progress, let's simplemindedly collapse these and other aspects of attraction into a single dimension.

Suppose these 2,000 people live in a monogamous society and each woman is engaged to marry the man who shares her ranking. She'd like to marry a higher-ranking man, but they're all taken by competitors who outrank her. The men too would like to marry up, but for the same reason can't. Now, before any of these engaged couples gets married, let's legalize polygyny and magically banish its stigma. And let's suppose that at least one woman who is mildly more desirable than average—a quite attractive but not overly bright woman with a ranking of, say, 400—dumps her fiancé (male #400, a shoe salesman) and agrees to become the second wife of a successful lawyer (male #40). This isn't wildly implausible—forsaking a family income of around $40,000 a year, some of which she would have to earn herself by working part-time at a Pizza Hut, for maybe $100,000 a year and no job requirement (not to mention the fact that male #40 is a better dancer than male #400).[6]

Even this first trickle of polygynous upward mobility makes most women better off and most men worse off. All 600 women who ranked below the deserter move up one notch to fill the vacuum; they still get a husband all to themselves, and a better husband at that. Meanwhile 599 men wind up with a wife slightly inferior to their former fiancées—and one man now gets no wife at all. Granted: in real life, the women wouldn't move up in lockstep. Very early in the process, you'd find a woman who, pondering the various intangibles of attraction, would stand by her man. But in real life, you'd probably have more than a trickle of upward mobility in the first place. The basic point stands: many, many women, even many women *who will choose not to share a husband*, have their options expanded when all women are free to share a husband.[7] By the same token, many, many men can suffer at the hands of polygyny.

All told, then, institutionalized monogamy, though often viewed as a big victory for egalitarianism and for women, is emphatically not egalitarian in its effects on women. Polygyny would much more evenly distribute the assets of males among them. It is easy—and wise—for beautiful, vivacious wives of charming, athletic corporate titans to dismiss polygyny as a violation of the basic rights of women. But married women living in poverty—or women without a husband or child, and desirous of both—could be excused for wondering just which women's rights are protected by monogamy. The only under-privileged citizens who should favor monogamy are men. It is what gives them access to a supply of women that would otherwise drift up the social scale.

So neither gender, as a whole, belongs on either side of the im-aginary bargaining table that yielded the tradition of monogamy. Monogamy is neither a minus for men collectively nor a plus for women collectively; within both sexes, interests naturally collide. More plausibly, the grand, historic compromise was cut between more fortunate and less fortunate men. For them, the institution of monogamy does represent a genuine compromise: the most fortunate men still get the most desirable women, but they have to limit them-selves to one apiece. This explanation of monogamy—as a divvying up of sexual property among men—has the virtue of consistency with the fact that opened this chapter: namely, that it is men who usually control sheerly political power, and men who, historically, have cut most of the big political deals.

This is not to say, of course, that men ever sat down and ham-mered out the one-woman-per-man compromise. The idea, rather, is that polygyny has tended to disappear in response to egalitarian values—not values of equality between the sexes, but of equality among men. And maybe "egalitarian values" is too polite a way of putting it. As political power became distributed more evenly, the hoarding of women by upper-class men simply became untenable. Few things are more anxiety-producing for an elite governing class than gobs of sex-starved and childless men with at least a modicum of political power.

This thesis remains only a thesis.[8] But reality at least loosely fits it. Laura Betzig has shown that in preindustrial societies, extreme polygyny often goes hand in hand with extreme political hierarchy, and reaches its zenith under the most despotic regimes. (Among the Zulu, whose king might monopolize more than a hundred women, coughing, spitting, or sneezing at his dining table was punishable by death.) And the allocation of sexual resources by political status has often been fine-grained and explicit. In Inca society, the four political offices from petty chief to chief were allotted ceilings of seven, eight, fifteen, and thirty women, respectively.[9] It stands to reason that as political power became more widely disbursed, so did wives. And the ultimate widths are one-man-one-vote and one-man-one-wife. Both characterize most of today's industrialized nations.

Right or wrong, this theory of the origin of modern institutionalized monogamy is an example of what Darwinism has to offer historians. Darwinism does not, of course, explain history *as* evolution; natural selection doesn't work nearly fast enough to drive ongoing change at the level of culture and politics. But natural selection did shape the minds that *do* drive cultural and political change. And understanding how it shaped those minds may afford fresh insight into the forces of history. In 1985 the eminent social historian Lawrence Stone published an essay that stressed the epic significance of the early Christian emphasis on the fidelity of husbands and the permanence of marriage. After reviewing a couple of theories as to how this cultural innovation spread, he concluded that the answer "remains obscure."[10] Perhaps a Darwinian explanation—that, given human nature, monogamy is a straightforward expression of political equality among men—deserved at least a mention. It may be no accident that Christianity, which served as a vehicle for monogamy politically as well as intellectually, has often pitched its message to poor and powerless men.[11]

WHAT'S WRONG WITH POLYGYNY?

This Darwinian analysis of marriage complicates the choice between monogamy and polygyny. For it shows that the choice isn't between equality and inequality. The choice is between equality among men and equality among women. A tough call.

There are several conceivable reasons to vote for equality among men (that is, monogamy). One is to dodge the wrath of the various feminists who will not be convinced that polygyny liberates down-trodden women. Another is that monogamy is the only system that, theoretically at least, can provide a mate for just about everyone. But the most powerful reason is that leaving lots of men without wives and children is not just inegalitarian; it's dangerous.

The ultimate source of the danger is sexual selection among males. Men have long competed for access to the scarcer sexual resource, women. And the costs of losing the contest are so high (genetic oblivion) that natural selection has inclined them to compete with special ferocity. In all cultures, men wreak more violence, including murder, than women. (Indeed, across the animal kingdom, males are the more belligerent sex, *except* in those species, such as phalaropes, where male parental investment is so high females can reproduce more often than males.) Even when the violence isn't against a sexual rival, it often boils down to sexual competition. A trivial dustup may es-calate until one man kills another to "save face"—to earn the sort of raw respect that, in the ancestral environment, could have raised status and brought sexual rewards.[12]

Fortunately, male violence can be dampened by circumstance. And one circumstance is a mate. We would expect womanless men to compete with special ferocity, and they do. An unmarried man between twenty-four and thirty-five years of age is about three times as likely to murder another male as is a married man the same age. Some of this difference no doubt reflects the kinds of men that do and don't get married to begin with, but Martin Daly and Margo Wilson have argued cogently that a good part of the difference may lie in "the pacifying effect of marriage."[13]

Murder isn't the only thing an "unpacified" man is more likely to do. He is also more likely to incur various risks—committing robbery, for example—to gain the resources that may attract women. He is more likely to rape. More diffusely, a high-risk, criminal life often entails the abuse of drugs and alcohol, which may then com-pound the problem by further diminishing his chances of ever earning enough money to attract women by legitimate means.[14]

This is perhaps the best argument for monogamous marriage, with its egalitarian effects on men: inequality among males is more socially destructive—in ways that harm women *and* men—than inequality among women. A polygynous nation, in which large numbers of low-income men remain mateless, is not the kind of country many of us would want to live in.

Unfortunately, this is the sort of country we already live in. The United States is no longer a nation of institutionalized monogamy. It is a nation of serial monogamy. And serial monogamy in some ways amounts to polygyny.[15] Johnny Carson, like many wealthy, high-status males, spent his career monopolizing long stretches of the reproductive years of a series of young women. Somewhere out there is a man who wanted a family and a beautiful wife and, if it hadn't been for Johnny Carson, would have married one of these women. And if this man has managed to find another woman, she was similarly snatched from the jaws of some other man. And so on—a domino effect: a scarcity of fertile females trickles down the social scale.

As abstractly theoretical as this sounds, it really can't help but happen. There are only about twenty-five years of fertility per woman. When some men dominate more than twenty-five years' worth of fertility, some man, somewhere, must do with less. And when, on top of all the serial husbands, you add the young men who live with a woman for five years before deciding not to marry her, and then do it again (perhaps finally, at age thirty-five, marrying a twenty-eight-year-old), the net effect could be significant. Whereas in 1960 the fraction of the population age forty or older that had never married was about the same for men and women, by 1990 the fraction was markedly larger for men than for women.[16]

It is not crazy to think that there are homeless alcoholics and rapists who, had they come of age in a pre-1960s social climate, amid more equally distributed female resources, would have early on found a wife and adopted a lower-risk, less destructive lifestyle. Anyway, you don't have to buy this illustration to buy the point itself: if polygyny would indeed have pernicious effects on society's less fortunate men, and indirectly on the rest of us, then it isn't enough to just oppose legalized polygyny. (Legalized polygyny wasn't a looming political threat last time I checked, anyway.) We have to worry about the de facto polygyny that already exists. We have to ask not whether monogamy can be saved, but whether it can be restored. And we might be enthusiastically joined in this inquiry not only by discontented wifeless men, but by a large number of discontented former wives—especially the ones who had the bad fortune to marry someone less wealthy than Johnny Carson.

DARWINISM AND MORAL IDEALS

This view of marriage is a textbook example of how Darwinism can and can't legitimately enter moral discourse. What it can't do is furnish us with basic moral values. Whether, for example, we want to live in an egalitarian society is a choice for us to make; natural selection's indifference to the suffering of the weak is not something we need emulate. Nor should we care whether murder, robbery, and rape are in some sense "natural." It is for us to decide how abhorrent we find such things and how hard we want to fight them.

But once we've made such choices, once we *have* moral ideals, Darwinism can help us figure out which social institutions best serve them. In this case, a Darwinian outlook shows the prevailing marital institution, serial monogamy, to be in many ways equivalent to polygyny. As such, this institution is seen to have inegalitarian effects on men, working against the disadvantaged. Darwinism also highlights the costs of this inequality—violence, theft, rape.

In this light, old moral debates assume a new cast. For example, the tendency of political conservatives to monopolize the argument for "family values" starts to look odd. Liberals, concerned about the destitute, and about the "root causes" of crime and poverty, might logically develop a certain fondness for "family values." For a drop in the divorce rate, by making more young women accessible to low-income men, might keep an appreciable number of men from falling into crime, drug addiction, and, sometimes, homelessness.

Of course, given the material opportunities that polygyny (even de facto polygyny) may afford poor women, one can also imagine a liberal argument *against* monogamy. One can even imagine a liberal *feminist* argument against monogamy. And, in any event, one can see that a Darwinian feminism will be a more complicated feminism.

Viewed in Darwinian terms, "women" are not a naturally coherent interest group; there is no single sisterhood.[17]

There is one other kind of fallout from current marital norms that comes into focus through the new paradigm: the toll taken on children. Martin Daly and Margo Wilson have written, "Perhaps the most obvious prediction from a Darwinian view of parental motives is this: Substitute parents will generally tend to care less profoundly for children than natural parents." Thus, "children reared by people other than their natural parents will be more often exploited and otherwise at risk. Parental investment is a precious resource, and selection must favor those parental psyches that do not squander it on nonrelatives."[18]

To some Darwinians, this expectation might seem so strong as to render its verification a waste of time. But Daly and Wilson took the trouble. What they found surprised even them. In America in 1976, a child living with one or more substitute parents was about one hundred times more likely to be fatally abused than a child living with natural parents. In a Canadian city in the 1980s, a child two years of age or younger was seventy times more likely to be killed by a parent if living with a stepparent and natural parent than if living with two natural parents. Of course, murdered children are a tiny fraction of the children living with stepparents; the divorce and re-marriage of a mother is hardly a child's death warrant. But consider the more common problem of nonfatal abuse. Children under ten were, depending on their age and the particular study in question, between three and forty times more likely to suffer parental abuse if living with a stepparent and a natural parent than if living with two natural parents.[19]

It is fair to infer that many less dramatic, undocumented forms of parental indifference follow this rough pattern. After all, the whole reason natural selection *invented* paternal love was to bestow benefits on offspring. Though biologists call these benefits "investment," that doesn't mean they're strictly material, wholly sustainable through monthly checks. Fathers give their children all kinds of tutelage and guidance (more, often, than either father or child realizes) and guard them against all kinds of threats. A mother alone simply can't pick up the slack. A stepfather almost surely won't pick up much, if any of it. In Darwinian terms, a young stepchild is an obstacle to fitness, a drain on resources.

There are ways to fool mother nature, to induce parents to love children that aren't theirs. (Hence cuckoldry.) After all, people can't telepathically sense that a child is carrying their genes. Instead, they rely on cues that, in the ancestral environment, would have signified as much. If a woman feeds and cuddles an infant day after day, she may grow to love the child, and so may a man who has been sleeping with her for years. This sort of bonding is what makes adopted children lovable and nannies loving. But both theory and casual observation suggest that, the older a child is when first seen by the substitute parent, the less likely deep attachment is. And a large majority of children who acquire stepfathers are past infancy.

One can imagine arguments among reasonable and humane people over whether a strongly monogamous society is better than a strongly polygynous one. But this much seems less controversial: whenever marital institutions—in either kind of society—are allowed to dissolve, so that divorce and unwed motherhood are rampant, and many children no longer live with both natural parents, there will ensue a massive waste of the most precious evolutionary resource: love. Whatever the relative merits of monogamy and polygyny, what we have now—serial monogamy, de facto polygyny—is, in an important sense, the worst of all worlds.

PURSUING MORAL IDEALS

Obviously, Darwinism won't always simplify moral and political debate. In this case, by stressing the tension between equality among men and among women, it actually complicates the question of which marital institutions best serve our ideals. Still, the tension was always there; at least now it's in the open, and debate can proceed in stronger light. Further, once we *have* decided, with help from the new paradigm, which institutions best serve our moral ideals, Darwinism can make its second kind of contribution to moral discourse: it can help us figure out what sorts of forces—which moral norms, which social policies—help nourish those institutions.

And here comes another irony in the "family values" debate: conservatives may be surprised to hear that one of the best ways to strengthen monogamous marriage is to more equally distribute income.[20] Young single women will feel less inclined to tempt husband A away from wife A if bachelor B has just as much money. And husband A, if he's not drawing flirtatious looks from young women, may feel more content with wife A, and less inclined to notice her wrinkles. This dynamic presumably helps explain why monogamous marriage has often taken root in societies with little economic stratification.

One standard conservative argument against antipoverty policies is their cost: taxes burden the affluent and, by reducing their incentive to work, lower overall economic output. But if one goal of the policy is to bolster monogamy, then making the affluent less affluent is a welcome side effect. Monogamy is threatened not just by poverty in an absolute sense but also by the relative wealth of the wealthy. That reducing this wealth cuts overall economic output may, of course, still be regrettable; but once we add more stable marriages to the benefits of income redistribution, the regret should lose a bit of its sting.

One might imagine that this whole analysis is steadily losing its relevance. After all, as more women enter the workforce, they can better afford to premise their marital decisions on something other than the man's income. But remember: we're dealing with women's deep romantic attractions, not just their conscious calculation, and these feelings were forged in a different environment. To judge by hunter-gatherer societies, males during human evolution controlled most of the material resources. And even in the poorest of these societies, where disparities in male wealth are hard to detect, a father's social status often translates subtly into advantages for offspring, material and otherwise, in ways that a mother's social status doesn't.[21] Though a modern woman can of course reflect on her wealth and her independently earned status, and try to gauge marital decisions accordingly, that doesn't mean she can easily override the deep aesthetic impulses that had such value in the ancestral environment. In fact, modern women manifestly do not override them. Evolutionary psychologists have shown that the tendency of women to place greater emphasis than men on a mate's financial prospects persists regardless of the income or expected income of the women in question.[22]

So long as a society remains economically stratified, the challenge of reconciling lifelong monogamy with human nature will be large. Incentives and disincentives (moral and/or legal) may be necessary. One way to see what sorts of incentives can work is to look at an economically stratified society where they worked. Say, for example, Victorian England. To search for the peculiarities of Victorian morality that helped marriages succeed (at least in the minimal sense of not dissolving) isn't to say we should adopt those peculiarities ourselves. One can see the "wisdom" of some moral tenet—see how it achieves certain goals by implicitly recognizing deep truths about human nature—without finding it, on balance, worth the side effects. But seeing the wisdom is still a good way to appreciate the contours of the challenge it met. Looking at a Victorian marriage—Charles and Emma Darwin's—from a Darwinian point of view is worth the effort.

Before we return to Darwin's life, one caution is in order. So far we've been analyzing the human mind in the abstract; we've talked about "species-typical" adaptations designed to maximize fitness. When we shift our focus from the whole species to any one individual, we should *not* expect that person to chronically maximize fitness, to optimally convey his or her genes to future generations. And the reason goes beyond the one that has so far been stressed: that most human beings don't live in an environment much like the one for which their minds were designed. Environments—even the environments for which organisms *are* designed—are unpredictable. That is why behavioral flexibility evolved in the first place. And unpredictability, by its nature, cannot be mastered. As John Tooby and Leda Cosmides have put it, "Natural selection cannot directly 'see' an individual organism in a specific situation and cause behavior to be adaptively tailored."[23]

The best that natural selection can do is give us adaptations—"mental organs" or "mental modules"—that play the odds. It can give males a "love of offspring" module, and make that module sensitive to the likelihood that the offspring in question is indeed the man's. But the adaptation cannot be foolproof. Natural selection can give women an "attracted to muscles" module, or an "attracted to status" module, and, what's more, it can make the strength of those attractions depend on all kinds of germane factors; but even a highly flexible module can't guarantee that these attractions translate into viable and prolific offspring.

As Tooby and Cosmides say, human beings aren't general purpose "fitness maximizers." They are "adaptation executers."[24] The adaptations may or may not bring good results in any given case, and success is especially spotty in environments other than a small hunter-gatherer village. So as we look at Charles Darwin, the question isn't: Can we conceive of things he could have done to have more viable, prolific offspring than he had? The question is: Is his behavior intelligible as the product of a mind that consists of a bundle of adaptations?

Chapter 5: **DARWIN'S MARRIAGE**

*Like a child that has something it loves beyond measure, I long
to dwell on the words* my own *dear Emma. . . . My own dear
Emma, I kiss the hands with all humbleness and gratitude, which
have so filled up for me the cup of happiness . . . but do dear
Emma, remember life is short, and two months is the sixth part
of the year.*

—Darwin in November 1838, in a letter to his fiancée,
urging an early wedding

Sexual desire makes saliva to flow . . . curious association.

—Darwin in his scientific notebook, same month, same year[1]

In the decade of Darwin's marriage, the 1830s, the number of British
couples filing for divorce averaged four per year. This is in some
ways a misleading statistic. It probably reflects, in part, the tendency
of men back then to die before reaching the climax of their midlife
crises. (A misnomer; often the wife's middle age, more than the man's,
brings on the crisis.) And it certainly reflects the fact that getting a
divorce required—literally—an act of Parliament. Marriages also
ended in other ways, especially through privately arranged separations. Still, there's no denying that marriage back then was, by and
large, for keeps, especially within Darwin's upper-middle-class social
stratum. And marriage stayed that way for half a century after 1857,
the year the Divorce Act made breaking up easier to do.[2] There was

something about Victorian morality conducive to staying married.

There is no telling how much misery was generated in Victorian England by unhappy, unendable marriages. But it may well not exceed the misery generated by modern marital dissolution.[3] In any case, we do know of some Victorian marriages that seem to have been successful. Among these is the marriage of Charles and Emma Darwin. Their devotion was mutual and seems, if anything, to have strengthened with time. They created seven children who lived to adulthood, and none of them penned nasty memoirs about tyrannical parents ("Darwin Dearest"). Their daughter Henrietta called their marriage a "perfect union."[4] Their son Francis wrote of his father: "In his relationship towards my mother, his tender and sympathetic nature was shown in its most beautiful aspect. In her presence he found his happiness, and through her, his life,—which might have been overshadowed by gloom,—became one of content and quiet gladness."[5] Viewed from today, the marriage of Charles and Emma Darwin appears almost idyllic in its geniality, tranquility, and sheer durability.

DARWIN'S PROSPECTS

On the Victorian marriage market, Darwin must have been a fairly desirable commodity. He had a winning disposition, a respectable education, a family tradition that augured well for his career, and, in any event, his looming inheritance. He wasn't notably handsome, but so what? The Victorians were very clear about their division of aesthetic labor, and it was consistent with evolutionary psychology: good financial prospects made for an attractive husband, and good looks made for an attractive wife. In the large correspondence between Darwin and his sisters—while he was at college, or, later, on the *Beagle*—there is much talk of romance, as his sisters report gossip and relay any reconnaissance work they've done on his behalf. Almost invariably, men are gauged by their ability to provide materially for a woman, while women are seen as providing a pleasant visual and auditory environment for a man. Newly betrothed women, and the women who qualify as prospects for Charles, are "pretty" or "charming" or, at the very least, "pleasant." "I am sure you would like her," Charles's sister Catherine wrote of one candidate. "She is so very merry and pleasant, and I think very pretty." Newly betrothed men, on the other hand, are either of means or not of means. Susan Darwin wrote to her brother during his voyage: "Your charming Cousin Lucy Galton is engaged to marry Mr. Moilliet: the eldest son of a *very fat* Mrs. Moilliet. . . . The young Gentleman has a good fortune, so of course the match gives great satisfaction."[6]

The *Beagle*'s voyage lasted longer than expected, and Darwin wound up spending five years—the heart of his twenties—away from England. But, like undistinguished looks, advancing age was something men didn't have to worry greatly about. Women of Darwin's class often spent their early twenties on prominent display, hoping to catch a man while in their prime. Men often spent their twenties as Darwin did—singlemindedly pursuing the sort of professional stature (and/or money) that might later attract a woman in her prime. There was no rush. It was considered natural for a woman to marry a markedly older man, whereas a Victorian man who married a much older woman was cause for dismay. While Darwin was aboard the *Beagle*, his sister Catherine reported that cousin Robert Wedgwood, who was around Darwin's age, had "fallen vehemently and desperately in love with Miss Crewe, who is 50 years old, and blind of one eye." His sister Susan chimed in sarcastically, "just 20 years difference in their ages!" And sister Caroline: "a woman more than old enough to be his Mother." Catherine had a theory: "She is a clever woman, and must have entrapped him by her artifices; & she has the remains of great beauty to help her."[7] In other words: the man's age-detection system was functioning as designed, but he happened upon a woman whose enduring beauty—that is, youthful appearance—fooled it.

The realm within which the young Darwin was likely to find a mate was not vast. From adolescence on, the likely candidates came from two well-to-do families not far from the Darwin home in Shrewsbury. There was the ever-popular Fanny Owen—"the prettiest, plumpest, Charming" Fanny Owen, as Darwin described her during college.[8] And then there were the three youngest daughters of Josiah Wedgwood II, Darwin's maternal uncle: Charlotte, Fanny, and Emma.[9]

As of the *Beagle*'s departure, no one seems to have picked Emma as the frontrunner for Charles's affections—though his sister Caroline, in a letter sent to him around then, did note in passing that "Emma is looking very pretty & chats very pleasantly."[10] (What more could a man ask for?) As fate would have it, the other three candidates dropped out of the running in short order.

First to go was Emma's sister Charlotte. In January of 1832, she wrote Darwin to announce her unexpected engagement to a man who, she admitted, had "only a very small income now" but stood to inherit much upon his grandmother's death, and, anyway, had "high principles & kind nature which gives me a feeling of security. . . ."[11] (Translation: resources in the offing, and a reliable willingness to invest them parentally.) Charlotte had, in truth, probably been a dark horse as far as Charles was concerned. Though she had impressed both him and his brother Erasmus—they referred to her as "the incomparable"—she was more than a decade older than Charles; Erasmus was probably more smitten by her (as he seems to have been by a series of women, none of whom he managed to marry).

More unsettling than Charlotte's fate, probably, was the almost simultaneous news that the beguiling Fanny Owen was also to take the plunge. Fanny's father wrote to Charles of the news, plainly disappointed that the groom was "now not very rich & indeed probably never will be."[12] On the other hand, her husband was a man of status, serving briefly in Parliament.

Darwin, responding to all this matrimonial news in a letter to his sister Caroline, made no pretense of happiness. "Well it may be all very delightful to those concerned, but as I like unmarried women better than those in the blessed state, I vote it a bore."[13]

The vision that Darwin's sisters had of his future—becoming a country parson and settling down with a good wife—was not growing more likely as potential wives fell by the wayside. Catherine surveyed the remains, Emma and Fanny Wedgwood, and gave the nod to Fanny. She wrote to Charles that she hoped Fanny would still be single when he returned—"a nice little invaluable Wife she would be."[14] We'll never know. She fell ill and died within a month, at age twenty-six. With three of the four contenders either married or buried, the odds shifted decisively in Emma's favor.

If Charles had long-standing designs on Emma, he hid them well. He had predicted, as Catherine recalled, that upon his return he would find Erasmus "tied neck and heels to Emma Wedgwood, and heartily sick of her." In 1832, Catherine wrote to Charles that "I am much amused at your prophecy, and I think it may possibly have a good effect, and prevent its own fulfillment."[15] Erasmus continued to show interest in Emma, but she was still available when the *Beagle* returned to England in 1836. In fact, you might say she was emphatically available. She had been a carefree twenty-three when the *Beagle* set sail, and over the next couple of years had gotten several marriage proposals. But now she was a year and a half from thirty and spending much time at home caring for her invalid mother; she wasn't getting quite the exposure she had once gotten.[16] In preparation for Darwin's arrival, she wrote to her sister-in-law, she was reading a book on South America "to get up a little knowledge for him."[17]

There was cause to wonder whether "a little knowledge" would be enough to keep Charles's attention focused on childhood friends. Upon returning, he possessed something that women in all cultures and at all times seem to have prized in men: status. He had always had high social standing by virtue of his family's rank, but now he had a prominence all his own. From the *Beagle*, Darwin had sent back fossils, organic specimens, and acute observations on geology that had won a large scientific audience. He now rubbed elbows with the great naturalists of the day. By the spring of 1837, he had settled into London, in bachelor's quarters a few doors down from his brother Erasmus, and was in social demand.

A person of greater vanity and less certain purpose might have been drawn into a time-sapping social whirl—a corruption that the gregarious Erasmus would have been delighted to abet. Certainly Darwin was aware of his growing stature ("I was quite a lion there," he reported of a visit to Cambridge). But he was too measured and earnest a man to be a mingler by nature. As often as not, he saw fit to forgo large gatherings. He would, he told his mentor, Professor John Henslow, "prefer paying you a quiet visit to meeting all the world at a great Dinner." A note of demurral to Charles Babbage, the mathematician who designed the "analytical engine," forerunner of the digital computer, began, "My dear Mr. Babbage, I am very much obliged to you for sending me cards for your parties, but I am afraid of accepting them, for I should meet some people there, to whom I have sworn by all the saints in Heaven, I never go out. . . ."[18]

With the time thus saved, Darwin embarked on a remarkable burst of accomplishment. Within two years of his return to England, he: (1) edited his shipboard journal into a publishable volume (which reads nicely, sold well, and is today in print, further abridged, under the title *Voyage of the Beagle*); (2) skillfully extracted a thousand-pound subsidy from the Chancellor of the Exchequer for publication of *The Zoology of the Voyage of H.M.S. Beagle* and lined up contributors to it; (3) consolidated his place in British science by presenting half a dozen papers, ranging from a sketch of a new species of American ostrich (named *Rhea darwinii* by the Zoological Society of London) to a new theory about the formation of topsoil ("every particle of earth forming the bed from which the turf in old pasture land springs, has passed through the intestines of worms");[19] (4) went on a geological expedition to Scotland; (5) hobnobbed with eminences at the exclusive men's club the Athenaeum; (6) was elected secretary of the Geological Society of London (a post he accepted reluctantly, fearing its demand on his time); (7) compiled scientific notebooks—on subjects ranging from the "species question" to religion to human moral faculties—of such high intellectual density that they would serve as the basis for his largest works in the ensuing four decades; and (8) thought up the theory of natural selection.

CHOOSING MARRIAGE

It was toward the end of this phase—a few months before natural selection dawned on him—that Darwin decided to marry. Not necessarily anyone in particular; it isn't clear that he had Emma Wedgwood even remotely in mind, and one common view is that she wasn't at the center of his thinking on the subject. In a remarkable deliberative memorandum, apparently composed around July of 1838, he decided the matter of marriage in the abstract.

The document has two columns, one labeled *Marry*, one labeled *Not Marry*, and above them, circled, are the words "This is the Question." On the pro-marriage side of the equation were "Children—(if it Please God)—Constant companion, (&friend in old age) who will feel interested in one,—object to be beloved & played with."

After reflection of unknown length, he modified the foregoing sentence with "better than a dog anyhow." He continued: "Home, & someone to take care of house— Charms of music & female chit-chat— These things good for one's health.— *but terrible loss of time*." Without warning, Darwin had, from the pro-marriage column, swerved uncontrollably into a major anti-marriage factor, so major that he underlined it. This issue—the infringement of marriage on his time, especially his work time—was addressed at greater length in the appropriate, *Not Marry* column. Not marrying, he wrote, would preserve "Freedom to go where one like—choice of Society & *little of it*.—Conversation of clever men at clubs—not forced to visit relatives, & to bend in every trifle—to have the expense & anxiety of children—Perhaps quarelling—*Loss of time*.—cannot read in the Evenings—fatness & idleness—Anxiety & responsibility— less money for books &c—if many children forced to gain one's bread."

Yet the pro-marriage forces carried the day, with this train of thought at the end of the *Marry* column: "My God, it is intolerable to think of spending ones whole life, like a neuter bee, working, working, & nothing after all.—No, no won't do.—Imagine living all one's day solitarily in smoky dirty London House.—Only picture to yourself a nice soft wife on a sofa with good fire, & books & music perhaps." After recording these images he wrote: "Marry-Mary[sic]-Marry Q.E.D."

Darwin's decision had to survive one more wave of doubt. The backlash began innocently enough, as Darwin wrote, "It being proved necessary to Marry, When? Soon or Late." But this question incited that final spasm of panic with which many grooms are familiar. Brides are familiar with it too, of course, but their doubt seems more often to be whether their choice of a lifelong mate is the right one. For men, as Darwin's memo attests, the panic isn't essentially related to any particular prospective mate; it is the *concept* of a lifelong mate that is at some level frightening. For—in a monogamous society, at least—it dampens the prospects for intimacy with all those other women that a man's genes urge him to find and get to know (however briefly).

This isn't to say that the premarital panic fixes itself coarsely on images of would-be sex partners; the subconscious can be more subtle than that. Still, there is, somewhat reliably among men who are about to pledge themselves to one woman for life, a dread of impending entrapment, a sense that the days of adventure are over. "Eheu!!" Darwin wrote, with one final shudder in the face of lifelong commitment. "I never should know French,—or see the Continent—or go to America, or go up in a Balloon, or take solitary trip in Wales—poor slave—you will be worse than a negro." But then, fatefully, he mustered the necessary resolve. "Never mind my boy—Cheer up—One cannot live this solitary life, with groggy old age, friendless & cold, & childless staring one in ones face, already beginning to wrinkle.—Never mind, trust to chance—keep a sharp look out—There is many a happy slave—" End of document.[20]

CHOOSING EMMA

Darwin had written an earlier deliberative memorandum, probably in April, in which he rambled on about career paths—teach at Cambridge? in geology? in zoology? or "Work at transmission of Species"?—and pondered the marriage question inconclusively.[21] There is no way of knowing what drove him to reopen the question and this time settle it. But it's intriguing that, of the six entries made between April and July in his sporadically kept personal journal, two say he was feeling "unwell." Unwellness was to become a way of life for Darwin, a fact he may already have suspected. It is ironic that hints of mortality can draw a man into marriage, for often it is these same hints, much later, that drive him out, to seek fresh proof of his virility. But the irony dissolves when reduced to ultimate cause: both the impulses to profess lifelong love to a woman and to wander lie within a man by virtue of how often, in his ancestors, they led to progeny. In that sense, both are an apt antidote to mortality, though in the end futile (except from the genes' point of view), and, in the latter case—wandering—often destructive as well.

Anyway, on a less philosophical plane: Darwin may have sensed that he would before long need a devoted helpmate and nurse. And perhaps he even had a glimmer of spending many years working, in patient and needy solitude, on a big book about evolution. As Darwin's health had gotten worse, his grasp of that subject had gotten better. He opened his first notebook on "transmutation of Species" in June or July of 1837, and his second in early 1838.[22] By the time he was rigorously mulling marriage, he had gone some of the way toward natural selection. He believed that one key to evolution lay in initially slight hereditary difference; that when a species is divided into two populations by, say, a body of water, what are at first merely two variants of the species grow apart until they qualify as new and distinct species.[23] All that remained—the hard part—was to figure out what guided that divergence. In July of 1838, he finished his second species notebook and opened his third, the one that would bring him the answer. And he may, as he penned his fateful marriage memorandum that same month, have had a sense of impending success.

In late September, the solution came. Darwin had just read Thomas Malthus's famous essay on population, which noted that a human population's natural rate of growth will tend to outstrip the food supply unless checked. Darwin recalled in his autobiography: "[B]eing well prepared to appreciate the struggle for existence which everywhere goes on from long-continued observation of the habits of animals and plants, it at once struck me that under these circumstances favourable variations would tend to be preserved, and unfavourable ones to be destroyed. The result of this would be the formation of new species. Here, then, I had at last got a theory by which to work."[24] Under the heading of September 28, Darwin jotted in his notebook some lines about Malthus and then, without explicitly describing natural selection, surveyed its effects: "One may say there is a force like a hundred thousand wedges trying [to] force every kind of adapted structure into the gaps in the economy of Nature, or rather forming gaps by thrusting out weaker ones. The final cause of all this wedgings, must be to sort out proper structure & adapt it to change."[25]

The direction of Darwin's professional life was set, and now he fixed the course of his personal life. Six weeks after writing this passage, on Sunday, November 11 ("The Day of Days!" he wrote in his personal journal), he proposed to Emma Wedgwood.

Viewed in the simplest Darwinian terms, Darwin's attraction to Emma seems strange. He was now a high-status and well-to-do man in his late twenties. Presumably he could have had a young and beautiful wife. Emma was a year older than he and, though not unattractive (at least in the eyes of her portrait painter), was not thought beautiful. Why would Darwin do anything so maladaptive as to marry a plain woman who had already exhausted more than a decade of her reproductive potential?

First of all, this simple equation—rich, high-status male equals young, beautiful wife—is a bit crude. There are many factors that make for a genetically auspicious mate, including intelligence, trust-worthiness, and various sorts of compatibility.[26] Moreover, the selection of a spouse is also the selection of a parent for one's offspring. Emma's sturdiness of character foreshadowed the attentiveness she would bring to her children. One of her daughters recalled: "Her sympathy, and the serenity of her temper, made her children feel absolutely at their ease with her, and sure of comfort in every trouble great or small, whilst her unselfishness made them know that she would never find anything a burden, and that they could go to her with all the many little needs of a child for help or explanation."[27]

Besides, if the issue is how "valuable" a wife Darwin should be expected to seek, the question isn't, strictly speaking, how marketable a mate he was, but how marketable he had been given the impression he was. By adolescence, if not earlier, people are getting feedback about their market value, feedback that shapes their self-esteem and thus affects how high they aim their sights. Darwin doesn't seem to have emerged from adolescence feeling like an alpha male. Though large, he was meek, not much of a fighter. And, as one of his daughters noted, he considered his face "repellently plain."[28]

Of course, all of this was rendered less relevant by later achievement. Darwin may not have had high status as a teenager, but he got it later, and status per se can compensate, in the eyes of women, for mediocre looks and a lack of brute strength. Yet his insecurity seems to have persisted—as, indeed, insecurities formed in adolescence often do. The question is why.

Maybe the developmental mechanism that fine-tuned Darwin's insecurity was a vestige of evolution, an adaptation that would have tended to raise fitness in the ancestral environment but no longer does. In many hunter-gatherer societies, the male dominance hierarchy is fairly firm by early adulthood; submissive, low-status men don't go to college, assiduously climb career ladders, and then wow ladies with their newly reached station. So in the ancestral environment, a self-esteem that began to harden shortly after adolescence may have been a reliable guide to one's enduring value on the marriage market; maybe it has become a faulty indicator only in a more modern environment.

Then again, maybe a stubbornly low opinion of oneself can be adaptive in almost any environment. Wives do cheat on husbands, after all. And folk wisdom, at least, has them often cheating with handsome, athletic men. Thus Darwin's low opinion of his animal magnetism may have kept him from marrying the kind of stunning woman who would draw overtures from world-class philanderers whom she would then find sexier than him.

EMMA SAYS YES

Emma accepted Darwin's proposal, leaving him with feelings of "hearty gratitude to her for accepting such a one as myself." She was pleased, she later reported, to find that he had been unsure of her answer.[29] It is everyone's desire not to be taken for granted by a mate, and naturally so, since this would bode ill for future devotion.

Emma shows no signs of having equivocated. She clearly admired Darwin's intelligence and, in explaining her assent, she also stressed his honesty, his affection toward his family, and his "sweet tempered" nature.[30] (Translation: he probably has some good genes and he seems likely to be a generous and considerate parental investor.) And it could not have escaped her attention that he came from a wealthy family and was of high and rising professional stature (that he would have plenty of resources, material and social, to invest).

To be sure, Emma came from a wealthier family. Her grandfather was an innovative and immensely successful potter whose name lives on in the form of Wedgwood china. She could have married a pauper without fear that her children would grow up deprived. But, as we've seen, attraction to mates who command social and material resources may have been so consistently conducive to the fitness of women during evolution that it has become a fairly rigid part of their minds.

Even if Emma Wedgwood could have *bought* her way into London's high society—through philanthropy, say—Darwin's social status probably would have wowed her. At any rate, it did. During her engagement to Charles, the couple was entertained by the Cambridge University geologist Adam Sedgwick. "What an honour for the great Sedgwick to invite me to his house," marveled Emma. "*Me* only think of it! I feel a greater person already for it & how my head will stand it when I am really Mrs. D . . . I can't tell."[31]

Men, of course, are hardly oblivious to the status and wealth of a mate. But if the importance of these things was indeed sexually asymmetrical during most of evolution, the male attraction to a moneyed or socially prominent woman may be less a matter of raw appeal and more a matter of conscious calculation. In the July marriage memorandum, when Darwin was fretting about the twin evils of marriage—the "loss of time every day" and the "horrid poverty" that could result, he slyly followed each of those phrases with a parenthetic qualifier: "without one's wife was an angel, & made one keep industrious," he added after the first worry. And after the second: "without one's wife was better than an angel & had money."

Regardless of how much Darwin did or didn't know about his future health and career, he had just given a composite sketch of the ideal wife for a chronically ill man who, while not affiliated with any university, is trying to write the most important scientific book of the century. And, regardless of whether he then had some inkling of who his wife would be, he had just given a roughly accurate portrait of Emma Wedgwood.[32] Between her father's wealth, Darwin's father's wealth, Darwin's book royalties, and his knack for sound investment, there would be ample money in the Darwin household.[33] And although Emma may not have "made" Darwin stay industrious, she certainly encouraged him to, nursing him faithfully and shielding him from distraction. Darwin, in characteristically oblique fashion, made this assignment clear from the beginning. Three weeks into the engagement, he wrote to her about the reaction of an acquaintance to the news: "She says 'so Mr. Darwin is going to be married; I suppose he will be buried in the country and lost to geology'. She little knows, what a good strict wife, I am going to be married to, who will send me to my lessons, and make me better, I trust, in every respect. . . ."[34]

DARWIN GETS EXCITED

To say that Darwin carefully and rationally selected his wife isn't to say that he wasn't in love with her. By the time of their wedding day, his letters to Emma were so laden with emotion as to raise a question: How did his feelings accelerate so rapidly? As of July, he is, depending on your interpretation of the evidence, either (a) not even dreaming of marrying her in particular; or (b) dithering violently over whether to marry her. In late July, he pays her a visit, and they have a long talk. On his next visit, three and a half months later, he pops the question. Now, suddenly, he is in ecstasy, writing florid letters about how he waits anxiously for the day's mail in the hope it will hold a letter from her; how he lies awake at night thinking of their future together; how "I long for the day when we shall enter the house together; how glorious it will be to see you seated by the fire of our own house."[35] What has happened to this man?

At the risk of seeming to harp on a single theme, I direct your attention again to the subject of genes. In particular: the differing genetic interests of a man and a woman who have never had sex with one another. Pre-sex, a woman's genes often call for wary evaluation. Affection should not too quickly become overwhelming passion. The male's genetic interest, meanwhile, often lies in speeding things up, saying things that will melt the woman's reserve. High on the list of these things are intimations of deep affection and eternal devotion. And nothing produces more convincing intimations then *feelings* of affection and devotion.

This logic may be amplified by various circumstances, and one of them is how much sex the man has been getting up until now. As Martin Daly and Margo Wilson have observed, "any creature that is recognizably on track toward complete reproductive failure" should, in theory, try with increasing intensity to change this trajectory.[36] That is: natural selection probably would not have been kind to the genes of men whose quest for sex wasn't accelerated by its prolonged absence. So far as anyone knows, Darwin went through his bachelor years without ever having sexual intercourse.[37] How little does it take to arouse a man who has been so long deprived? When the *Beagle* docked in Peru, Darwin saw elegant ladies shrouded in veils that exposed only one eye. "But then," he wrote, "that one eye is so black and brilliant and has such powers of motion and expression that its effect is very powerful."[38] It is not surprising that when Emma Wedgwood was placed within reach—her whole face visible, and her body soon to be his—Darwin began to salivate. (Literally, it would seem. See Darwin's journal excerpt at the head of this chapter.)

It is hard to estimate the exact ratio of love to lust in Darwin's heart as the wedding approached; their relative reproductive value during our evolutionary past has varied widely from moment to moment (as it still does) and from millennium to millennium. A few weeks before the wedding, Darwin mused in one of his scientific notebooks, "What passes in a man's mind when he says he loves a person . . . it is blind feeling, something like sexual feelings—love being an emotion does it regard—is it influenced by—other emotions?"[39] Like many passages in Darwin's notebooks, this is cryptic, but in mentioning love and sexual feelings in the same breath, and in suggesting that love may be subterraneanly rooted in other feelings, it seems headed in the general direction of a modern Darwinian view of human psychology. And it suggests (as does his mention of salivation) that he was, at that point, experiencing more than one kind of feeling toward Emma.

What was Emma feeling? If indeed the male's intense interest in impending sex is often matched by lingering female wariness, she might be expected to feel somewhat less ardor than Darwin. There are all kinds of factors that could change things in any one case, of course, but this is the generic expectation: more female than male ambivalence about consummation. Thus the Victorian postponement of sex until marriage should theoretically have shifted power toward women during the engagement. While the man had cause to be eager for the wedding day (compared to today's men, at least), the woman had cause to pause and reflect (compared to today's women, at least).

Emma complied with theory. Weeks into the engagement, she suggested that the wedding be put off until spring, whereas Darwin was pushing for winter. She cited the feelings of her sister Sarah Elizabeth, who, fifteen years her senior and still unwed, had mixed emotions about the event. But Emma added candidly, in a letter to Darwin's sister Catherine, "besides which I should wish it myself." She urged, "Do dear Catty clog the wheels a little slow."[40]

Darwin, enlisting some lush prose ("hope deferred does make my heart quite sick to call you in truth my wife"), kept the honeymoon from receding. But even after the wedding date was firm, he seems to have been left a bit insecure by Emma's reluctance, and perhaps by her overall tenor; her letters are warm, but they're far from effusive. Darwin wrote: "I earnestly pray, you may never regret the great, & I will add very good, deed, you are to perform on *the* Tuesday." Emma tried to reassure him, but she wasn't under the same magical spell he was under: "You need not fear my own dear Charles that I shall not be quite as happy as you are & I shall always look upon the event of the 29th as a most happy one on my part though perhaps not so great or so good as you do."[41] Ouch.

Now, all of this may reflect wholly on the peculiar dynamics between Charles and Emma, and not at all on the Victorian linkage of marriage with consummation. Emma was never an overly sentimental woman.[42] And, anyway, she may have begun having doubts about Charles's health, doubts that would have been warranted. Still, the basic point is probably valid in the aggregate: if it is harder to drag men to the altar today than it used to be, one reason is that they don't have to stop there on the way to the bedroom.

AFTER THE HONEYMOON

Consummation can alter the balance of affection. Though the average woman is more selective in unleashing her ardor than the average man, she should, in theory, be less inclined to rein it in once it has left the gate. Having deemed a man worthy of joining in her epic parental investment, she typically has a strong genetic interest in keeping him involved. Again, Emma's behavior matches expectations. Within the first few months of their marriage she wrote: "I cannot tell him how happy he makes me and how dearly I love him and thank him for all his affection which makes the happiness of my life more and more every day."[43]

Whether a man's devotion will be nourished by consummation is a less certain matter. Maybe his professions of affection were self-delusion; maybe, once his mate is pregnant, a better deal will come along. But in Darwin's case, the early signs were good. Months after his wedding day (and weeks after the conception of his first child), Darwin, writing in his notebook, groped for an evolutionary explanation of why a man's acts of "kindness to wife and children would give him pleasure, without any regard to his own interest," which suggests that his affection for Emma still felt deep.[44]

Perhaps this is not surprising. The tactical value of a woman's sexual reserve isn't just that men desperately want sex and to get it may say anything, even *believe* anything—including "I want to spend my whole life with you." If a Madonna-whore switch is indeed built into the male brain, then a woman's early reticence can lastingly affect a man's view of her. He is more likely to respect her in the morning—and perhaps for many years to come—if she doesn't weaken under his advances. He may *say* "I love you" to various women he yearns for, and he may mean it; but he may be more likely to keep meaning it if he doesn't get them right away. There may have been a bit of wisdom in the Victorian disapproval of premarital sex.

Even beyond this disapproval, Victorian culture was finely calibrated to excite the "Madonna" part of a man's mind and numb the "whore" part. The Victorians themselves called their attitude toward females "woman worship." The woman was a redeemer—innocence and purity incarnate; she could tame the animal in a man and rescue his spirit from the deadening world of work. But she could only do this in a domestic context, under the blessing of marriage, and after a long, chaste courtship. The secret was to have, as the title of one Victorian poem put it, an "Angel in the House."[45]

The idea wasn't just that men were supposed to at some point quit sowing wild oats, get hitched, and worship their wives. They were supposed to not sow the wild oats in the first place. Though the double standard for promiscuity prevailed in nineteenth-century Britain as elsewhere, it was battled by the more austere guardians of

Victorian morality (including Dr. Acton), who preached not just extramarital, but also premarital, abstinence for men. In *The Victorian Frame of Mind* Walter Houghton writes, "To keep body and mind untainted, the boy was taught to view women as objects of the greatest respect and even awe." Though he was supposed to give all women this respect, a certain kind of woman warranted something more. "He was to consider nice women (like his sister and his mother, like his future bride) as creatures more like angels than human beings— an image wonderfully calculated not only to dissociate love from sex, but to turn love into worship, and worship of purity."[46]

When Houghton says "calculated," he means it. One author in 1850 expressed the virtue of male premarital chastity as follows: "Where should we find that reverence for the female sex, that tenderness towards the feelings, that deep devotion of the heart to them, which is the beautiful and purifying part of love? Is it not certain that all of [the] delicate and chivalric which still pervades our sentiments towards women, may be traced to *repressed*, and therefore hallowed and elevated passion? . . . And what, in these days, can preserve chastity, save some relic of chivalrous devotion? Are we not all aware that a young man can have no safeguard against sensuality and low intrigue, like an early, virtuous, and passionate attachment?"[47]

Aside from the word *repressed*, which probably mischaracterizes the psychodynamics, this passage is plausible enough. It implies that a man's passion can be "hallowed and elevated" if not quenched too readily—that a chaste courtship, in other words, helps move a woman into the "Madonna" part of his mind.

This is not the only reason that a chaste courtship may encourage marriage. Recall how different the ancestral environment was from the modern environment. In particular: there were no condoms, diaphragms, or birth-control pills. So if an adult couple paired up, slept together for a year or two, and produced no baby, the chances were good that one of them wasn't fertile. No way of telling which one, of course; but for both of them there was little to lose and much to gain by dissolving the partnership and finding a new mate. The adaptation expected to arise from this logic is a "mate-ejection module"—a mental mechanism, in both male and female, that would encourage souring on a mate after lots of sex without issue.[48]

This is a quite speculative theory, but it has some circumstantial evidence on its side. In cultures around the world, barren marriages are among the most likely to break up.[49] (Though cases where barrenness is cited as the cause of breakup don't quite get at the crux of this theory: the *unconsciously* motivated alienation from a mate.) And, as many husbands and wives can attest, the birth of a child often cements a marital bond, if obliquely; the love of spouse is partly diverted to the child and then refracted diffusely onto the family as a whole, mate included. It is a different *kind* of love for the spouse, but it's sturdy in its own way. In the absence of this roundabout recharge, love of spouse may tend to disappear entirely—by design.

Darwin once worried that contraceptive technology would "spread to unmarried women & would destroy chastity on which the family bond depends; & the weakening of this bond would be the greatest of all possible evils to mankind."[50] He surely didn't grasp all the plausible Darwinian reasons that contraception and the attendant premarital sex might indeed discourage marriage. He didn't suspect the deep basis of the Madonna-whore dichotomy or the possible existence of a "mate-ejection module." And even today, we're far from certain about these things. (The established correlations between premarital sex and divorce, and between premarital cohabitation and divorce, are suggestive but ambiguous.)[51] Still, it is harder now than it would have been thirty years ago to dismiss Darwin's fear as the rantings of an aging Victorian.

Contraception isn't the only technology that may affect the structure of family life. Women who breast-feed often report a weakened sex drive—and with good Darwinian cause, since they're usually incapable of conception. Husbands, meanwhile, sometimes fail to find a breast-feeding wife sexually exciting, presumably for the same ultimate reason. Thus bottle-feeding may make wives both more lustful and more attractive. Whether this is, on balance, good for family cohesion is hard to say. (Does it more often tempt wives into extramarital affairs or distract husbands from having them?) In any event, this logic may make sense of Dr. Acton's otherwise comical-sounding claim that "the best mothers, wives, and managers of households, know little or nothing of sexual indulgences. Love of home, children, and domestic duties, are the only passions they feel." In Victorian England, when so many wives spent so many of their fertile years either pregnant or nursing, their passion may indeed have spent much time in abeyance.[52]

Even if a succession of babies helps keep both partners devoted, the interests of husband and wife may diverge as time goes by. The older the children (the less urgently needful of paternal investment), and the older the wife, the less support a man's devotion gets from his evolutionary heritage. More and more of the harvest has been reaped; the ground is less and less fertile; it may be time to move on.[53] Of course, whether the husband feels this impulse strongly may depend on how likely it is to bear fruit. A dashing and wealthy man may get the kinds of glances from women that fuel it; a poor and disfigured man may not. Still, the strength of the impulse will tend to be greater in the husband than in the wife.

Though the shifting balance of attraction between husband and wife is seldom described this explicitly, it is often reflected more obliquely—in novels, in aphorisms, in the folk wisdom offered up as advice to bride and groom. Professor Henslow, a fifteen-year veteran of the blessed state, wrote to Darwin shortly before his marriage: "All the advice, which I need not give you, is, to remember that as you take your wife for better for worse, be careful to value the better & care nothing for the worse." He added: "It is the neglect of this little particular which makes the marriage state of so many men worse than their single blessedness."[54] In other words, just remember one simple rule: don't stop loving your wife, as men seem inclined to do.

Emma, meanwhile, was getting advice not about overlooking Charles's flaws, but about concealing any flaws of her own, especially those that make a woman look old and haggard. An aunt (perhaps mindful of Emma's noted lack of fashion consciousness) wrote: "If you do pay a little more, be always dressed in good taste; do not despise those little cares which give everyone more pleasing looks, because you know you have married a man who is above caring for such little things. No man is above caring for them. . . . I have seen it even in my half-blind husband."[55]

The logic of male intolerance typically remains opaque to all concerned. A man souring on a mate doesn't think, "My reproductive potential is best served by getting out of this marriage, so for entirely selfish reasons I'll do so." Awareness of his selfishness would only impede its pursuit. It's much simpler for the feelings that got him into the marriage to simply stage a slow but massive retreat.

The increasingly severe view that a restless husband may take of an aging wife was well illustrated by Charles Dickens, one of the few upper-class Victorians who actually got out of a marriage (by separation, not divorce). Dickens, who was elected to membership in London's Athenaeum Club on the same day in 1838 as Darwin, had then been married for two years to the woman he called his "better half." Two decades later—now much more famous, and thus commanding the attention of many young women—he was having trouble seeing her brighter side. It now seemed to him that she lived in a "fatal atmosphere which slays every one to whom she should be dearest." Dickens wrote to a friend: "I believe that no two people were ever created with such an impossibility of interest, sympathy, confidence, sentiment, tender union of any kind between them, as there is between my wife and me." (If so, mightn't he have discussed this with her *before* she bore him ten children?) "To his eyes," a chronicler of their marriage has written, his wife "had become unresponsive, grudging, inert, close to inhuman."[56]

Emma Darwin, like Catherine Dickens, grew old and shapeless. And Charles Darwin, like Charles Dickens, rose markedly in stature after his wedding. But there's no evidence that Darwin ever saw Emma as close to inhuman. What accounts for the difference?

Chapter 6: **THE DARWIN PLAN FOR MARITAL BLISS**

She has been my greatest blessing, and I can declare that in my whole life I have never heard her utter one word which I had rather have been unsaid. . . . She has been my wise adviser and cheerful comforter throughout life, which without her would have been during a very long period a miserable one from ill-health. She has earned the love and admiration of every soul near her.

—*Autobiography* (1876)[1]

In his pursuit of a lasting and fulfilling marriage, Charles Darwin possessed several distinct advantages.

To begin with, there was his chronic ill health. Nine years into marriage, while visiting his ailing father, and while ailing himself, he wrote to Emma of how he "yearned" for her, as "without you, when sick I feel most desolate." He closed the letter: "I do long to be with you & under your protection for then I feel safe."[2] After three decades of marriage, Emma would observe that "nothing marries one so completely as sickness."[3] This reflection may have been more bitter than sweet; Darwin's illness was a lifelong burden for her, and she couldn't have grasped its full weight until well after the wedding. But, whether or not it gave her second thoughts about the marriage, it meant that, for much of Darwin's married life, he wasn't a very marketable commodity. And in marriage, an unmarketable commodity—male or female—is often a contented one, with little if any sexual restlessness.

A complementary asset that Darwin brought to his marriage was hearty subscription to the Victorian ideal of woman as spiritual salvation. In his premarital deliberative soliloquies, he had imagined an "angel" who would keep him industrious yet not let him suffocate in his work. He got that, and a nurse too. And, for good measure, the chasteness of the courtship may have helped keep Emma filed under "Madonna" in Darwin's mind. Something did. "I marvel at my good fortune," he wrote toward the end of his life, "that she, so infinitely my superior in every single moral quality, consented to be my wife."[4]

A third advantage was residential geography. The Darwins lived, gibbon-like, on an eighteen-acre parcel, two hours by coach from London and its young female distractions. Male sexual fantasies tend to be essentially visual in nature, whereas female fantasies more often include tender touching, soft murmurs, and other hints of future investment. Not surprisingly, male fantasies, and male sexual arousal, are also more easily activated by sheerly visual cues, by the mere sight of anonymous flesh.[5] So visual isolation is an especially good way to keep a man from thinking the thoughts that could lead to marital discontent, infidelity, or both.

Isolation is hard to come by these days, and not just because attractive young women no longer stay in their homes, barefoot and pregnant. Images of beautiful women are everywhere we look. The fact that they're two-dimensional doesn't mean they're inconsequential. Natural selection had no way of "anticipating" the invention of photography. In the ancestral environment, distinct images of many beautiful young women would have signified a (genetically) profitable alternative to monogamy, and it would have been adaptive for feelings to shift accordingly. One evolutionary psychologist has found that men shown pictures of *Playboy* models later describe themselves as less in love with their wives than do men shown other images. (Women shown pictures from *Playgirl* felt no such attitude adjustment toward spouses.)[6]

The Darwins also had the blessing of fecundity. A marriage that produces a steady stream of children, and has the resources to care for them, may dampen the wanderlust of both women and men. Wandering takes time and energy, both of which can be well invested in those endearing little vehicles of genetic transmission. That divorce does grow less likely as more children are born is sometimes taken to mean that couples choose to endure the pain of matrimony "for the sake of children." No doubt this happens. But it's at least possible that evolution has inclined us to love a mate more deeply when marriage proves fruitful.[7] In either event, couples who say they'll stay married but won't have children may well prove wrong on one count or the other.[8]

We can now roughly sketch a Charles Darwin plan for marital bliss: have a chaste courtship, marry an angel, move to the country not long after the wedding, have tons of kids, and sink into a deeply debilitating illness. A heartfelt commitment to your work probably helps too, especially when the work doesn't entail business trips.

MARRIAGE TIPS FOR MEN

From the point of view of the average, late-twentieth-century man, the Darwin plan doesn't get high marks for feasibility. Perhaps some more practicable keys to lifelong monogamy can be gleaned from Darwin's life. Let's start with his three-step approach to marriage: (1) decide, rationally and systematically, to get married; (2) find someone who in most practical ways meets your needs; (3) marry her.

One biographer has chastised Darwin for this formulaic approach, complaining that "there is an emotional emptiness about his ponderings on marriage."[9] Maybe so. But it's worth noting that Darwin was a loving husband and father for around half a century. Any men who would like to fill this role might profit from looking closely at the "emotional emptiness" of Darwin's ruminations on marriage. They may hold a lesson transplantable to modern times.

Namely: lasting love is something a person has to *decide* to experience. Lifelong monogamous devotion is just not natural—not for women even, and emphatically not for men. It requires what, for lack of a better term, we can call an act of will. Hence the aptness of Darwin's apparent separation of the marriage question from the marriage-partner question. That he made up his mind—firmly, in the end—to get married and to make the most of his marriage was as important as his choice of mate.

This isn't to say that a young man can't hope to be seized by love. Darwin himself got fairly worked up by his wedding day. But whether the sheer fury of a man's feelings accurately gauges their likely endurance is another question. The ardor will surely fade, sooner or later, and the marriage will then live or die on respect, practical compatibility, simple affection, and (these days, especially) determination. With the help of these things, something worthy of the label "love" can last until death. But it will be a different kind of love from the kind that began the marriage. Will it be a richer love, a deeper love, a more spiritual love? Opinions vary. But it's certainly a more impressive love.

A corollary of the above is that marriages aren't made in heaven. One great spur to divorce is the belief of many men (and no few women) that somehow they just married the "wrong" person and next time they'll get it "right." Not likely. Divorce statistics support Samuel Johnson's characterization of a man's decision to remarry as "the triumph of hope over experience."[10]

John Stuart Mill held a similarly sober view. Mill insisted on *tolerance* of moral diversity, and stressed the long-term value of experimentation by society's nonconformists, but he didn't recommend moral adventurism as a lifestyle. Beneath the radicalism of *On Liberty* lay Mill's belief in keeping our impulses under firm cerebral control. "Most persons have but a very moderate capacity of happiness," he wrote in a letter. "Expecting . . . in marriage a far greater degree of happiness than they commonly find: and knowing not that the fault is in their own scanty capabilities of happiness—they fancy they should have been happier with some one else." His advice to the unhappy: sit still until the feeling passes. "[I]f they remain united, the feeling of disappointment after a time goes off, and they pass their lives together with fully as much happiness as they could find either singly or in any other union, without having undergone the wearing of repeated and unsuccessful experiments."[11]

Many men—and some, but fewer, women—would enjoy the opening stages of those experiments. But in the end they might find that the glimpse of lasting joy the second time around was just another delusion sponsored by their genes, whose primary goal, remember, is to make us prolific, not lastingly happy (and which, anyway, aren't operating in the environment of our design; in a modern society, where polygamy is illegal, a polygamous impulse can do more emotional damage to all concerned—notably offspring—than natural selection "intended"). The question then becomes whether the fleeting fun of greener pastures outweighs the pain caused by leaving the golden-brown ones. This isn't a simple question, much less a question whose answer is easy to impose on one's yearnings. But more often than many people (men in particular) care to admit, the answer is no.

And, anyway, there is debate over whether a minute-by-minute summation of pleasure and pain should settle the issue. Maybe the cumulative coherence of a life counts for something. Men from many generations have testified that over the long haul, a life shared with another person and several little people, for all its diverse frustrations, brings rewards of a sort unattainable through other means. Of course, we shouldn't give infinite weight to the testimony of old married men. For every one of them who claims to have had a fulfilling life, there is at least one bachelor claiming to be enjoying his series of conquests. But it's noteworthy that a number of these old men went through early phases of sexual liberty and concede that they enjoyed them. None of those making the other side of the argument can say they know what it's like to create a family and stay with it until the end.

John Stuart Mill made this point in a larger context. Even Mill, who, as the foremost publicist of utilitarianism, insisted that "pleasure, and freedom from pain, are the only things desirable as ends," didn't mean that the way it sounds. He believed that the pleasure and pain of all people affected by your actions (emphatically including any people your marriage created) belong in your moral calculus. Further, Mill stressed not just quantity of pleasure but quality, attaching special value to pleasures involving the "higher faculties." He wrote: "Few human creatures would consent to be changed into any of the lower animals, for a promise of the fullest allowance of a beast's pleasures. . . . It is better to be a human being dissatisfied than a pig satisfied; better to be Socrates dissatisfied than a fool satisfied. And if the fool, or the pig, is of a different opinion, it is because they only know their own side of the question. The other party to the comparison knows both sides."[12]

DIVORCE THEN AND NOW

Since Darwin's day, the incentive structure surrounding marriage has been transformed—indeed, inverted. Back then men had several good reasons to get married (sex, love, and societal pressure) and a good reason to stay married (they had no choice). Today an unmarried man can get sex, with or without love, regularly and respectably. And if for some reason he does stumble into matrimony, there's no cause for alarm; when the thrill is gone, he can just move out of the house and resume an active sex life without raising local eyebrows. The ensuing divorce is fairly simple. Whereas Victorian marriage was enticing and ultimately entrapping, modern marriage is unnecessary and eminently escapable.

This change had begun by the turn of the century, and it reached dramatic proportions after midcentury. The American divorce rate, which was level during the 1950s and early 1960s, doubled between 1966 and 1978, reaching its present level. Meanwhile, as escape from marriage became simple and commonplace, the incentive for men (and, in a probably less dramatic way, for women) to enter one was being dulled. Between 1970 and 1988, though the average age of a woman upon (first) marriage was rising, the number of eighteen-year-old girls who reported having had intercourse grew from 39 to 70 percent. For fifteen-year-olds it went from one in twenty to one in four.[13] The number of unmarried couples living together in the United States grew from half a million in 1970 to nearly three million in 1990.

Hence the double whammy: as easy divorce creates a growing population of formerly married women, easy sex creates a growing population of never-married women. Between 1970 and 1990, the number of American women aged thirty-five to thirty-nine who had never been married rose from one in twenty to one in ten.[14] And of women in that age group who *have* gotten married, about a third have also gotten divorced.[15]

For men, the figures are even more severe. One in seven men aged thirty-five to thirty-nine has never married; as we've seen, serial monogamy tends to leave more men than women in that condition.[16] Still, women may be the bigger losers here. They are more likely than men to want children, and a forty-year-old unmarried childless woman, unlike her male counterpart, is watching her chances of parenthood head briskly toward zero. And as for the relative fortunes of *formerly* married men and women: in the United States, divorce brings the average man a marked increase in standard of living, while his wife, along with her children, suffers the opposite.[17]

The Divorce Act of 1857, which helped legitimize marital breakup in England, was welcomed by many feminists. Among them was John Stuart Mill's wife, Harriet Taylor Mill, who, until the death of her first husband, had been trapped in a marriage she detested. Mrs. Mill, who seems never to have been a great enthusiast for sexual intercourse, had painfully come to believe "that all men, with the exception of a few lofty minded, are sensualists more or less" and that "women on the contrary are quite exempt from this trait." To any wife who shared her distaste for sex, Victorian marriage could seem like a series of rapes punctuated by dread. She favored divorce on demand for the sake of women.

Mill too favored divorce on demand (*assuming* the couple had no children). But his view of the matter differed from hers. He saw wedding vows as a constraint less on the wife than on the husband. The strict marriage laws of the day, Mill observed, with great insight into the likely origins of institutionalized monogamy, had been written "*by* sensualists, *for* sensualists, and *to bind* sensualists."[18] He was not alone in this view. Behind opposition to the Divorce Act of 1857 was a fear that it would turn men into serial monogamists. Gladstone opposed the bill because, he said, it "would lead to the degradation of woman."[19] (Or, as an Irish woman would put it more than a century later: "A woman voting for divorce is like a turkey voting for Christmas.")[20] The effects of easy divorce have been complex, but in many ways the evidence supports Gladstone. Divorce is very often a raw deal for women.

There's no point in trying to turn back the clock, no point in trying to sustain marriages by making the alternative virtually illegal. Studies show that the one thing harder on children than divorce is for parents to stay together even though locked in mortal combat.

But surely there shouldn't be a financial *incentive* for a man to get divorced; divorce shouldn't *raise* his personal standard of living, as it now usually does. In fact, it seems only fair to lower his standard of living—not necessarily to punish him, but because that's often the only way to keep the living standards of his wife and children from plunging, given the inefficiencies of two households compared to one. If financially secure, women can often be happy rearing children without a man—happier than they were with him, sometimes, and happier even than he'll be after he's gotten used to the grass on the other side of the fence.

R·E·S·P·E·C·T

There is a difference of opinion over how much "respect" women get in the modern moral climate. Men think they get lots. The portion of American men saying women are better respected than in the past went from 40 percent in 1970 to 62 percent in 1990. Women disagree. In a 1970 survey, they were most likely to call men "basically kind, gentle, and thoughtful," but a 1990 survey conducted by the same pollster found women most likely to describe men as valuing only their own opinions, trying to keep women down, preoccupied with getting women into bed, and not paying attention to household affairs.[21]

Respect is an ambiguous word. Maybe those men who consider women well "respected" mean that women have been accepted at work as worthy colleagues. And maybe women are indeed getting more of this kind of respect. But if by *respect* you mean what the Victorians meant when they urged respect for women—not treating them as objects of sexual conquest—then respect has probably dropped since 1970 (and it certainly has since 1960). One interpretation of the above numbers is that women would like to have more of this second kind of respect.

There's no clear reason for a sharp trade-off between the two; no reason that feminists of the late 1960s and early 1970s, in insisting on the first kind of respect, had to undermine the second (which, actually, they said they also wanted). But, as things happened, they did. They preached the innate symmetry of the sexes in all major arenas, including sex. Many young women took the doctrine of symmetry to mean they could follow their sexual attractions and disregard any vague visceral wariness: sleep with any man they liked, without fear that his sexual interest didn't signify comparable affection, without fear that sex might be more emotionally entangling for her than for him. (Some feminists practiced casual sex almost out of a sense of ideological commitment.) Men, for their part, used the doctrine of symmetry to ease themselves off the moral hook. Now they could sleep around without worrying about the emotional fallout; women were just like them, so no special consideration was necessary. In this they were, and are, aided by women who actively resist special moral consideration as patronizing (which it sometimes is, and certainly was in Victorian England).

Lawmakers, meanwhile, took sexual symmetry to mean that women needed no special legal protection.[22] In many states, the 1970s brought "no-fault" divorce and the automatically equal division of a couple's assets—even if one spouse, usually the wife, hasn't been on a career track and thus faces bleaker prospects. The lifelong alimony that a divorced woman could once expect may now be replaced by a few years of "rehabilitative maintenance" payments, which are supposed to give her time to launch her career recovery—a recovery that, in fact, will extend beyond a few years if she has a few children to tend. In trying to get a more equitable deal, it won't help to point out that the cause of the breakup was her husband's rampant philandering, or his sudden, brutal intolerance. These things, after all, are nobody's fault. The no-fault philosophy is one reason divorce is literally a profitable enterprise for men. (The other reason is lax enforcement of the man's financial obligations.) The height of the no-fault vogue has now passed, and state legislatures have undone some of the damage, but not all.

The feminist doctrine of innate sexual symmetry wasn't the only culprit, or even, initially, the main one. Sexual and marital norms had been changing for a long time, for many reasons, ranging from contraceptive technology to communications technology, from residential patterns to recreational trends. So why dwell on feminism?

Partly because of the sheer irony that (perfectly laudable) attempts to stop one kind of exploitation of women aided another kind. Partly, too, because, though feminists didn't single-handedly create the problem, some of them have helped sustain it. Until very recently, fear of feminist backlash was far and away the main obstacle to an honest discussion of differences between the sexes. Feminists have written articles and books denouncing "biological determinism" without bothering to understand biology or determinism. And the increasing, if belated, feminist discussion of sex differences is sometimes vague and disingenuous; there is a tendency to describe differences that are plausibly explicable in Darwinian terms while dodging the question of whether they are innate.[23]

UNHAPPILY MARRIED WOMEN

The "Darwin plan" for staying married—and, indeed, the general thrust of this chapter so far—may seem to presuppose a simple picture: women love marriage, men don't. Obviously, life is more complicated than that. Some women don't want to get married, and many more, once married, are far from bliss. If this chapter has stressed the *male* mind's incongruence with monogamous marriage (and it has), it's not because I think the female mind is a perpetual font of adulation and fidelity. It's because I think the male mind is the largest single obstacle to lifelong monogamy—and certainly the largest such obstacle that emerges distinctly from the new Darwinian paradigm.

The incongruence between the *female* mind and modern marriage is less simple and clear-cut (and, in the end, less disruptive). The clash isn't so much with monogamy itself as with the social and economic setting of modern monogamy. In the typical hunter-gatherer society, women have both a working life and a home life, and reconciling the two isn't hard. When they go out to gather food, child care is barely an issue; their children may go with them or, instead, stay with aunts, uncles, grandparents, cousins. And when mothers, back from work, do care for children, the context is social, even communal. The anthropologist Marjorie Shostak, after living in an African hunter-gatherer village, wrote: "The isolated mother burdened with bored small children is not a scene that has parallels in !Kung daily life."[24]

Most modern mothers seem to find themselves well to one side or the other of the (reasonably) happy medium that a hunter-gatherer woman naturally strikes. They may work forty or fifty hours a week, worry about the quality of day care, and feel vaguely guilty. Or they may be full-time housewives who rear their children alone and are driven nearly mad by monotony. Some housewives, of course, manage to build a solid social infrastructure even amid the transience and anonymity of the typical modern neighborhood. But the unhappiness of the many women who don't is virtually inevitable. It's no surprise that modern feminism gathered such momentum during the 1960s, after post–World War II suburbanization (and so much else) had diluted the sense of neighborhood community and pulled the extended family apart; women weren't designed to be suburban housewives.

The generic suburban habitat of the fifties was more "natural" for men. Like many hunter-gatherer fathers, vintage suburban husbands spent a little time with their children and a lot of time out bonding with males, in work, play, or ritual.[25] For that matter, many Victorian men (though not Darwin) had the same setup. Although lifelong monogamy per se is less natural for men than for women, one form that monogamous marriage has often taken, and still often takes, may be harder on women than on men.

But that's not the same as saying that the female mind threatens modern monogamy more than, or as much as, the male mind. A mother's discontent seems not to translate into breakup as naturally as a father's. The ultimate reason is that, in the ancestral environment, seeking a new husband once there were children was seldom a genetically winning proposition.

To make modern monogamy "work"—in the sense both of enduring and of leaving husband and wife fairly happy—is a challenge of overwhelming complexity. A successful overhaul might well entail tampering with the very structure of modern residential and vocational life. Any aspiring tamperers would do well to ponder the social environment in which human beings evolved.

People weren't, of course, designed to be relentlessly happy in the ancestral environment; there, as here, anxiety was a chronic motivator, and happiness was the always pursued, often receding, goal. Still, people *were* designed not to go crazy in the ancestral environment.

THE EMMA PLAN

Notwithstanding the discontents of modern marriage, many women aspire to find a lifelong mate and have children. Given that the current climate doesn't favor this goal, what are they to do? We've talked about how men might behave if they want marriage to be a sturdier institution. But giving men marriage tips is a little like offering Vikings a free booklet titled "How Not to Pillage." If women are closer than men to being naturally monogamous, and often suffer from divorce, maybe they are a more logical locus for reform. As George Williams and Robert Trivers discovered, much of human sexual psychology flows from the scarceness of eggs relative to sperm. This scarcity gives women more power—in individual relationships, and in shaping the moral fabric—than they sometimes realize.

Sometimes they do realize it. Women who would like a husband and children have been known to try the Emma Wedgwood plan for landing a man. In its most extreme form, the plan runs as follows: if you want to hear vows of eternal devotion right up to your wedding day—and if you want to make sure there *is* a wedding day—don't sleep with your man until the honeymoon.

The idea here isn't just that, as the saying goes, a man won't buy the cow if he can get the milk for free. If the Madonna-whore dichotomy is indeed rooted firmly in the male mind, then early sex with a woman may tend to stifle any budding feelings of love for her. And, if there are such things as "mate-ejection modules" in the human mind, sustained sex without issue may bring—in the man *or* the woman—a cooling toward the other.

Many women find the Emma strategy abhorrent. To "trap" a man, some say, is beneath their dignity; if he has to be coerced into marriage, they'd rather do without him. Others say the Emma approach is reactionary and sexist, a revival of the hoary demand that women carry the moral burden of self-control for the sake of the social order. Still others say this approach seems to presume that sexual restraint for a woman is easy, which it often is not. These are all valid reactions.

There is one other common complaint about the Emma strategy: it doesn't work. These days much sex is available to men for little commitment. Not as much as a few years ago, maybe, but enough so that, if any one woman cuts off the supply, alternatives abound. Prim women sit at home alone and bask in their purity. Around Valentine's Day of 1992, the *New York Times* quoted a twenty-eight-year-old single woman who "lamented the lack of romance and courtship." She said, "Guys still figure if you don't come across, someone else will. It's like there's no incentive to wait until you get to know each other better."[26]

This, too, is a valid point, and a good reason why a one-woman austerity crusade isn't likely to bring great rewards. Still, some women have found that a move *of some distance* toward austerity may make sense.[27] If a man isn't interested enough in a woman, as a human being, to endure, say, two months of merely affectionate contact before graduating to lust, he's unlikely to stick around for long in any event. Some women have decided not to waste the time—which, needless to say, is more precious for them than for men.

This mild version of the Emma approach can be self-reinforcing. As more women discover the value of a short cooling-off period, it becomes easier for each of them to impose a longer one. If an eight-week wait is common, then a ten-week wait won't put a woman at much of a competitive disadvantage. Don't expect this trend to reach Victorian extremes. Women, after all, do like sex. But expect the trend, which seems already to be under way, to continue. Much of today's incipiently conservative sexual climate may come from a fear of sexually transmitted diseases; but, to judge by the increasingly clear opinion of many women that men are basically pigs, a part of the new climate may come from women rationally pursuing self-interest in recognition of harsh truths about human nature. And one generally safe bet is that people will continue to pursue their self-interest as they see it. In this case, evolutionary psychology helps them see it.

A THEORY OF MORAL CHANGE

There is another reason trends in sexual morality—whether toward or away from sexual reserve—may be self-sustaining. If men and women are indeed designed to tailor their sexual strategies to local market conditions, the norm among each sex depends on the norm among the other. We've already seen evidence, from David Buss and others, that when men deem a woman promiscuous, they treat her accordingly—as a short-term conquest, not a long-term prize. We've also seen evidence, from Elizabeth Cashdan, that women who perceive males to be generally pursuing short-term strategies are themselves more likely to look and act promiscuous: to wear sexy clothes and have sex often.[28] One can imagine these two tendencies getting locked in a spiral of positive feedback, leading to what the Victorians would have called ongoing moral decline. A proliferation of low-cut dresses and come-hither looks might send the visual cues that discourage male commitment; and as men, thus discouraged, grow less deferential toward women, and more overtly sexual, low-cut dresses might further proliferate. (Even come-hither looks that show up on billboards, or in the pages of *Playboy*, might have some effect.)[29]

If things should for some reason begin moving in the other direction, *toward* male parental investment, the trend could be sustained by the same dynamic of mutual reinforcement. The more Madonnaish the women, the more daddish and less caddish the men, and thus the more Madonnaish the women, and so on.

To call this theory speculative would almost be an understatement. It has the added disadvantage of being (like many theories of cultural change) hard to test directly. But it does rest on theories of individual psychology that are themselves testable. The Buss and Cashdan studies amount to a preliminary test, and so far the two pillars of the theory hold up. The theory also has the virtue of helping to explain why trends in sexual morality persist for so long. Just as Victorian prudishness, at its high-water mark, was the culmination of a century-long trend, it seems to have then receded for a long, long time.

Why does the long, slow swing of the pendulum ever reverse? The possible reasons range from changes in technology (contraceptive, for example) to changes in demographics.[30] It's possible too that the pendulum may tend to reverse when a large fraction of one sex or the other (or both) finds its deepest interests not being served and begins to consciously reevaluate its lifestyle. In 1977 Lawrence Stone observed, "The historical record suggests that the likelihood of this period of extreme sexual permissiveness continuing for very long without generating a strong backlash is not very great. It is an ironic thought that just at the moment when some thinkers are heralding the advent of the perfect marriage based on full satisfaction of the sexual, emotional and creative needs of both husband and wife, the proportion of marital breakdowns, as measured by the divorce rate, is rising rapidly."[31] Since he wrote, women, who have much leverage over sexual morality, have, in apparently growing numbers, asked basic questions about the wisdom of highly casual sex. Whether we are entering a period of long growth in moral conservatism is impossible to say. But modern society doesn't exude an overwhelming sense of satisfaction with the status quo.

VICTORIA'S SECRET

Many judgments have been rendered about Victorian sexual morality. One is that it was horribly and painfully repressive. Another is that it was well suited to the task of preserving marriage. Darwinism affirms these two judgments and unites them. Once you have seen the odds against lifelong monogamous marriage, especially in an economically stratified society—in other words, once you have seen human nature—it is hard to imagine anything short of harsh repression preserving the institution.

But Victorianism went well beyond simple, general repression. Its particular inhibitions were strikingly well-tailored to the task at hand.

Perhaps the greatest threat to lasting marriage—the temptation of aging, affluent, or high-status men to desert their wives for a younger model—was met with great social firepower. Though Charles Dickens did manage, amid great controversy and at real social cost, to leave his wife, he forever confined contact with his mistress to secret meetings. To admit that his desertion was, in fact, a desertion would have been to draw a censure he wasn't willing to face.

It's true that some husbands spent time in one or another of London's many brothels (and housemaids sometimes served as sexual outlets for men of the upper classes). But it's also true that male infidelity may not threaten marriage so long as it doesn't lead to desertion; women, more easily than men, can reconcile themselves to living with a mate who has cheated. And one way to ensure that male infidelity doesn't lead to desertion is to confine it to, well, whores. We can safely bet that few Victorian men sat at the breakfast table daydreaming about leaving their wives for the prostitute they had enjoyed the night before; and we can speculate with some confidence that part of the reason is a Madonna-whore dichotomy planted deeply in the male psyche.

If a Victorian man did more directly threaten the institution of monogamy, if he committed adultery with "respectable" women, the risk was great. Darwin's physician, Edward Lane, was accused in court by a patient's husband of committing adultery with the patient. In those days, this sort of case was so scandalous that the *Times* of London carried daily coverage. Darwin followed it closely. Perhaps conveniently, he doubted Lane's guilt ("I never heard a sensual expression from him"), and he worried about Lane's future: "I fear it will ruin him."[32] It probably would have, if the judge hadn't exonerated him.

Of course, in keeping with the double standard, female adulterers drew even stronger censure than their male counterparts. Both Lane and his patient were married, yet her diary recounted a post-tryst conversation between them that allocated blame in the following proportions. "I entreated him to believe that since my marriage I had never before in the smallest degree transgressed. He consoled me for what I had done, and conjured me to forgive myself."[33] (Lane's lawyer convinced the court that her diary was a mad fantasy, but even if so, it reflects prevailing morality.)

The double standard may not be fair, but it does have a kind of rationale. Adultery per se is a greater threat to monogamy when the wife commits it. (Again: the average man will have much more trouble than the average woman continuing a marriage with a mate known to have been unfaithful.) And if the husband of an adulterer does for some reason stay in the marriage, he may start treating the children less warmly, now that doubts about their paternity have arisen.

Conducting this sort of brisk, clinical appraisal of Victorian morality is risky. People are inclined to misunderstand. So let's be clear: clinical isn't the same as prescriptive; this is *not* an argument on behalf of the double standard, or any other particular aspect of Victorian morality.

Indeed, whatever contribution the double standard may have once made to marital stability by offering a vent for male lust, times have changed. These days a high-powered businessman doesn't confine his extramarital affairs to prostitutes, or to maids and secretaries whose cultural backgrounds make them unlikely wives. With women more widely in the workplace, he will meet young single women at the office, or on a business trip, who are exactly the sort he might marry if he had it to do all over again; and he *can* do it all over again. Whereas extramarital activity in the nineteenth century, and often in the 1950s, was a sheerly sexual outlet for an otherwise committed husband, today it is often a slippery slope toward desertion. The double standard may have once bolstered monogamy, but these days it brings divorce.

Even aside from the question of whether Victorian morality would "work" today, there is the question of whether any benefits would justify its many peculiar costs. Some Victorian men and women felt desperately trapped by marriage. (Although when marriages seem inescapable, almost literally unthinkable, people may dwell less on shortcomings.) And prevailing morality made it hard for some women to guiltlessly enjoy even marital sex—not to mention the fact that Victorian men weren't known for their sexual sensitivity. Life was also hard for women who wanted to be more than ornaments, more than an "Angel in the House." The Darwin sisters reported to Charles with some concern about brother Erasmus's budding, if ambiguous, friendship with the author Harriet Martineau, who didn't fit meekly into the feminine mold. Darwin, upon meeting her, had this impression: "She was very agreeable and managed to talk on a most wonderful number of subjects, considering the limited time. I was astonished to find how little ugly she is, but as it appears to me, she is overwhelmed with her own projects, her own thoughts and own abilities. Erasmus palliated all this, by maintaining one ought not to look at her as a woman."[34] This sort of remark is one of many reasons we shouldn't try to recreate Victorian sexual morality wholesale.

There no doubt are other moral systems that could succeed in sustaining monogamous marriage. But it seems likely that any such system will, like Victorianism, entail real costs. And although we can certainly strive for a morality that distributes costs *evenly* between men and women (and evenly among men and among women), distributing the costs *identically* is less likely. Men and women are different, and the threats that their evolved minds pose to marriage are different. The sanctions with which an efficient morality combats these threats will thus be different for the two sexes.

If we are really serious about restoring the institution of monogamy, *combat*, it seems, will indeed be the operative word. In 1966, one American scholar, looking back at the sense of shame surrounding the sexual impulse among Victorian men, discerned "a pitiable alienation on the part of a whole class of men from their own sexuality."[35] He's certainly right about the alienation. But the "pitiable" part is another question. At the other end of the spectrum from "alienation" is "indulgence"—obedience to our sexual impulses as if they were the voice of the Noble Savage, a voice that could restore us to some state of primitive bliss which, in fact, never was. A quarter-century of indulging these impulses has helped bring a world featuring, among other things: lots of fatherless children; lots of embittered women; lots of complaints about date rape and sexual harassment; and the frequent sight of lonely men renting X-rated videotapes while lonely women abound. It seems harder these days to declare the Victorian war against male lust "pitiable." Pitiable compared to what? Samuel Smiles may seem to have been asking a lot when he talked about spending one's life "armed against the temptation of low indulgences," but the alternative isn't obviously preferable.

WHERE DO MORAL CODES COME FROM?

The intermittently moralistic tone of this chapter is in a sense ironic. Yes, on the one hand, the new Darwinian paradigm does suggest that any institution as "unnatural" as monogamous marriage may be hard to sustain without a strong (that is, repressive) moral code. But the new paradigm also has a countervailing effect: nourishing a certain moral relativism—if not, indeed, an outright cynicism about moral codes in general.

The closest thing to a generic Darwinian view of how moral codes arise is this: people tend to pass the sorts of moral judgments that help move their genes into the next generation (or, at least, the kinds of judgments that would have furthered that cause in the environment of our evolution). Thus a moral code is an informal compromise among competing spheres of genetic self-interest, each acting to mold the code to its own ends, using any levers at its disposal.[36]

Consider the sexual double standard. The most obvious Darwinian explanation is that men were designed, on the one hand, to be sexually loose themselves yet, on the other, to relegate sexually loose women ("whores") to low moral status—even, remarkably, as those same men encourage those same women to be sexually loose. Thus, to the extent that men shape the moral code, it may include a double standard. Yet on closer inspection, this quintessentially male judgment is seen to draw natural support from other circles: the parents of young, pretty girls, who encourage their daughters to save their favors for Mr. Right (that is, to remain attractive targets for male parental investment), and who tell their daughters it's "wrong" to do otherwise; the daughters themselves, who, while saving their virtue for a high bidder, self-servingly and moralistically disparage the competing, low-rent alternatives; happily married women who consider an atmosphere of promiscuity a clear and present danger to their marriage (that is, to continued high investment in their offspring). There is a virtual genetic conspiracy to depict sexually loose women as evil. Meanwhile, there is relative tolerance for male philandering, and not only because some males (especially attractive or rich ones) may themselves like the idea. Wives, too, by finding a husband's desertion more shattering than his mere infidelity, reinforce the double standard.

If you buy this way of looking at moral codes, you won't expect them to serve the interests of society at large. They emerge from an informal political process that presumably gives extra weight to powerful people; they are quite unlikely to represent everyone's interests equally (though more likely to do so, perhaps, in a society with free speech and economic equality). And there's definitely no reason to assume that existing moral codes reflect some higher truth apprehended via divine inspiration or detached philosophical inquiry.

Indeed, Darwinism can help highlight the contrast between the moral codes we have and the sort that a detached philosopher might arrive at. For example: though the double standard's harsh treatment of female promiscuity may be a natural by-product of human nature, an ethical philosopher might well argue that sexual license is more often *morally* dubious in the case of the man. Consider an unmarried man and an unmarried woman on their first date. The man is more likely than the woman to exaggerate emotional commitment (consciously or unconsciously) and obtain sex under these false pretenses. And, if he does so, his warmth is then more likely than hers to fade. This is far, far from a hard and fast rule; human behavior is very complex, situations and individuals vary greatly, and members of both sexes get emotionally chewed up in all kinds of ways. Still, as a gross generalization, it is probably fair to say that single men cause more pain to partners of short duration through dishonesty than single women do. So long as women don't sleep with already mated men, their sexual looseness typically harms others obliquely and diffusely, if at all. Thus, if you believe, as most people seem to, that it is immoral to cause others pain by implicitly or explicitly misleading them, you might be more inclined to condemn the sexual looseness of men than of women.

That, at any rate, would be my inclination. If in this chapter I seemed to suggest that women practice sexual restraint, the advice wasn't meant to carry any overtone of obligation. It was self-help, not moral philosophy.

This may sound paradoxical: one can, from a Darwinian vantage point, advise sexual restraint for women, roughly echoing traditional moral exhortation, while at the same time decrying the moral censure of women who don't take the advice. But you might as well get used to the paradox, for it's part of a more general Darwinian slant on morality.

On the one hand, a Darwinian may treat existing morality with suspicion. On the other hand, traditional morality often embodies a certain utilitarian wisdom. After all, the pursuit of genetic interest sometimes, though not always, coincides with the pursuit of happiness. Those mothers who urge their daughters to "save themselves" may at one level be counseling ruthless genetic self-interest, but they are, on another level, concerned for the long-term happiness of their daughters. So too for the daughters who follow mother's advice, believing it will help them become lastingly married and have children: yes, the reason they want childen is because their genes "want" them to want children; nonetheless, the fact remains that they *do* want children, and may well, in fact, have more fulfilling lives if they get them. Though there's nothing inherently good about genetic self-interest, there's nothing inherently wrong with it either. When it does conduce to happiness (which it won't always), and doesn't gravely hurt anyone else, why fight it?

For the Darwinian inclined toward moral philosophy, then, the object of the game is to examine traditional morality under the assumption that it is laden with practical, life-enhancing wisdom, yet is also laced with self-serving and philosophically indefensible pronouncements about the absolute "immorality" of this or that. Mothers may be wise to counsel restraint in their daughters—and, for that matter, wise to condemn competing girls who aren't so restrained. But the claim that these condemnations have *moral* force may be just a bit of genetically orchestrated sophistry.

Extricating the wisdom from the sophistry will be the great and hard task of moral philosophers in the decades to come, assuming that more than a few of them ever get around to appreciating the new paradigm. It is a task, in any event, to which we'll return toward the end of this book, after the origins of the most fundamental moral impulses have become apparent.

SUGARCOATED SCIENCE

One common reaction to discussions of morality in light of the new Darwinism is: Aren't we getting a little ahead of the game here? Evolutionary psychology is just getting started. It has produced some theories with powerful support (an innate difference in male and female jealousy); some with fair-to-middling support (the Madonna-whore dichotomy); and many more that are sheer, if plausible, speculation (the "mate ejection" module). Is this body of theories really capable of supporting sweeping pronouncements about Victorian, or any other, morality?

Philip Kitcher, a philosopher who in the 1980s established himself as sociobiology's preeminent critic, has carried this doubt a step further. He believes Darwinians should tread carefully not just in making moral or political extensions from their inchoate science (extensions most of them avoid anyway, thanks to the scorching a few received in the 1970s), but in making the science in the first place. After all, even if they don't cross the line between science and values, someone else will; theories about human nature will inevitably be used to support this or that doctrine of morality or social policy. And if the theories turn out to be wrong, they may have done a lot of damage in the meanwhile. Social science, Kitcher notes, is different from physics or chemistry. If we embrace "an incorrect view of the origins of a distant galaxy," then "the mistake will not prove tragic. By contrast, if we are wrong about the bases of human social behavior, if we abandon the goal of a fair distribution of the benefits and burdens of society because we accept faulty hypotheses about ourselves and our evolutionary history, then the consequences of a scientific mistake may be grave indeed." Thus, "When scientific claims bear on matters of social policy, the standards of evidence and of self-criticism must be extremely high."[37]

There are two problems here. First, "self-criticism" per se is not an essential part of science. Criticism from colleagues—a kind of *collective* self-criticism—is. It is what keeps the "standards of evidence" high. And this collective self-criticism can't even begin until a hypothesis is put forward. Presumably Kitcher isn't suggesting that we short-circuit this algorithm of scientific progress by refraining from the proposal of weak hypotheses; the way weak hypotheses get strong is by being proposed and then mercilessly scrutinized. And if Kitcher is suggesting only that we label speculative hypotheses as such, no one has any objection to that. Indeed, thanks to people like Kitcher (and this isn't meant sarcastically), many Darwinians are now masters of careful qualification.

Which brings us to the second problem with Kitcher's argument: the suggestion that Darwinian social scientists, but not social scientists generally, should proceed with great caution. The unspoken assumption is that incorrect Darwinian theories about behavior will tend to be more pernicious than incorrect *non*-Darwinian theories about behavior. But why should that be so? One long-standard and utterly non-Darwinian doctrine of psychology—that there are no important innate mental differences between men and women bearing on courtship and sex—seems to have caused a fair amount of suffering over the past few decades. And it depended on the lowest imaginable "standards of evidence"—no real evidence whatsoever, not to mention the blatant and arrogant disregard of folk wisdom in every culture on the planet. For some reason, though, Kitcher isn't upset about this; he seems to think that theories involving genes can have bad effects but theories not involving genes can't.

A more reliable generalization would seem to be that incorrect theories are more likely than correct theories to have bad effects. And if, as is often the case, we don't know for sure which theories are right and which are wrong, our best bet is to go with the ones that seem most likely to be right. The premise of this book is that evolutionary psychology, in spite of its youth, is now far and away the most likely source of theories about the human mind that will turn out to be right—and that, indeed, many of its specific theories already have fairly firm grounding.

Not all threats to the honest exploration of human nature come from the enemies of Darwinism. Within the new paradigm, truth sometimes gets sugarcoated. It is often tempting, for example, to downplay differences between men and women. Regarding the more polygamous nature of men, politically sensitive Darwinian social scientists may say things like: "Remember, these are only statistical generalizations, and any one person may diverge greatly from the norm for his or her sex." Well, yes, but few of those divergences are very close to the other sex's norm (and half of the divergences, re-member, are *farther* than average from the other sex's norm). Or: "Remember, behavior is influenced by the local environment and conscious choice. Men don't *have* to philander." True—and crucially important. But many of our impulses are, by design, very strong, so any force that is to stifle them may have to be pretty harsh. It is grossly misleading to talk as if self-restraint is as easy as punching a channel on the remote control.

It's dangerous too. George Williams, perhaps the closest thing there is to a single founding father of the new paradigm, may be going too far when he says that natural selection is "evil." After all, it created everything benign in human nature as well as everything destructive. But surely it's true that the roots of all evil can be seen in natural selection, and are expressed (along with much that is good) in human nature. The enemy of justice and decency does indeed lie in our genes. If in this book I seem to depart from the public-relations strategy practiced by some Darwinians, and stress the bad in human nature more than the good, it is because I think we are more in danger of underestimating the enemy than overestimating it.

Part Two: **SOCIAL CEMENT**

Chapter 7: FAMILIES

[W]ith the working ant we have an insect differing greatly from its parents, yet absolutely sterile; so that it could never have transmitted successively acquired modifications of structure or instinct to its progeny. It may well be asked how is it possible to reconcile this case with the theory of natural selection?

—*The Origin of Species* (1859)

[Yesterday] Doddy [Darwin's son William] was generous enough to give Anny the last mouthful of his gingerbread & today . . . he again put his last crumb on the sofa for Anny to run to & then cried in rather a vain-glorious tone "oh kind Doddy" "kind Doddy."

—Observations of Darwin's children (1842)[1]

We all like to think of ourselves as selfless. And on occasion we are. But we are pigs compared to the social insects. Bees die for their fellow bees, disemboweling themselves upon stinging an intruder. Some ants, also in defense of the colony, detonate themselves. Other ants spend their lives as doors, keeping out insects that lack security clearance, or as food sacks, hanging bloated from the ceiling in case of scarcity.[2] These pieces of furniture have no offspring.

Darwin spent more than a decade wondering how natural selection could have produced whole castes of ants that create no descendants. Meanwhile, he was creating plenty of descendants himself. The problem of insect sterility had gotten his attention by the time his fourth child, Henrietta, was born in late 1843, and he still had not solved it by the birth of his tenth and last child, Charles, in 1856. For all those years, he kept the theory of natural selection secret, and one reason may be the seemingly blatant contradiction of it by ants. The paradox seemed "insuperable, and actually fatal to my whole theory."[3]

Darwin probably didn't suspect, as he pondered the insect puzzle, that the solution to it could explain, as well, the texture of his growing family's everyday life: why his children showed affection for one another, why they sometimes fought; why he felt compelled to teach them the virtues of kindness, why they sometimes resisted; even why he and Emma grieved more deeply the loss of one of their children than another. Understanding self-sacrifice among insects would un-lock the dynamics of family life among mammals, including people.

Though Darwin finally conceived, at least vaguely, the correct explanation of insect sterility, and suspected that it might have rel-evance to human behavior, he came nowhere near seeing the breadth and diversity of the relevance. Neither did anyone else until a century later.

One reason for this may be that Darwin's explanation, as he phrased it, was hard to grasp. In *The Origin of Species* he wrote that the paradox of evolved sterility "is lessened, or, as I believe, disap-pears, when it is remembered that selection may be applied to the family, as well as to the individual, and may thus gain the desired end. Thus, a well-flavoured vegetable is cooked, and the individual is destroyed; but the horticulturist sows seeds of the same stock, and confidently expects to get nearly the same variety; breeders of cattle wish the flesh and fat to be well marbled together; the animal has been slaughtered, but the breeder goes with confidence to the same family."[4]

However strange it may seem to bring plant and animal breeders into the picture, this made perfect sense after 1963, when a young British biologist named William D. Hamilton sketched out the theory of kin selection.[5] Hamilton's theory is an articulation and extension of Darwin's insight in the language of genetics, a language that didn't exist in Darwin's day.

The term *kin selection* itself suggests a link with Darwin's assertion that "selection may be applied to the family," and not just to the individual organism. But this suggestion, while true, is misleading. The beauty of Hamilton's theory is that it sees selection as taking place not so much at the level of the individual *or* the family, but, in an important sense, at the level of the gene. Hamilton was the first to clearly sound this central theme of the new Darwinian paradigm: looking at survival from the gene's point of view.

Consider a young ground squirrel that has not yet produced any offspring and that, upon sighting a predator, gets up on its hind legs and delivers a loud alarm call, which may attract the predator's attention and bring sudden death. If you look at natural selection the way almost all biologists looked at it through the mid-twentieth century—a process concerned with the survival and reproduction of animals, and of their offspring—this warning call doesn't make sense. If the ground squirrel giving it has no offspring to save, then the warning call is evolutionary suicide. Right? This is the question that was momentously answered in the negative by Hamilton.

In the Hamiltonian view, attention shifts from the ground squirrel that is sounding the alarm to the gene (or, in real life, the series of genes) responsible for the alarm. After all, ground squirrels don't live forever, and neither do any other animals. The only potentially immortal organic entity is a gene (or, strictly speaking, the pattern of information encoded in the gene, since the physical gene itself will pass away after conveying the pattern through replication). So, in an evolutionary time frame, over hundreds or thousands or millions of generations, the question isn't how individual animals fare; we all know the finally grim answer to that one. The question is how individual *genes* fare. Some will pass away and some will thrive, and which do which is a matter of consequence. How will a "suicidal warning call" gene fare?

The somewhat surprising answer, which lay at the core of Hamilton's theory, is: quite well, under the right circumstances. The reason is that the ground squirrel containing the gene may have some nearby relatives who will be saved by the alarm call, and some of those relatives probably carry the same gene. Half of all brothers and sisters, for example, can be assumed to possess the gene (unless they're half-siblings, in which case the fraction is a still nontrivial one-fourth).

If the warning call saves the lives of four full siblings that would otherwise die, two of which carry the gene responsible for it, then the gene has done well for itself, even if the sentry containing it pays the ultimate sacrifice. This superficially selfless gene will do much better over the ages than a superficially selfish gene that induced its carrier to scurry to safety while four siblings—and two copies of the gene, on average—perished.* The same is true if the gene saves only one full sibling, while giving the sentry a one-in-four chance of dying. Over the long run, there will be two genes saved for every gene lost.

GENES FOR BROTHERLY LOVE

There is nothing mystical going on here. Genes don't magically sense the presence of copies of themselves in other organisms and try to save them. Genes aren't clairvoyant, or even conscious; they don't "try" to do anything. But should a gene appear that *happens* to make its vehicle behave in ways that help the survival or reproductive prospects of other vehicles likely to contain a copy of that gene, then the gene may thrive, even if prospects for *its* vehicle are lowered in the process. This is kin selection.

This logic could apply, as in this case, to a gene that inclines a mammal to produce a warning call when it sees a threat to its home burrow, where relatives reside. The logic could also apply to a gene that leads an insect to be sterile, so long as the insect spends its life helping fertile relatives (who contain the gene in "unexpressed" form) to survive or reproduce. And the logic could apply to genes inclining human beings to sense early on who their siblings are and thereafter share food with them, give guidance to them, defend them, and so on—genes, in other words, leading to sympathy, empathy, compassion: genes for love.

A failure to appreciate familial love had helped keep the principle of kin selection from clear view before Hamilton's day. In 1955, in

---

*Actually, a ground squirrel (or a person) shares much more than half of his genes with a sibling—and, indeed, with other members of his species. But fairly *novel* genes, genes that have just appeared within a population, will, on average, reside in half of an organism's full siblings. And novel genes are the ones that matter when we're talking about the evolution of new traits.

a popular article, the British biologist J.B.S. Haldane had noted that a gene inclining you to jump in a river and save a drowning child, taking a one-in-ten chance of dying, could flourish so long as the child were your offspring or your brother or sister; the gene could even spread, at a slower rate, if that child were your first cousin, since first cousins share, on average, one eighth of your genes. But rather than sustain this train of thought, he cut it short by observing that in emergencies people don't have time to make mathematical calculations; and surely, he said, our Paleolithic ancestors hadn't run around calculating their degree of relatedness to each other. So Haldane concluded that genes for heroism would spread only "in rather small populations where most of the children were fairly near relatives of the man who risked his life."[6] In other words: an indiscriminate heroism, reflecting the *average* degree of relatedness to people in the general vicinity, could evolve if that average were fairly high.

For all Haldane's insight in looking at things from the gene's, rather than the individual's, point of view, his failure to follow this logic to its end is odd, to say the least. It's as if he thought natural selection realizes its calculations by having organisms consciously repeat them, rather than by filling organisms with feelings that, in their fine contours, are proxies for calculation. Hadn't Haldane noticed that people tend to have the warmest feelings for the people who share the largest fraction of their genes? And that people are more inclined to risk their lives for the people they feel warmly toward? Why should it matter that Paleolithic men weren't math whizzes? They were animals; they had feelings.

Technically speaking, Haldane was right insofar as he went. Within a small, closely related population, an indiscriminate altruism could indeed evolve. And that's true even though some of the altruism would get spent on people who weren't relatives. After all, even if you channel your altruism precisely toward siblings, some of it is wasted, in evolutionary terms, since siblings don't share all your genes, and any given sibling may not carry the gene responsible for the altruism. What matters, in both cases, is that the altruism gene *tends* to improve prospects for vehicles that will *tend* to carry copies of itself; what matters is that the gene does more good than harm, in the long run, to its own proliferation. Behavior always takes place amid uncertainty, and all natural selection can do is play the odds. In the Haldane scenario the way to play the odds is to instill a mild and generalized altruism, the exact strength depending on the average extent of kinship with people regularly in the vicinity. This is conceivable.

But as Hamilton noted in 1964, natural selection will, given the opportunity, improve the odds by minimizing uncertainty. Any genes that sharpen the precision with which altruism is channeled will thrive. A gene that leads a chimpanzee to give two ounces of meat to a sibling will eventually prevail over a gene that leads it to give an ounce to a sibling and an ounce to an unrelated chimp. So unless identifying kin is very hard, evolution should produce a strong and well-targeted strain of benevolence, not a weak and diffuse strain. And that is what has happened. It has happened, at least to some extent, with ground squirrels, which are more likely to deliver warning calls in the presence of close kin.[7] It has happened, to some extent, with chimpanzees and other nonhuman primates, which often have uniquely supportive sibling relationships. And it has happened, to a great extent, with us.

Maybe the world would be a better place if it hadn't. Brotherly love in the literal sense comes at the expense of brotherly love in the biblical sense; the more precisely we bestow unconditional kindness on relatives, the less of it is left over for others. (This, some believe, is what kept Haldane, a Marxist, from facing the truth.) But, for better or worse, the literal kind of brotherly love is the kind we have.

Many social insects recognize their kin with the help of chemical signals called pheromones. It is less clear how humans and other mammals figure out (consciously or unconsciously) who their kin are. Surely seeing our mother feed and care for a child day after day is one conspicuous cue. We may also, by observing our mother's social affiliations, develop a sense for, say, who her sister is, and hence who her sister's offspring are. Besides, since the advent of language, mothers have been able to tell us who's who—instruction it is in their genetic interest to give and in our genetic interest to heed. (That is to say, genes inclining the mother to help children identify kin would thrive, as would genes inclining children to pay attention.) It's hard to say what other kin-recognition mechanisms, if any, are at work, since experiments that might settle the question involve unethical things like removing children from families.[8]

What's clear is that mechanisms exist. Anyone with siblings—anyone in any culture—is familiar with the empathy for a sibling in great need, the sense of fulfillment at giving aid, the guilt at not giving it. Anyone who has endured a sibling's death is familiar with grief. These people know what love is, and they have kin selection to thank for it.

That goes double for males, who, in the absence of kin selection, might never have felt deep love at all. Back before our species became high in male parental investment, there was no reason for males to be intensely altruistic toward offspring. That sort of affection was the exclusive province of females, in part because only they could be sure who their offspring were. But males could be pretty sure who their brothers and sisters were, so love crept into their psyches via kin selection. Had males not thus acquired the capacity for sibling love, they might not have been so readily steered toward high male parental investment, and the even deeper love it brings. Evolution can only work with the raw materials that happen to be lying around; if love for certain kinds of children—siblings—hadn't been part of males' minds several million years ago, the path to loving their own children—the path to high MPI—might have been too tortuous.

THE NEW MATH

With Hamilton's theory in hand, it's easier to appreciate the connection Darwin saw between a cow that has "well marbled" beef, and gets slaughtered and eaten, and an ant that works hard all its life without issue. The cow gene responsible for the good marbling, to be sure, has done nothing for its vehicle, which is now slaughtered, and may do nothing for the direct genetic legacy of its vehicle; dead cows can't have more offspring. But the gene will still do much for the indirect genetic legacy of its vehicle, for by producing the marbling, it prompts a farmer to feed and breed the vehicle's close relatives, some of which contain copies of the gene. So too with the sterile ant. The ant has no direct legacy, but the genes responsible for this fact do just fine, thank you, so long as the time and energy that would have been devoted to reproduction are profitably spent helping close relatives be prolific. Though the gene for sterility lies dormant in these relatives, it is there, and passes to the next generation, where it again produces gobs of sterile altruists devoted to its transmission. This is the exact sense in which worker bees and tasty cattle are alike: some genes, by impeding their transmission through one conduit, lubricate their transmission through others, and the net result is more transmission.

That Darwin, working with no knowledge of genes, with no sound understanding of the nature of heredity, should sense this parallel a century before Hamilton is one of the higher tributes to the care and precision of his thought.

Still, let there be no doubt about the superiority of Hamilton's version of kin selection to Darwin's. It is accurate enough to say, as Darwin did, that sometimes (as with insect sterility) natural selection operates on the family and sometimes on the individual organism. But why not keep things simple? Why not just say that in both of these cases, the ultimate unit of selection is the gene? Why not make a single brief statement that encompasses all forms of natural selection? Namely: those genes that are conducive to the survival and reproduction *of copies of themselves* are the genes that win. They may do this straightforwardly, by prompting their vehicle to survive, beget offspring, and equip the offspring for survival and reproduction. Or they may do this circuitously—by, say, prompting their vehicle to labor tirelessly, sterilely, and "selflessly," so that a queen ant can have lots of offspring containing them. However the genes get the job done, it is selfish from *their* point of view, even if it seems altruistic at the level of the organism. Hence the title of Richard Dawkins's book, *The Selfish Gene.* (The title has caught flack from people who note that genes don't have intentions, and so can't be "selfish." True, of course, but the phrase wasn't meant literally.)

Naturally, the level of the organism is of primary concern to human beings; human beings are organisms. But it's of secondary importance to natural selection. If there is a sense in which natural selection "cares" about anything—and there is, metaphorically—that thing isn't us; it's the information in our sex cells, our eggs and our sperm. Of course, natural selection "wants" us to behave in certain ways. But, so long as we comply, it doesn't care whether we are made happy or sad in the process, whether we get physically mangled, even whether we die. The only thing natural selection ultimately "wants" to keep in good shape is the information in our genes, and it will countenance any suffering on our part that serves this purpose.

This was the philosophical import of the simple point Hamilton made abstractly, skeletally, in a 1963 letter to the editors of the journal *The American Naturalist*. He imagined a gene G that causes an altruistic behavior and noted: "Despite the principle of 'survival of the fittest' the ultimate criterion which determines whether G will spread is not whether the behavior is to the benefit of the behaver but whether it is to the benefit of the gene G; and this will be the case if the average net result of the behavior is to add to the gene-pool a handful of genes containing G in higher concentration than does the gene-pool itself."[9]

Hamilton gave flesh to this observation the next year with his paper "The Genetical Evolution of Social Behaviour" in *The Journal of Theoretical Biology*. The paper, after going underappreciated for years, has become one of the most widely cited works in the history of Darwinian thought, and has revolutionized the mathematics of evolutionary biology. Before the theory of kin selection, it was common to talk as if the final arbiter in evolution were "fitness," whose ultimate manifestation, it seemed, was the sum total of the organism's direct biological legacy. Genes that made an organism fitter—that maximized the number of offspring, grand-offspring, and so on—would be the genes that flourished. Now the final arbiter of evolution is thought of as "inclusive fitness," which takes into account also the genes' indirect legacy, realized via siblings, cousins, and so on. Hamilton wrote in 1964: "Here then we have discovered a quantity, inclusive fitness, which under the conditions of the model tends to maximize in much the same way that fitness tends to maximize in the simpler classical model."

Hamiltonian math contains a potent symbol—r—introduced earlier by the biologist Sewall Wright but now given new consequence; r represents the degree of relatedness among organisms. Among full siblings, r is ½, among half-siblings, nieces, nephews, aunts, and uncles, it is ¼, and among first cousins it is ⅛. The new math says that genes for sacrificial behavior will thrive so long as the cost to the altruist (in terms of impact on future reproductive success) is less than the benefit to the recipient (ditto) times the degree of relatedness between the two. That is, so long as c is less than br.

When Hamilton introduced the theory of kin selection, he used as his example the very group of organisms that had perplexed Darwin. Like Darwin, he had been struck by the extraordinary self-sacrifice among many insects of the order Hymenoptera, notably the highly social ants, bees, and wasps. Why is this intensity of altruism, and its attendant social cohesion, found in so few other parts of the insect world? There may be several evolutionary reasons, but Hamilton put his finger on what seems a central one. He noted that, thanks to a bizarre form of reproduction, these species feature an unusually large r. Sister ants share ¾ of their genes by common descent, not just ½. So altruism of extraordinary magnitude is justified in the eyes of natural selection.

When r is even larger than ¾, the evolutionary argument for altruism, and social solidarity, grows even stronger. Consider the cellular slime mold, which is so tightly interwoven that it has inspired reasoned debate as to whether it is best thought of as a society of cells or a single organism. Because slime-mold cells reproduce asexually, the r among them is 1; they are all identical twins. From the point of view of the gene, then, there is *no* difference between the fate of its own cell and the fate of a nearby cell. It's not surprising that many slime-mold cells fail to reproduce, and devote themselves instead to buffering fertile fellow cells from the elements. Their neighbors' welfare, in evolutionary terms, is identical to their own. *That's* altruism.

So too with human beings—not groups of human beings, but the groups of cells that *are* human beings. At some point hundreds of millions of years ago, multicellular life arose. Societies of cells became so highly integrated as to qualify for the title "organism," and these organisms eventually begat us. But as the cellular slime mold attests, the line between society and organism is unclear. It is fair, technically speaking, to consider even so coherent an organism as a human being a tight-knit community of single-celled organisms. These cells exhibit a kind of cooperation and self-sacrifice that makes even the machine-like efficiency of an insect colony look ragged by comparison. Almost all of the cells in the human body are sterile. Only the sex cells— our "queen bees"—get to make copies of themselves for posterity. That the zillions of sterile cells act as if they were perfectly content with this arrangement is doubtless grounded in the fact that the r between them and the sex cells is 1; genes in sterile cells are transmitted to future generations as assuredly via sperm or egg as they would be if their particular cellular vehicles were doing the transmitting. Again: when r is 1, altruism is ultimate.

THE LIMITS OF LOVE

The reverse side of this coin is that when r isn't 1, altruism isn't ultimate. Even pure sibling love—brotherly love—isn't total love. J.B.S. Haldane is said to have remarked once that he would never give his life for a brother—but, rather, for "two brothers or eight cousins." Presumably he was joking—parodying, perhaps, what he wrongly considered the overly fine extension of Darwinian logic. But his joke captures a basic truth. To define the degree of commitment to any relative is to define the degree of indifference and, potentially, antagonism; the cup of common interest between siblings is half-empty as well as half-full. While it makes genetic sense to help a brother or sister, even at great expense, that expense is not unlimited.

Thus, on the one hand, no modern Darwinian would expect a child to monopolize the food supply while a brother or sister grew weak from hunger. But neither should we expect that, given two siblings and one sandwich, the question of its allotment will be amicably resolved. It may not be hard to teach children to share with brothers and sisters (at least in some circumstances), but it is hard to teach them to share *equally*, for this runs against their genetic interest. That, at any rate, is what natural selection implies. We can leave it for veteran parents to say whether the prediction is borne out.

The divergence of genetic interests between siblings creates an exasperating, if sometimes charming, paradox. They fiercely compete for the affection and attention of their parents, with all the resources that can bring, and in the process display jealousy so petty that it's hard to credit them with love; but let one of them become truly needy, or seem genuinely endangered, and love will surface. Darwin saw one such shift in attitude on the part of his son Willy, then nearly five years old, toward younger sister Annie. "Whenever she hurts herself when we are present Willy appears not to mind, & sometimes makes a great noise as if to distract our attention," Darwin wrote. But one day Annie hurt herself with no adults in view, so Willy couldn't assume that any real danger was being addressed. Then his reaction "was quite different. He first attempted to comfort her very nicely & then said he would call Bessy & she not being in sight his fortitude gave way & he began to cry also."[10] Darwin didn't explain this, or any instances of sibling love, in terms of kin, or what he called "family," selection; he seems never to have seen the connection between insect self-sacrifice and mammalian affection.[11]

The biologist who first emphasized the partial emptiness of the cup of common genetic interest is Robert Trivers. He has noted, in particular, that a child's genetic interest diverges not only from a brother's or sister's, but from a parent's. Each child should, in theory, see itself as twice as valuable as its sibling, while the parent, being equally related to the two, values them equally. Hence another Darwinian prediction: not only will siblings have to be taught to share equally; parents will, in fact, try to teach them.

In 1974, Trivers dissected parent-offspring conflict in a paper by that name. By way of illustration, he discussed the contentious mammalian issue of when a suckling should get off the teat. A caribou calf, he observed, will continue to suckle long after milk has ceased to be essential to its survival, even though this prevents the mother from conceiving another calf that will share some of its genes. After all: "the calf is completely related to himself but only partially related to his future siblings. . . ."[12] The time will come when the nutritional rewards from suckling are so marginal that genetic interest favors another calf over milk. But the mother, valuing (implicitly) the two offspring equally, reaches that point sooner. So the theory of natural selection, stated in terms of inclusive fitness, implies that conflict over weaning will be a regular part of mammalian life—as it seems to be. The conflict can last for several weeks and become pretty wild, as infants shriek for milk and even strike their mother. Veteran baboon watchers know that a good way to find a baboon troop is to listen each morning for the sound of mother-offspring strife.[13]

In the battle over resources, expect children to use any tools at their disposal, including dishonesty. The dishonesty may be crude and directed at other siblings. ("Willy sometimes tries a little ruse to prevent Annie wishing to have his apple. . . .'Yours is larger than mine Annie.' ") But the ruse may be more subtle, and directed to a larger audience, including parents. One good way to short-circuit parental demands for greater sacrifice is to exaggerate—or, shall we say, selectively highlight—sacrifices already made. An example appears at the head of this chapter: Willy, then two years old, and nicknamed Doddy, had given his younger sister his last bit of gingerbread and then exclaimed, for all to hear, "Oh kind Doddy, kind Doddy."[14] Many parents are familiar with this sort of conspicuous nonconsumption.

Another ploy children use to extract resources from parents is to embellish their needs. Emma Darwin recorded three-year-old son Leonard's actions after "he scraped 2 little bits of skin off his wrist": "He thought Papa did not pity him enough & nodded emphatically at him. 'The skin's come off—& its lost—& the bleed's coming out.' " A year later Leonard was heard to say, "Papa, I have coughed awfully—many times awfully—five awfullys—and more, too—so mayn't I have some black stuff [licorice]?"[15]

To further bolster an image of entitlement, youngsters may stress the cruelty and injustice being inflicted by parents. At peak intensity, this emphasis is known as a temper tantrum—part of young life not just in our species, but also among chimpanzees, baboons, and other primates. Many a young outraged chimp has been known, as one primatologist put it half a century ago, to "glance furtively at its mother or the caretaker as if to discover whether its action was attracting attention."[16]

Fortunately for young primates, parents are ripe for exploitation. Attention to a child's crying and complaining is in the interest of the parent's genes, since cries and complaints may signal real needs felt by a vehicle containing copies of them. In other words: parents love their children and can be blinded by that love.

Still, the idea of a temper tantrum as manipulation won't strike most parents as a revolutionary insight, and that's evidence that they aren't entirely blind. Though natural selection made parents open to manipulation in the first place, it should, in theory, have equipped them thereafter with antimanipulation devices, such as a discerning suspicion of childhood whining. But once this discernment exists, natural selection should equip the children with counterdiscernment technology in the form of more heartfelt whining. The arms race goes on forever.

As Trivers stressed in his 1974 paper, the gene's-eye view implies that parents are themselves dishonestly manipulative. They want— or, at least, their genes "want" them—to extract from the child more kin-directed altruism and sacrifice, and thus to instill in the child more love, than is in the child's genetic interest. This is true not just of sibling love, but of love for uncles, aunts, and cousins, all of whom (on average) have twice as many of the parent's genes as of the child's. Hence the general dearth of disputes in which a parent demands that a child be *less* considerate of the parent's siblings, nieces, and nephews.

Children are biologically vulnerable to a parent's propaganda campaign, just as the parent is vulnerable to the child's. The reason is that it so often makes Darwinian sense to do what parents say. Although the genetic interests of parent and child diverge, there is a 50 percent overlap, so no one has a stronger genetic incentive than a parent to fill a child's head with useful facts and adages. Thus there is no one to whom the child should pay more attention. A child's genes should "want" the child to tap into the uniquely devoted data-banks housed in its parents.

And plainly, the genes get their way. When young, we are filled with awe and credulity in the presence of our parents. One of Darwin's daughters recalled that "Whatever he said was absolute truth and law to us." Surely she exaggerates. (When Darwin found five-year-old Leonard jumping on the sofa and told him that was against the rules, Leonard replied, "Well then I *advise* you to go out of the room.")[17] Still, young children do have a basic, if not entire, trust in their parents, and parents should, in theory, abuse it.

In particular, parents should do what Trivers called "molding" in the guise of "teaching." He wrote: "Since teaching (as opposed to molding) is expected to be recognized by offspring as being in their own self-interest, parents would be expected to overemphasize their role as teachers in order to minimize resistance in their young."[18] Trivers might view with cynicism one of Darwin's recollections of his mother: "I remember her saying 'if she did ask me to do something . . . it was solely for my good.' "[19]

Parents have a second, more specific, advantage in trying to (partly) frustrate the genes of their children. Kin selection has ensured that the conscience looks out for siblings, generating guilt after any major neglect of them. So parents have a guilt-center to play on, and natural selection should make them good at playing on it. On the other hand, Trivers noted, natural selection should then turn around and give the children antiexploitation equipment—perhaps, for example, a pointed skepticism of parental claims about fraternal duty. Another arms race.

The result of all this is a full-fledged battle for every child's soul. Trivers has written: "The personality and conscience of the child is formed in an arena of conflict."[20]

Trivers sees the prevailing view of child-rearing—as a process of "enculturation," in which parents dutifully equip children with vital skills—as hopelessly naïve. "One is not permitted to assume that parents who attempt to impart such virtues as responsibility, decency, honesty, trustworthiness, generosity, and self-denial are merely providing the offspring with useful information on appropriate behavior in the local culture, for all such virtues are likely to affect the amount of altruistic and egoistic behavior impinging on the parent's kin, and parent and offspring are expected to view such behavior differently." It almost seems that Trivers sees the very prevalence of the notion of "enculturation" as an unspoken conspiracy among oppressors. "The prevailing concept of socialization," he notes, "is to some extent a view one would expect adults to entertain and disseminate."[21]

This hints at a sense in which Darwinism, long stereotyped as a right-wing worldview, can have emanations of another sort. Seen through the new paradigm, moral and ideological discourse can look like a constant power struggle, in which the powerful often prevail and the weak are often exploited. "The ruling ideas of each age," wrote Karl Marx and Friedrich Engels, "have ever been the ideas of its ruling class."[22]

MOM ALWAYS LIKED YOU BEST

So far we have been looking at the no-frills models of kin selection and parent-offspring conflict, relying on convenient but in some cases dubious assumptions. One such assumption is that during human evolution siblings have had the same father as well as mother. To the extent that this premise is flawed—and it surely is, to some extent—then the "natural" ratio of altruism among siblings isn't two to one in favor of the self, but somewhere between two and four to one. (This amendment may ease the concerns of parents who find their offspring more mutually antagonistic than Hamilton's math seems to deem "natural.") Of course, it is possible that offspring actually (unconsciously) gauge the odds that their siblings share their father as well as their mother, and then treat them accordingly. It would be interesting, for example, to see if siblings with two homebody parents are more generous toward one another than siblings whose parents are often apart.

Another oversimplification has been the idea that r by itself—your degree of relatedness to other people—defines your genetically optimal attitude toward them. The mathematical question William Hamilton raised—Is c less than br?—has two other variables: the cost of your altruism to you (c) and the benefit to the recipient (b). Both are stated in terms of Darwinian fitness: how much the chances of producing viable, reproductively successful offspring will drop for you because of the altruism and how much they'll rise for the recipient. Both of these things, obviously, depend on what those chances were to begin with—on how much, if any, reproductive potential you and the other person have. And reproductive potential is a much-changing thing, both from one relative to the next and, for any one relative, from one decade to the next.

For example: a big, strong, smart, handsome, ambitious brother has greater likely reproductive success than a withdrawn, dull, inept brother. And this would have been especially true in the social environment of human evolution, where high-status males might have more than one wife—or, failing that, might be widely adulterous. In theory, parents should (consciously or unconsciously) pay attention to such differences. They should dole out investments in their various children with all the discernment of a Wall Street portfolio manager, the goal always being to maximize overall reproductive return on each increment of investment. Thus, there may be an evolutionary basis for the complaint that "Mom [or Dad] always liked you best." The Smothers Brothers comedy team made that line famous during the 1960s, and it was always the dull-witted, pie-faced Tommy making the complaint to his sharper, more dynamic brother Dick.[23]

The relative reproductive potential of two offspring may depend on more than the offspring themselves. It may depend on a family's social standing. For a poor family with a pretty girl and a handsome but not otherwise especially gifted boy, the daughter is more likely to produce children who begin life with material advantages; girls more often "marry up" on the socioeconomic scale than boys.[24] For rich, high-status families, it's the sons who, all other things being equal, have higher reproductive potential; a man, unlike a woman, can use wealth and status to produce scores of offspring.

Are human beings programmed to execute this unsettling logic? Do parents who find themselves resource-rich or high in status unconsciously decide to lavish attention on their sons at the expense of daughters, since sons can (or could, during evolution) more efficiently convert status or material resources into offspring? Do parents who find themselves poor do the reverse? Sounds creepy, but that doesn't mean it doesn't happen.

The logic is grounded in a more general point that Robert Trivers made in 1973 in a paper coauthored with the mathematician Dan E. Willard.[25] In any polygynous species, some males mate prolifically and others fail entirely to reproduce. So mothers in poor physical condition might profit (genetically) from treating daughters as more valuable assets than sons. For, assuming the mother's poor health leads—through skimpy milk, say—to frail offspring, it will bode especially ill for the sons. Malnourished males may be shut out of reproductive competition altogether, whereas a fertile female in almost any condition can usually attract a sex partner.

Some nonhuman mammals seem to comply with this logic. Florida pack rat mothers, if fed poorly, will force sons off the teat, even letting them starve to death, while daughters nurse freely. In other species, even the *birth ratio* of males to females is affected, with mothers in the most auspicious condition having mostly sons and less advantaged mothers having mostly daughters.[26]

In our species, somewhat polygynous through much of its evolution, wealth and status can be as important as health. Both are weapons with which men compete for women—and, in the case of status, at least, this has been the case for millions of years. So, for parents who find themselves socially and materially advantaged, to invest more in sons than daughters would make (Darwinian) sense. This is a good example of logic that often strikes people as somehow too Machiavellian to be part of human nature. To a Darwinian, the Machiavellian coldness, if anything, adds credibility. (As Thomas Huxley said after Darwin proposed a particularly seamy hypothesis about reproduction in jellyfish: "The indecency of the process is to a certain extent in favour of its probability.")[27] So far, there is evidence that the Darwinians are right, and much less evidence to the contrary.

In the late 1970s, the anthropologist Mildred Dickemann, after studying nineteenth-century India and China and medieval Europe, concluded that female infanticide—the killing of newborn daughters because they are daughters—has been most intensively practiced among the upper classes.[28] There is also the well-known tendency in many cultures, including Darwin's, for wealthy families to pass their largest assets down to their sons, not their daughters. (A relative of Darwin's, the early twentieth-century economist Josiah Wedgwood, noted in a study of inheritance that "It appeared to be usual among the wealthier predecessors in my sample for the sons to receive a larger share than daughters. In the case of the smaller estates, equal division is much more common.")[29] Biases toward sons or daughters can assume subtler form. The anthropologists Laura Betzig and Paul Turke, working in Micronesia, found that high-status parents spend more time with sons and low-status parents with daughters.[30] All these findings are consistent with the logic of Trivers and Willard: for families at the upper end of the socioeconomic scale, sons are a better investment than daughters.[31]

What may be the most intriguing support for the Trivers-Willard hypothesis is among the most recent. A study of North American families found pronounced differences in how indulgent the parents of different social classes are of boys and girls. More than half of the daughters born to low-income women were breast-fed, while fewer than half of the sons were; around 60 percent of the daughters born to affluent women were breast-fed, and nearly 90 percent of the sons. And, more dramatically, low-income women, on average, had another child within 3.5 years of the birth of a son and within 4.3 years of the birth of a daughter. In other words: in the contest over how soon to create a sibling, low-income mothers are inclined to let a daughter win; they wait longer to produce a competing target for investment. For affluent women, the opposite was true: daughters had a rival sibling within 3.2 years of birth, sons within 3.9 years.[32] Presumably few if any mothers in the study knew how social status can affect the reproductive success of males and females (or, strictly speaking, how it would have done so in the environment of our evolution). This is another reminder that natural selection tends to work underground, by shaping human feelings, not by making humans conscious of its logic.[33]

Though these studies all focused on *parental* investment, the same logic would apply to sibling investment. If you're poor, you should, in theory, direct more altruism toward a sister than a brother, and if you're rich, vice versa. It is certainly true that, in Darwin's well-to-do family, his sisters spent much time worrying about and catering to their brothers. But that trend may have been just as pronounced among the lower classes back when the subservience of women was a social ideal (a reminder that culture can push our behavior against the grain of Darwinian logic).

Besides, extraordinary female helpfulness has other Darwinian explanations. Reproductive potential changes over the life cycle, and changes differently for males and females. In his 1964 paper, Hamilton noted abstractly that, "the behaviour of a post-reproductive animal may be expected to be entirely altruistic."[34] After all, once the vehicle that genes inhabit can't transmit them to the next generation, they are well advised to direct all energies toward vehicles that can. Since only women spend much of their lives in post-reproductive mode, the implication is that older women, much more than older men, will shower attention on kin. And they do. The single aunt who devotes her life to relatives is a much more common sight than the single uncle who does the same. Darwin's sister Susan and his brother Erasmus were both middle-aged and unmarried when their sister Marianne died, but it was Susan who adopted her children.[35]

PATTERNS OF GRIEF

Even for a male, reproductive potential changes *somewhat* over time. In fact, it changes every year for everyone. A fifty-year-old of either sex has, on average, many fewer potential offspring in his or her future than he or she did at thirty—at which point, in turn, the potential was less than it had been at fifteen. On the other hand, the average fifteen-year-old has more future offspring than the average one-year-old, since the one-year-old may yet die before adolescence—a fairly common thing during much of human evolution.

Here lies another oversimplification in the no-frills model of kin selection. Since reproductive potential is embedded in both the cost and benefit sides of the altruism equation, the age of both the giver and the recipient help determine whether the altruism will tend to raise inclusive fitness, and thus be favored by natural selection. In other words, how warm and generous we feel toward kin depends, theoretically, both on our age and on the kin's age. There should, for example, be continuous change, throughout a child's life, in how dear that life seems to parents.[36]

Specifically, parental devotion should grow until around early adolescence, when reproductive potential peaks, and then begin to drop. Just as a horse breeder is more disappointed by the death of a thoroughbred the day before its first race than the day after its birth, a parent should be more heartbroken by the death of an adolescent than by the death of an infant. Both the adolescent and the mature racehorse are assets on the brink of bringing rewards, and in both cases it will take much time and effort, starting from scratch, to get another asset to that point. (This isn't to say a parent will never feel more tender or protective toward an infant than toward an adolescent. If, say, a marauding band approaches, a mother's natural impulse will be to grab the infant before fleeing, thus leaving the adolescent to fend for itself; but this impulse exists because adolescents *can* fend for themselves, not because they're less precious than infants.)

As predicted, parents do grieve more over the death of an adolescent than of a three-month-old—or, also in keeping with theory, of a forty-year-old. It is tempting to dismiss such results: *of course* we regret a young man's death more than an older man's; it's obviously tragic to die with so much of life unlived. To which Darwinians reply: Yes, but remember—the very "obviousness" of the pattern may be a product of the same genes that, we propose, created it. The way natural selection has worked its will is to make some things seem "obvious" and "right" and "desirable" and others "absurd" and "wrong" and "abhorrent." We should probe our commonsense reactions to evolutionary theories carefully before concluding that the common sense itself isn't a cognitive distortion created by evolution.

In this case, we should ask: If an adolescent boy's vast unlived life is what makes his death seem so sad, why doesn't the death of an infant seem even sadder? One answer is that we've had more time to get to know the adolescent and can thus see the unlived life more clearly. But what a coincidence that the countervailing changes in these two quantities—the growing intimacy with a person over time and the shrinking size of that person's unlived life—happen to reach some sort of maximum combined grief value right around adolescence, when that person's reproductive potential is highest. Why doesn't the peak come, say, at age twenty-five, when the contours of the unlived life are *really* clear? Or at age five, when there's so *much* unlived life?

The evidence so far is that grief does comply exquisitely with Darwinian expectations. In a 1989 Canadian study, adults were asked to imagine the death of children of various ages and estimate which deaths would create the greatest sense of loss in a parent. The results, plotted on a graph, show grief growing until just before adolescence and then beginning to drop. When this curve was compared with a curve showing changes in reproductive potential over the life cycle (a pattern calculated from Canadian demographic data), the correlation was fairly strong. But much stronger—nearly perfect, in fact—was the correlation between the grief curve of these modern Canadians and the reproductive-potential curve of a hunter-gatherer people, the !Kung of Africa. In other words, the pattern of changing grief was almost exactly what a Darwinian would predict, given demographic realities in the ancestral environment.[37]

In theory, and in fact, the dearness of parents to children also changes over time. In the pitiless eyes of natural selection, the utility of our parents to us declines, after a certain point, even faster than ours to them. As we pass through adolescence, they are less and less critical databanks, providers, and protectors. And as they pass through middle age, they are less and less likely to further promulgate our genes. By the time they are old and infirm, we have little if any genetic use for them. Even as we attend to their needs (or pay someone else to), we may feel traces of impatience and resentment. Our parents, in the end, are as dependent on us as we once were on them, yet we don't look after their needs with quite the same gusto they brought to ours.

The ever-shifting but almost perennially uneven balance of affection and obligation between parent and child is one of life's deepest and most bittersweet experiences. And it illustrates how imprecise the genes can be in turning on and off our emotional spigots. Though there seems to be no good Darwinian reason to spend time and energy on an old, dying father, few of us would, or could, turn our backs. The stubborn core of familial love persists beyond its evolutionary usefulness. Most of us, presumably, are glad for this crudeness of genetic control—although, of course, there's no way of knowing what our opinion would be if the controls were more precise.

DARWIN'S GRIEF

Darwin had many occasions to grieve, including the deaths of three of his ten children and of his father. His behavior generally matches theory.

The death of the Darwins' third child, Mary Eleanor, came only three weeks after her birth in 1842. Charles and Emma were undeniably saddened, and the funeral was hard on Charles, but there are no signs of overwhelming or lasting grief. Emma wrote that, "Our sorrow is nothing to what it would have been if she had lived longer and suffered more," assuring her sister-in-law that, with two other children to distract her and Charles, "you need not fear that our sorrow will last long."[38]

The death of the last Darwin child, Charles Waring, should also, in theory, have been a glancing blow. He was young—a year and a half—and was retarded. One of the most straightforward Darwinian predictions is that parents will care relatively little for children who are so defective as to have negligible reproductive value. (In many preindustrial societies, infants with obvious defects have been routinely killed, and even in industrial societies, handicapped children are especially prone to abuse.)[39] Darwin wrote a short memorial to his dead son, but it was, in places, clinically detached ("He often made strange grimaces & shivered, when excited . . .") and was nearly devoid of anguish.[40] One of the Darwin daughters later said of the baby: "Both my father and mother were infinitely tender towards him, but, when he died in the summer of 1858, after their first sorrow they could only feel thankful."[41]

Nor should the death of Darwin's father in 1848 have been devastating. Charles was by now easily self-sufficient, and his father, at age eighty-two, had spent his reproductive potential. Darwin did show signs of deep grief in the days after the death, and of course there's no way of being sure he didn't keep suffering for months. But in his letters he never got more effusive than to note that "no one, who did not know him, would believe that a man above 83 years old [sic], could have retained so tender & affectionate a disposition, with all his sagacity unclouded to the last." He wrote three months after the death, "[W]hen last I saw him he was very comfortable & his expression which I have now in my mind's eye serene & cheerful."[42]

Clearly distinct from all three of these cases was the death of the Darwins' daughter Annie in 1851, after a periodic illness that had begun the year before. She was ten years old, her reproductive potential just a few years from its peak.

In the days leading up to her death, there is an anguished and poignant exchange of letters between Charles, who had traveled with her to a doctor, and Emma. A few days after the death, Darwin composed a memorial to Annie that is strikingly different in tone from the later memorial to Charles Waring. "Her joyousness and animal spirits radiated from her whole countenance, and rendered every movement elastic and full of life and vigour. It was delightful and cheerful to behold her. Her dear face now rises before me, as she used sometimes to come running downstairs with a stolen pinch of snuff for me, her whole form radiant with the pleasure of giving pleasure. . . . In the last short illness, her conduct in simple truth was angelic. She never once complained; never became fretful; was ever considerate of others, and was thankful in the most gentle, pathetic manner for everything done for her. . . . When I gave her some water, she said, 'I quite thank you;' and these, I believe, were the last precious words ever addressed by her dear lips to me." He wrote in closing, "We have lost the joy of the household, and the solace of our old age. She must have known how we loved her. Oh, that she could now know how deeply, how tenderly, we do still and shall ever love her dear joyous face! Blessings on her!"[43]

It is possible to inject this analysis of Darwin's grief with (believe it or not) a bit more cynicism. Annie, it seems, was the Darwins' favorite child. She was bright and talented ("a second Mozart," Darwin once said)—assets that would have raised her value on the marriage market, and hence her reproductive potential. And she was an exemplary child, a model of generosity, morals, and manners.[44] Or, as Trivers might put it: Emma and Charles had successfully conned her into pursuing their inclusive fitness at the expense of hers. Perhaps an analysis of "favorite children" would confirm that they tend to possess these sorts of valuable attributes—valuable from the perspective of the parents' genes, which may or may not imply value from the perspective of the child's.

Only months after his father's death, Darwin had declared his grieving at an end, referring in a letter to "my dear Father about whom it is now to me the sweetest pleasure to think."[45] In the case of Annie, no such point was reached for either Emma or Charles. Another of their daughters, Henrietta, would later write that "it may almost be said that my mother never really recovered from this grief. She very rarely spoke of Annie, but when she did the sense of loss was always there unhealed. My father could not bear to reopen his sorrow, and he never, to my knowledge, spoke of her." Twenty-five years after Annie's death, he wrote in his autobiography that thinking of her still brought tears to his eyes. Her death, he wrote, had been the "only one very severe grief" the family had suffered.[46]

In 1881, after Darwin's brother Erasmus had died, and less than a year, in fact, before Darwin's own death, he was moved to remark, in a letter to his friend Joseph Hooker, on the difference between "the death of the old and young." He wrote, "Death in the latter case, when there is a bright future ahead, causes grief never to be wholly obliterated."[47]

Chapter 8: **DARWIN AND THE SAVAGES**

Mr. J. S. Mill speaks, in his celebrated work, "Utilitarianism,"
of the social feelings as a "powerful natural sentiment," and as
"the natural basis of sentiment for utilitarian morality;" but on
the previous page he says, "if, as is my own belief, the moral
feelings are not innate, but acquired, they are not for that reason
less natural." It is with hesitation that I venture to differ from
so profound a thinker, but it can hardly be disputed that the
social feelings are instinctive or innate in the lower animals; and
why should they not be so in man?

—*The Descent of Man* (1871)[1]

When Darwin first encountered a primitive society, he reacted roughly as you would expect a nineteenth-century English gentleman to react. As the *Beagle* sailed into a bay in Tierra del Fuego, he saw a group of Indians who yelled and "threw their arms wildly round their heads." With "their long hair streaming," he wrote to his mentor, John Henslow, "they seemed the troubled spirits of another world." Closer inspection reinforced the impression of barbarism. Their language, "according to our notions, scarcely deserves to be called articulate"; their houses "are like what children make in summer, with boughs of trees." Nor were these homes graced by affection between husband and wife, "unless indeed the treatment of a master to a laborious slave can be considered as such."[2]

To top it all off, the Fuegians seemed to have a habit of eating old women when food got scarce. Darwin reported grimly that a Fuegian boy, when asked why they didn't eat their dogs instead, had replied, "Dog catch otter—woman good for nothing—man very hungry." Darwin wrote to his sister Caroline: "Was ever any thing so atrocious heard of, to work them like slaves to procure food in the summer & occasionally in winter to eat them.—I feel quite a disgust at the very sound of the voices of these miserable savages."[3]

It turns out that the part about eating the women was apocryphal. But Darwin saw plenty of other examples of violence in the various preliterate societies he visited during the voyage. The savage, he wrote decades later in *The Descent of Man*, "delights to torture his enemies, offers up bloody sacrifices, practises infanticide without remorse."[4] So it's doubtful that, had Darwin known the Fuegians didn't in fact eat their senior citizens, he would have much altered the view of primitive peoples in his popular account of the *Beagle*'s voyage: "I could not have believed how wide was the difference, between savage and civilized man. It is greater than between a wild and domesticated animal. . . ."[5]

Nonetheless, Fuegian life did include some things that lay at the core of civilized life in Victorian England. For example: friendship, signified by mutual generosity and sealed by a ritual of solidarity. Darwin wrote of the Fuegians: "After we had presented them with some scarlet cloth, which they immediately tied round their necks, they became good friends. This was shown by the old man patting our breasts, and making a chuckling kind of noise, as people do when feeding chickens. I walked with the old man, and this demonstration of friendship was repeated several times; it was concluded by three hard slaps, which were given me on the breast and back at the same time. He then bared his bosom for me to return the compliment, which being done, he seemed highly pleased."[6]

Darwin's consciousness of the savage's humanity was further raised by an experiment in cross-culturalization. On a previous voyage, Captain FitzRoy had brought four Fuegians to England, and now three of them were being returned to their native land, freshly educated and civilized (complete with respectable clothing), to help spread enlightenment and Christian morality in the New World. The experiment failed in several respects, most ignominiously when one newly civilized Fuegian stole all the possessions of another newly civilized Fuegian and headed for another part of the continent under cover of darkness.[7] But the experiment did, at least, produce three English-speaking Fuegians, and thus gave Darwin a chance to do something with natives other than stare at them in disbelief. He later wrote: "The American aborigines, Negroes and Europeans are as different from each other in mind as any three races that can be named; yet I was incessantly struck, whilst living with the Fuegians on board the 'Beagle' with the many little traits of character, shewing how similar their minds were to ours and so it was with a full-blooded negro with whom I happened once to be intimate."[8]

This perception of a fundamental unity among human beings—a human nature—is the first step toward becoming an evolutionary psychologist. The second step—trying to explain parts of that nature in terms of natural selection—Darwin also took. In particular, he tried to explain parts of the human psyche that, to judge by some of his letters from the *Beagle*, you might have thought the Fuegians and other "savages" didn't possess at all: "the moral sense, which tells us what we ought to do, and . . . the conscience which reproves us if we disobey it . . ."[9]

Once again, as with the sterility of insects, Darwin had chosen to confront a major obstacle to his theory of evolution. The moral sentiments are hardly an obvious product of natural selection.

To a certain extent, Darwin's solution to the sterility problem *was* a solution to the morality problem. His concept of "family" selection, or kin selection, can explain altruism in mammals, and hence conscience. But kin selection accounts only for acts of conscience within the family. And human beings are amply able to feel sympathy for nonkin, to help them, and to feel guilty about failing to. Bronislaw Malinowski, in the early twentieth century, would note that the Trobriand Islanders have two words for "friend," depending on whether the friend is within one's own clan or from another clan. He translated the words as meaning "friend within the barrier" and "friend across the barrier."[10] Even the Fuegians, those "miserable savages," had been able to make friends with a young white-skinned man who came from across the ocean. The question remains, even after the theory of kin selection: Why do we have friends across the barrier?

The question is even larger than that. Human beings can feel sympathy for people "across the barrier" who aren't friends—people, indeed, whom they don't even know. Why is this? Why are there Good Samaritans? Why do most people have trouble walking past a beggar without at least a twinge of discomfort?

Darwin found an answer to these questions. His answer, it now seems, is misguided. But it is misguided in a very illuminating way. It rests on a particular kind of confusion that periodically afflicted biology until late this century, when it was finally swept away and the path was cleared for modern evolutionary psychology. What's more, Darwin's analysis of human morality, up to the point where he made his big mistake, is in some ways exemplary; in places it is a paragon of the method of evolutionary psychology even by today's standards.

MORALITY GENES?

The first problem facing anyone who seeks evolutionary insight into morality is its huge diversity. There is the prudishness and gentility of Victorian England, the morally sanctioned savagery of savages, and lots in between. Darwin wrote with some perplexity of the "absurd rules of conduct" reflected in, for example, the "horror felt by a Hindoo who breaks his caste" and the "shame of a Mahometan woman who exposes her face."[11]

If morality is grounded in human biology, how can moral codes differ so widely? Do Arabs and Africans and Englishmen have different "morality genes"?

This is not the explanation that modern evolutionary psychology favors, and not the one Darwin stressed. To be sure, he did believe the races had inborn mental differences, some of them morally relevant.[12] This belief was standard in the nineteenth century, an era when some scholars (Darwin not among them) argued strenuously that the races weren't races at all, but *species*. Still, Darwin believed the world's varied moral customs were rooted—in at least a general sense—in a common human nature.

To begin with, he noted the deep sensitivity of all human beings to public opinion. "The love of approbation and the dread of infamy, as well as the bestowal of praise or blame" are grounded in instinct, he asserted. A breach of norms can cause a man "agony," and the violation of some trivial bit of etiquette, when recalled even years after, can bring back a "burning sense of shame."[13] Thus, adherence to any moral rule has an innate basis. It is only the specific contents of moral codes that are not inborn.

Why do the contents vary so? Darwin believed different peoples have different rules because, for their own historical reasons, they judge different norms to be in the interest of the community.

Often, Darwin said, these judgments are in error, yielding patterns of behavior that are pointless, if not, indeed, "in complete opposition to the true welfare and happiness of mankind." One gets the impression that, from Darwin's point of view, it was England, or at least Europe, in which the fewest errors had been made. And savages, plainly, had made more than their share. They seemed to possess "insufficient powers of reasoning" to discern nonobvious connections between moral laws and public welfare, and they lacked, perhaps constitutionally, self-discipline; "their utter licentiousness, not to mention unnatural crimes, is something astounding."[14]

Still, Darwin believed that none of this savagery should distract us from the second universal element in human morality. Fuegians and Englishmen alike possessed the "social instincts," central among them sympathy for their fellow man. "[F]eelings of sympathy and kindness are common, especially during sickness, between the members of the same tribe. . . ." And "many instances have been recorded of barbarians, . . . not guided by any religious motive, who have deliberately as prisoners sacrificed their lives, rather than betray their comrades; and surely their conduct ought to be considered moral."[15]

True, barbarians had an unfortunate tendency to define everyone outside of their tribe as morally worthless, even to the point of deeming harm to outsiders an honorable endeavor. Indeed, "it has been recorded that an Indian Thug conscientiously regretted that he had not strangled and robbed as many travellers as did his father before him."[16] Still, this was a question of the scope of sympathy, not of its existence; so long as all peoples have a core capacity for moral concern, no people is beyond edification. In *The Voyage of the Beagle*, Darwin wrote about an island off the coast of Chile: "It is a pleasant thing to see the aborigines advanced to the same degree of civilization, however low that may be, which their white conquerors have attained."[17]

Any savages who feel flattered that Darwin accorded them full possession of sympathetic impulses and the underlying social instincts should be aware that he bestowed a similar honor on some nonhuman forms of life. He saw sympathy in reports of crows that dutifully fed their blind compatriots and of baboons that heroically saved their youngsters from a pack of dogs; and "Who can say what cows feel, when they surround and stare intently on a dying or dead companion?"[18] Darwin described signs of tenderness among two chimpanzees, relayed to him by a zookeeper who watched their first encounter: "They sat opposite, touching each other with their much protruded lips; and the one put his hand on the shoulder of the other. They then mutually folded each other in their arms. Afterwards they stood up, each with one arm on the shoulder of the other, lifted their heads, opened their mouths, and yelled with delight."[19]

Some of these examples may be cases of altruism among close relatives, in which case the simple explanation is kin selection. And, for that matter, the scene with the chimps getting acquainted may have been embellished by an anthropomorphizing zookeeper. But, for the record, chimpanzees do, in fact, form friendships, and this single fact is sufficient for the point Darwin was making: that however special we may consider our species, we are not unique in our capacity for sympathetic behavior, even beyond the confines of family.

Certainly, Darwin noted, human beings carry moral behavior to unique lengths. They can, via complex language, learn precisely what sort of conduct is expected of them in the name of the common good. And they can look back into the past, recall the ultimately painful result of allowing their "social instincts" to be overridden by baser instincts, and resolve to do better. Indeed, Darwin suggested, on grounds such as these, that the word *moral* itself be reserved for our species.[20] Still, at the root of this full-blown morality he saw a social instinct that long predates humanity, even if human evolution had enriched it.

In figuring out how evolution favored moral (or any other) impulses, it is critical to focus on the behaviors they bring. After all, behavior, not thought or emotion, is what natural selection passes judgment on; acts, not the feelings themselves, directly guide the transportation of genes. Darwin perfectly understood this principle. "It has often been assumed that animals were in the first place rendered social, and that they feel as a consequence uncomfortable when separated from each other, and comfortable whilst together; but it is a more probable view that these sensations were first developed, in order that those animals which would profit by living in society, should be induced to live together, . . . for with those animals which were benefited by living in close association, the individuals which took the greatest pleasure in society would best escape various dangers; whilst those that cared least for their comrades and lived solitary would perish in greater numbers."[21]

GROUP SELECTIONISM

In the course of his basically sound approach to evolutionary psychology, Darwin succumbed to a temptation known as group selectionism. Consider his central explanation for the evolution of the moral sense. In *The Descent of Man* he wrote that "an advancement in the standard of morality and an increase in the number of well-endowed men will certainly give an immense advantage to one tribe over another. There can be no doubt that a tribe including many members who, from possessing in a high degree the spirit of patriotism, fidelity, obedience, courage, and sympathy, were always ready to give aid to each other and to sacrifice themselves for the common good, would be victorious over most other tribes; and this would be natural selection."[22]

Yes, this would be natural selection, if it actually happened. But, while it isn't *impossible* for it to happen, the more you think about it, the less likely it seems. Darwin himself had seen the main snag only a few pages earlier: "It is extremely doubtful whether the offspring of the more sympathetic and benevolent parents, or of those which were the most faithful to their comrades, would be reared in greater number than the children of selfish and treacherous parents of the same tribe." On the contrary, the bravest, most self-sacrificial men "would on an average perish in larger number than other men." A noble man "would often leave no offspring to inherit his noble nature."[23]

Exactly. So even though a tribe full of selfless people would prevail over a tribe full of selfish people, it is hard to see how a tribe would get full of selfless people in the first place. Everyday prehistoric life, with its normal share of adversity, would presumably favor the genes of people who, say, hoarded food rather than share it, or let neighbors fight their own battles rather than risk injury; and this intratribal advantage would, if anything, have grown when the intertribal competition at the heart of Darwin's group-selectionist theory heated up, as during war or famine (unless, after wars, societies took exceedingly good care of the kin of dead war heroes). So there might never be a way for biologically based impulses of selflessness to pervade a group. Even if you magically intervened and implanted "sympathetic" genes in 90 percent of the population, these would steadily lose out to their less ennobling rival genes.

Granted, as Darwin said, the resulting rampant selfishness might mean that this tribe perished in competition with another tribe. But all tribes are subject to the same internal logic, so the victors presumably wouldn't be paragons of virtue themselves. And any meager quantity of selflessness they had put to good use should, in theory, be declining even as they savored the fruits of victory.

The problem with Darwin's theory is a common problem with group-selectionist theories: it is hard to imagine group selection spreading some trait that individual selection on its own wouldn't favor; it is hard to imagine natural selection resolving a direct conflict between group welfare and individual welfare in favor of the group. To be sure, one can dream up scenarios—with particular rates of migration among groups, and particular rates of group extinction—where group selection might favor individual sacrifice; and there are a few biologists who believe group selection did play an important role in human evolution.[24] Still, group-selectionist scenarios do tend to be a bit convoluted. Indeed, George Williams found them so generally onerous that he proposed, in *Adaptation and Natural Se-*

lection, an official bias against them: "One should postulate adaptation at no higher a level than is necessitated by the facts."[25] In other words: first look very hard for a way that genes underlying a trait could be favored in everyday, head-to-head competition. Only after failing should you resort to competition between separate populations, and then with great caution. This has become the unofficial credo of the new paradigm.

In the same book, Williams put his doctrine to vivid use. Without resort to group selection, he proposed what is now the accepted explanation for the human moral sentiments. Writing in the mid-sixties, just after Hamilton had explained the origin of altruism among kin, Williams suggested a way that evolution could extend altruism beyond the barrier of kinship.

Chapter 9: **FRIENDS**

[I]t is not a little remarkable that sympathy with the distresses of others should excite tears more freely than our own distress; and this certainly is the case. Many a man, from whose eyes no suffering of his own could wring a tear, has shed tears at the sufferings of a beloved friend.
—*The Expression of the Emotions in Man and Animals* (1872)[1]

Darwin, perhaps sensing the weakness of his main theory of the moral sentiments, threw in a second theory for good measure. During human evolution, he wrote in *The Descent of Man,* "as the reasoning powers and foresight . . . became improved, each man would soon learn from experience that if he aided his fellow-men, he would commonly receive aid in return. From this low motive he might acquire the habit of aiding his fellows; and the habit of performing benevolent actions certainly strengthens the feeling of sympathy, which gives the first impulse to benevolent actions. Habits, moreover, followed during many generations probably tend to be inherited."[2]

That last sentence, of course, is wrong. We now know that habits are passed from parent to child by instruction or example, not via the genes. In fact, no life experiences (except, say, exposure to radiation) affect the genes handed down to offspring. The very beauty of Darwin's theory of natural selection, in its strict form, was that it didn't require the inheritance of acquired traits, as had previous evolutionary theories, such as Jean-Baptiste de Lamarck's. Darwin saw this beauty, and stressed mainly the pure version of his theory. But he was willing, especially as he grew older, to invoke more dubious mechanisms to solve especially nettlesome issues, such as the origin of the moral sentiments.

In 1966, George Williams suggested a way to make Darwin's musings about the evolutionary value of mutual assistance more useful: take out not only the last sentence, but also the part about "reasoning" and "foresight" and "learning." In *Adaptation and Natural Selection*, Williams recalled Darwin's reference to the "low motive" of doing favors in hopes of reciprocation and wrote: "I see no reason why a conscious motive need be involved. It is necessary that help provided to others be occasionally reciprocated if it is to be favored by natural selection. It is not necessary that either the giver or the receiver be aware of this." He continued, "Simply stated, an individual who maximizes his friendships and minimizes his antagonisms will have an evolutionary advantage, and selection should favor those characters that promote the optimization of personal relationships."[3]

Williams's basic point (which Darwin certainly understood, and stressed in other contexts)[4] is one we've encountered before. Animals, including people, often execute evolutionary logic not via conscious calculation, but by following their feelings, which were designed as logic executers. In this case, Williams suggested, the feelings might include compassion and gratitude. Gratitude can get people to repay favors without giving much thought to the fact that that's what they're doing. And if compassion is felt more strongly for some kinds of people—people to whom we're grateful, for example—it can lead us, again with scarce consciousness of the fact, to repay kindness.

Williams's terse speculations were transmuted into a full-fledged theory by Robert Trivers. In 1971, exactly one hundred years after Darwin's allusion to reciprocal altruism appeared in *The Descent of Man*, Trivers published a paper titled "The Evolution of Reciprocal Altruism" in *The Quarterly Review of Biology*. In the paper's abstract, he wrote that "friendship, dislike, moralistic aggression, gratitude, sympathy, trust, suspicion, trustworthiness, aspects of guilt, and some forms of dishonesty and hypocrisy can be explained as important adaptations to regulate the altruistic system." Today, more than two decades after this nervy pronouncement, there is a diverse and still-growing body of evidence to support it.

GAME THEORY AND RECIPROCAL ALTRUISM

If Darwin were put on trial for not having conceived and developed the theory of reciprocal altruism, one defense would be that he came from an intellectually disadvantaged culture. Victorian England lacked two tools that together form a uniquely potent analytical medium: game theory and the computer.

Game theory was developed during the 1920s and thirties as a way to study decision making.[5] It has become popular in economics and other social sciences, but it suffers from a reputation for being a bit too, well, cute. Game theorists cleverly manage to make the study of human behavior neat and clean, but they pay a high price in realism. They sometimes assume that what people pursue in life can be tidily summarized in a single psychological currency—pleasure, or happiness, or "utility"; and they assume, further, that it is pursued with unwavering rationality. Any evolutionary psychologist can tell you that these assumptions are faulty. Humans aren't calculating machines; they're animals, guided somewhat by conscious reason but also by various other forces. And long-term happiness, however appealing they may find it, is not really what they're designed to maximize.

On the other hand, humans are designed *by* a calculating machine, a highly rational and coolly detached process. And that machine does design them to maximize a single currency—total genetic proliferation, inclusive fitness.[6]

Of course, the designs don't always work. Individual organisms often fail, for various reasons, to transmit their genes. (Some are bound to fail. That is the reason evolution so assuredly happens.) In the case of human beings, moreover, the design work was done in a social environment quite different from the current environment. We live in cities and suburbs and watch TV and drink beer, all the while being pushed and pulled by feelings designed to propagate our genes in a small hunter-gatherer population. It's no wonder that people often seem not to be pursuing any particular goal—happiness, inclusive fitness, whatever—very successfully.

Game theorists, then, may want to follow a few simple rules when applying their tools to human evolution. First, the object of the game should be to maximize genetic proliferation. Second, the context of the game should mirror reality in the ancestral environment, an environment roughly like a hunter-gatherer society. Third, once the optimal strategy has been found, the experiment isn't over. The final step—the payoff—is to figure out what feelings would lead human beings to pursue that strategy. Those feelings, in theory, should be part of human nature; they should have evolved through generations and generations of the evolutionary game.

Trivers, at the suggestion of William Hamilton, employed a classic game called the prisoner's dilemma. Two partners in crime are being interrogated separately and face a hard decision. The state lacks the evidence to convict them of the grave offense they committed but does have enough evidence to convict both on a lesser charge—with, say, a one-year prison term for each. The prosecutor, wanting a harsher sentence, pressures each man individually to confess and implicate the other. He says to each: If you confess but your partner doesn't, I'll let you off scot-free and use your testimony to put him away for ten years. The flip side of this offer is a threat: If you *don't* confess but your partner does, *you* go to prison for ten years. And if you confess and it turns out your partner confesses too, I'll put you both away, but only for three years.[7]

If you were in the shoes of either prisoner, and weighed your options one-by-one, you would almost certainly decide to confess—to "cheat" on your partner. Suppose, first of all, that your partner cheats on you. Then you're better off cheating: you get three years in prison, as opposed to the ten you'd get if you stayed mum while he confessed. Now, suppose he doesn't cheat on you. You're still better off cheating: by confessing while he stays mum, you go free, whereas you'd get one year if you too kept your silence. Thus, the logic seems irresistible: betray your partner.

Yet if both partners follow this nearly irresistible logic, and cheat on each other, they end up with three years in jail, whereas both could have gotten off with one year had they stayed mutually faithful and kept their mouths shut. If only they were allowed to communicate and reach an agreement—then cooperation could emerge, and both would be better off. But they aren't, so how can cooperation emerge?

The question roughly parallels the question of how dumb animals, which can't make promises of repayment, or, for that matter, grasp the concept of repayment, could evolve to be reciprocally altruistic. Betraying a partner in crime while he stays faithful is like an animal's benefiting from an altruistic act and never returning the favor. Mutual betrayal is like neither animal's extending a favor in the first place: though both might benefit from reciprocal altruism, neither will risk getting burned. Mutual fidelity is like a single successful round of reciprocal altruism—a favor is extended and returned. But again: Why extend the favor if there's no guarantee of return?

The match between model and reality isn't perfect.[8] With reciprocal altruism there is a time lag between the altruism and its reciprocation, whereas the players in a prisoner's dilemma commit themselves concurrently. But this is a distinction without much of a difference. Because the prisoners can't communicate about their concurrent decisions, each is in the situation faced by prospectively altruistic animals: unsure whether any friendly overture will be matched. Further, if you keep pitting the same players against one another, game after game after game—an "iterated prisoner's dilemma"—each can refer to the other's past behavior in deciding how to act toward him in the future. Thus each player may reap in the future what he has sown in the past—just as with reciprocal altruism.

All in all, the match between model and reality is quite good. The logic that would lead to cooperation in an iterated prisoner's dilemma is fairly precisely the logic that would lead to reciprocal altruism in nature. The essence of that logic, in both cases, is non-zero-sumness.

NON-ZERO-SUMNESS

Suppose you are a chimp that has just killed a young monkey and you give some meat to a fellow chimp that has been short of food lately. Let's say you give him five ounces, and let's call that a five-point loss for you. Now, in an important sense, the other chimp's gain is larger than your loss. He was, after all, in a period of unusual need, so the real value of food to him—in terms of its contribution to his genetic proliferation—was unusually high. Indeed, if he were human, and could think about his plight, and were forced to sign a binding contract, he might rationally agree to repay five ounces of meat with, say, six ounces of meat right after payday next Friday. So he gets six points in this exchange, even though it cost you only five.

This asymmetry is what makes the game non-zero-sum. One player's gain isn't canceled out by the other player's loss. The essential feature of non-zero-sumness is that, through cooperation, or reciprocation, *both* players can be better off.[9] If the other chimp repays you at a time when meat is bountiful for him and scarce for you, then he sacrifices five points and you get six points. Both of you have emerged from the exchange with a net benefit of one point. A series of tennis sets, or of innings, or of golf holes eventually produces only one winner. The prisoner's dilemma, being a non-zero-sum game, is different. Both players can win if they cooperate. If caveman A and caveman B combine to hunt game that one man alone can't kill, both cavemen's families get a big meal; if there's no such cooperation, neither family does.

Division of labor is a common source of non-zero-sumness: you become an expert hide-splicer and give me clothes, I carve wood and give you spears. The key here—and in the chimpanzee example above, as well as in much non-zero-sumness—is that one animal's surplus item can be another animal's rare and precious good. It happens all the time. Darwin, recalling an exchange of goods with the Fuegian Indians, wrote of "both parties laughing, wondering, gaping at each other; we pitying them, for giving us good fish and crabs for rags, &c.; they grasping at the chance of finding people so foolish as to exchange such splendid ornaments for a good supper."[10]

To judge by many hunter-gatherer societies, division of economic labor wasn't dramatic in the ancestral environment. The most common commodity of exchange, almost surely, was information. Knowing where a great stock of food has been found, or where someone encountered a poisonous snake, can be a matter of life or death. And knowing who is sleeping with whom, who is angry at whom, who cheated whom, and so on, can inform social maneuvering for sex and other vital resources. Indeed, the sorts of gossip that people in all cultures have an apparently inherent thirst for—tales of triumph, tragedy, bonanza, misfortune, extraordinary fidelity, wretched betrayal, and so on—match up well with the sorts of information conducive to fitness.[11] Trading gossip (the phrase couldn't be more apt) is one of the main things friends do, and it may be one of the main reasons friendship exists.

Unlike food or spears or hides, information is shared without being actually surrendered, a fact that can make the exchange radically non-zero-sum.[12] Of course, sometimes information is of value only if hoarded. But often that's not the case. One Darwin biographer has written that, after scientific discussions between Darwin and his friend Joseph Hooker, "each vied with the other in claiming that the benefits he had received . . . far outweighted whatever return he might have been able to make."[13]

Non-zero-sumness is, by itself, not enough to explain the evolution of reciprocal altruism. Even in a non-zero-sum game, cooperation doesn't *necessarily* make sense. In the food-sharing example, though you gain one point from a single round of reciprocal altruism, you gain *six* points by cheating—accepting generosity and never returning it. So the lesson seems to be: if you can spend your life exploiting people, by all means do; the value of cooperation pales by comparison. Further, if you can't find people to exploit, cooperation *still* may not be the best strategy. If you're surrounded by people who are always trying to exploit you, then reciprocal exploitation is the way to cut your losses. Whether non-zero-sumness actually fuels the evolution of reciprocal altruism depends heavily on the prevailing social environment. The prisoner's dilemma will have to do more than simply illustrate non-zero-sumness if it is to be of much use here.

Testing theories, of course, is a general problem for evolutionary biologists. Chemists and physicists test a theory with carefully controlled experiments that either work as predicted, corroborating the theory, or don't. Sometimes evolutionary biologists can do that. As we've seen, researchers have nutritionally deprived pack rat mothers to see if they would, as predicted, then favor female offspring. But biologists can't experiment with human beings the way they do with pack rats, and they can't conduct the ultimate experiment: rewind the tape and replay evolution.

Increasingly, though, biologists can replay approximations of evolution. When Trivers laid out the theory of reciprocal altruism in 1971, computers were still exotic machines used by specialists; the personal computer didn't even exist. Though Trivers put the prisoner's dilemma to good analytical use, he didn't talk about actually *animating* it—creating, inside a computer, a species whose members regularly confront the dilemma and may live or die by it, and then letting natural selection take its course.

During the late 1970s, Robert Axelrod, an American political scientist, devised such a computer world and then set about populating it. Without mentioning natural selection—which wasn't, initially, his interest—he invited experts in game theory to submit a computer program embodying a strategy for the iterated prisoner's dilemma: a rule by which the program decides whether to cooperate on each encounter with another program. He then flipped the switch and let these programs mingle. The context for the competition nicely mirrored the social context of human, and prehuman, evolution. There was a fairly small society—several dozen regularly interacting individuals. Each program could "remember" whether each other program had cooperated on previous encounters, and adjust its own behavior accordingly.

After every program had had 200 encounters with every other program, Axelrod added up their scores and declared a winner. Then he held a second generation of competition after a systematic culling: each program was represented in proportion to its first-generation success; the fittest had survived. And so the game proceeded, generation after generation. If the theory of reciprocal altruism is correct, you would expect reciprocal altruism to "evolve" inside Axelrod's computer, to gradually dominate the population.

It did. The winning program, designed by the Canadian game theorist Anatol Rapoport (who had once written a book called *Prisoner's Dilemma*), was named TIT FOR TAT.[14] TIT FOR TAT was guided by the simplest of rules—literally: its computer program was five lines long, the shortest submitted. (So if the strategies had been created by random computer mutation, rather than by design, it probably would have been among the first to appear.) TIT FOR TAT was just what its name implied. On the first encounter with any program, it would cooperate. Thereafter, it would do whatever the other program had done on the previous encounter. One good turn deserves another, as does one bad turn.

The virtues of this strategy are about as simple as the strategy itself. If a program demonstrates a tendency to cooperate, TIT FOR TAT immediately strikes up a friendship, and both enjoy the fruits of cooperation. If a program shows a tendency to cheat, TIT FOR TAT cuts its losses; by withholding cooperation until that program reforms, it avoids the high costs of being a sucker. So TIT FOR TAT never gets repeatedly victimized, as indiscriminately cooperative programs do. Yet TIT FOR TAT also avoids the fate of the indiscriminately *un*cooperative programs that try to exploit their fellow programs: getting locked into mutually costly chains of mutual betrayal with programs that would be perfectly willing to cooperate if only you did. Of course, TIT FOR TAT generally forgoes the large one-time gains that can be had through exploitation. But strategies geared toward exploitation, whether through relentless cheating or repeated "surprise" cheating, tended to lose out as the game wore on. Programs quit being nice to them, so they were denied both the large gains of exploitation and the more moderate gains of mutual cooperation. More than the steadily mean, more than the steadily nice, and more than various "clever" programs whose elaborate rules made them hard for other programs to read, the straightforwardly conditional TIT FOR TAT was, in the long run, self-serving.

HOW TIT FOR TAT FEELS

TIT FOR TAT's strategy—do unto others as they've done unto you—gives it much in common with the average human being. Yet it has no human foresight. It doesn't *understand* the value of reciprocation. It just reciprocates. In that sense it is perhaps more like *Australopithecus,* our small-brained forebears.

What feelings would natural selection have instilled in an australopithecine to make it employ the clever strategy of reciprocal altruism in spite of its dim-wittedness? The answer goes beyond the simple, indiscriminate "sympathy" that Darwin stressed. True, this kind of sympathy would come in handy at first, prompting TIT FOR TAT's initial overture of goodwill. But thereafter sympathy should be dished out selectively, and supplemented by other feelings. TIT FOR TAT's reliable return of favors might emerge from a sense of gratitude and obligation. The tendency to cut off largesse for mean australopithecines could be realized via anger and dislike. And the tendency to be nice toward erstwhile meanies who have mended their ways would come from a sense of forgiveness—an eraser of suddenly counterproductive hostility. All of these feelings are found in all human cultures.

In real life, cooperation isn't a matter of black and white. You don't run into an acquaintance, try to extract useful information, and either fail or succeed. More often, the two of you swap miscellaneous data, each providing something of possible use to the other, and the contributions don't exactly balance. So the human rules for reciprocal altruism are likely to be a bit less binary than TIT FOR TAT's. If person F has been distinctly nice on several occasions, you might lower your guard and do favors without constantly monitoring F, remaining alert only to gross signs of incipient meanness, and periodically reviewing—consciously or unconsciously—the cumulative account. Similarly, if person E has been mean for months, it's probably best to write him off. The sensations that would encourage you to behave in these time-and-energy-saving fashions are, respectively, affection and trust (which entail the concept of "friend"); and hostility and mistrust (along with the concept of "enemy").

Friendship, affection, trust—these are the things that, long before people signed contracts, long before they wrote down laws, held human societies together. Even today, these forces are one reason human societies vastly surpass ant colonies in size and complexity even though the degree of kinship among cooperatively interacting people is usually near zero. As you watch the kind but stern TIT FOR TAT spread through the population, you are seeing how the human species's uniquely subtle social cement could grow out of fortuitous genetic mutations.

More remarkable, perhaps, is that the fortuitous mutations thrive without "group selection." That was Williams's whole point back in 1966: altruism toward nonkin, though a critical ingredient in group cohesion, needn't have been created for the "good of the tribe," much less the "good of the species." It seems to have emerged from simple, day-to-day competition among individuals. Williams wrote in 1966: "There is theoretically no limit to the extent and complexity of group-related behavior that this factor could produce, and the immediate goal of such behavior would always be the well-being of some other individual, often genetically unrelated. Ultimately, however, this would not be an adaptation for group benefit. It would be developed by the differential survival of individuals and would be designed for the perpetuation of the genes of the individual providing the benefit to another."[15]

One key to this emergence of macroscopic harmony from microscopic selfishness is feedback between macro and micro. As the number of TIT FOR TAT creatures grows—that is, as the amount of social harmony grows—the fortunes of each individual TIT FOR TAT grow. The ideal neighbor for TIT FOR TAT, after all, is another TIT FOR TAT. The two settle quickly and painlessly into an enduringly fruitful relationship. Neither ever gets burned, and neither ever needs to dish out mutually costly punishment. Thus, the more social harmony, the better each TIT FOR TAT fares, and the more social harmony, and so on. Through natural selection, simple cooperation can actually feed on itself.

The person who pioneered the modern study of this sort of self-reinforcing social coherence, and also the evolutionary application of game theory, is John Maynard Smith. We've seen how he used the idea of "frequency-dependent" selection to show how two kinds of bluegill sunfish—drifters and fine, upstanding citizens—could exist in equilibrium: if the number of drifters grows relative to upstanding citizens, the drifters become less genetically prolific, and their number returns to normal. TIT FOR TAT is also subject to frequency-dependent selection, but here the dynamic works in the other direction, with feedback that is positive, not negative; the more TIT FOR TATs there are, the more successful TIT FOR TAT is. If negative feedback sometimes produces an "evolutionarily stable state"—a balance among different strategies—positive feedback can produce an "evolutionarily stable *strategy*": a strategy that, once it has pervaded a population, is impervious to small-scale invasion. There is no alternative strategy that, if introduced via a single mutant gene, can flourish. Axelrod, after watching TIT FOR TAT triumph and analyzing its success, concluded that it was evolutionarily stable.[16]

Cooperation can begin to feed on itself early in the game. If even a small chunk of the population employs TIT FOR TAT and all other creatures are steadfastly uncooperative, an expanding circle of cooperation will suffuse the population generation after generation. And the reverse isn't true. Even if several steadfast noncooperators arrive on the scene at once, they still can't subvert a population of TIT FOR TATs. Simple, conditional cooperation is more infectious than unmitigated meanness. Robert Axelrod and William Hamilton, in a jointly authored chapter of Axelrod's 1984 book *The Evolution of Cooperation*, wrote: "[T]he gear wheels of social evolution have a ratchet."[17]

Unfortunately, this ratchet doesn't kick in at the very beginning. If only *one* TIT FOR TAT creature enters a climate of pure meanness, it is doomed to extinction. Steadfast uncooperativeness, apparently, is itself an evolutionarily stable strategy; once it pervades a population, it is immune to invasion by a single mutant employing any other strategy, even though it is vulnerable to a small cluster of conditionally cooperative mutants.

In that sense, Axelrod's tournament gave TIT FOR TAT a head start. Though the strategy didn't at first enjoy the company of any exact clones, most of its neighbors were designed to cooperate under at least some circumstances, thus raising the value of its own good nature. Had TIT FOR TAT been tossed in with forty-nine steadfast meanies, there would have been a forty-nine-way tie for first place, and only one clear loser. However inexorable TIT FOR TAT's success appears on the computer screen, reciprocal altruism's triumph wasn't so obviously in the cards many millions of years ago, when meanness pervaded our evolutionary lineage.

How did reciprocal altruism get off the ground? If any new gene offering cooperation gets stomped into the dust, how did there ever arise the small population of reciprocal altruists needed to shift the odds in favor of cooperation?

The most appealing answer is one suggested by Hamilton and Axelrod: that kin selection gave reciprocal altruism a subtle boost. As we've seen, kin selection can favor any gene that raises the precision with which altruism flows toward relatives. Thus, a gene counseling apes to love other apes that suckled at their mother's breast—younger siblings, that is—might thrive. But what are the younger siblings supposed to do? They never see their older siblings suckle, so what cues can they go by?

One cue is altruism itself. Once genes directing altruism toward sucklers had taken hold by benefiting younger siblings, genes directing altruism toward altruists would benefit older siblings. These genes—reciprocal-altruism genes—would thus spread, at first via kin selection.

Any such imbalance of information between two relatives about their relatedness is fertile ground for a reciprocal-altruism gene. And such imbalances are quite likely to have existed in our past. Before the advent of language, aunts, uncles, and even fathers often had conspicuous cues about the identities of their younger relatives when the reverse wasn't true; so altruism would have flowed largely from older to younger relatives. That imbalance would itself have been a reliable cue for youngsters to use in steering altruism toward relatives—at least, it probably would have been more reliable than other simple cues, which is all that matters. A gene that repaid kindness with kindness could thus have spread through the extended family, and, by interbreeding, to other families, where it would thrive on the same logic.[18] At some point the TIT FOR TAT strategy would be widespread enough to keep flourishing even without the aid of kin selection. The ratchet of social evolution was now forged.

Kin selection probably paved the way for reciprocal altruism genes in a second way too: by placing handy psychological agents at their disposal. Long before our ancestors were reciprocal altruists, they were capable of familial affection and generosity, of trust (in kin) and of guilt (a reminder not to mistreat kin). These and other elements of altruism were part of the ape mind, ready to be wired together in a new way. That almost surely made things easier for natural selection, which often makes thrifty use of the materials at hand.

Given these likely links between kin selection and reciprocal altruism, one can view the two phases in evolution almost as a single creative thrust, in which natural selection crafted an ever-expanding web of affection, obligation, and trust out of ruthless genetic self-interest. The irony alone would make the process worth savoring, even if this web didn't include so many of the experiences that make life worthwhile.

BUT IS IT SCIENCE?

Game theory and computer simulation are neat and fun, but how much do they really add up to? Is the theory of reciprocal altruism genuine science? Does it succeed in explaining what it aims to explain?

One answer is: Compared to what? There isn't exactly a surplus of rival theories. Within biology, the only alternatives are group-selectionist theories, which tend to face the sort of problem Darwin's group-selectionist theory faced. And within the social sciences, this subject is a giant void.

To be sure, social scientists, going back at least to the turn-of-the-century anthropologist Edward Westermarck, have recognized that reciprocal altruism is fundamental to life in all cultures. There is a whole literature on "social exchange theory," in which the everyday swapping of sometimes intangible resources—information, social support—is gauged with care.[19] But because so many social scientists have resisted the very idea of an inherent human nature, reciprocation has often been seen as a cultural "norm" that just happens to be universal (presumably because distinct peoples independently discovered its utility). Few have noted that the daily life of every human society rests not just on reciprocity, but on a common foundation of feelings—sympathy, gratitude, affection, obligation, guilt, dislike, and so on. Even fewer have offered an ultimate explanation for this commonality. There must be *some* explanation. Does any one have an alternative to the theory of reciprocal altruism?

The theory thus wins by default. But it doesn't win *only* by default. Since Trivers published his paper in 1971, the theory has been tested and so far has fared well.[20]

The Axelrod tournament was one test. If uncooperative strategies had prevailed over cooperative ones, or if cooperative strategies had paid off only after they made up much of the population, things would have looked worse for the theory. But conditional niceness was shown to have the upper hand over meanness, and indeed to be a nearly inexorable evolutionary force once it gains even a small foothold.

The theory has also gotten support in the natural world: evidence that reciprocal altruism can evolve without a human's abstract comprehension of its logic, so long as the animals in question are smart enough to recognize individual neighbors and record their past deeds, whether consciously or unconsciously. Williams, in 1966, noted the existence of mutually supportive and long-lasting coalitions of rhesus monkeys. And he suggested that the mutually "solicitous" behavior of porpoises might be reciprocal—a suspicion later confirmed.[21]

Vampire bats, not mentioned by either Trivers or Williams, also turn out to be reciprocally altruistic. Any given bat has sporadic success in its nightly forays to suck blood from cattle, horses, and other victims. Since blood is highly perishable, and bats don't have refrigerators, scarcity faces individual bats pretty often. And periodic individual scarcity, as we've seen, invites non-zero-sum logic. Sure enough, bats that return to the roost empty-handed are often favored with regurgitated blood from other bats—and they tend to return the favor in the future. Some of the sharing is, not surprisingly, between kin, but much takes place within partnerships—two or more unrelated bats that recognize each other by distinctive "contact calls" and often groom each other.[22] Bat buddies.

The most vital zoological support for the evolution of reciprocal altruism in humans has come from our close relatives the chimpanzees. When Williams and Trivers first wrote about reciprocity, the social life of chimpanzees was just coming into clear view. There were few signs of how utterly reciprocal altruism permeates it. Now we know that chimpanzees share food reciprocally and form somewhat durable alliances. Friends groom each other and help each other confront or fend off enemies. They give reassuring caresses and hearty embraces. When one friend betrays another, seemingly heartfelt outrage may ensue.[23]

The theory of reciprocal altruism also passes a very basic, essentially aesthetic scientific test: the test of elegance, or parsimony. The simpler a theory, and the more varied and numerous the things it explains, the more "parsimonious" it is. It is hard to imagine anyone isolating a single and fairly simple evolutionary force that, like the force Williams and Trivers isolated, could plausibly account for things so diverse as sympathy, dislike, friendship, enmity, gratitude, a gnawing sense of obligation, acute sensitivity to betrayal, and so on.[24]

Reciprocal altruism has presumably shaped the texture not just of human emotion, but of human cognition. Leda Cosmides has shown that people are good at solving otherwise baffling logical puzzles when the puzzles are cast in the form of social exchange—in particular, when the object of the game is to figure out if someone is cheating. This suggests to Cosmides that a "cheater-detection" module is among the mental organs governing reciprocal altruism.[25] No doubt others remain to be discovered.

THE MEANING OF RECIPROCAL ALTRUISM

One common reaction to the theory of reciprocal altruism is discomfort. Some people are troubled by the idea that their noblest impulses spring from their genes' wiliest ploys. This is hardly a necessary response, but for those who choose it, full immersion is probably warranted. If indeed the genetically selfish roots of sympathy and benevolence are grounds for despair, then extreme despair is in order. For, the more you ponder reciprocal altruism's finer points, the more mercenary the genes seem.

Consider again the question of sympathy—in particular, its tendency to grow in proportion to the gravity of a person's plight. Why do we feel sadder for a starving man than for a slightly hungry man? Because the human spirit is a grand thing, devoted to allaying suffering? Guess again.

Trivers addressed this question by asking why gratitude itself varies according to the plight from which the grateful are rescued. Why are you lavishly thankful for a life-saving sandwich after three days in the wilderness and moderately thankful for a free dinner that evening? His answer is simple, credible, and not too startling: gratitude, by reflecting the value of the benefit received, calibrates the repayment that's in order. Gratitude is an I.O.U., so naturally it records what's owed.

For the benefactor, the moral of the story is clear: the more desperate the plight of the beneficiary, the larger the I.O.U. Exquisitely sensitive sympathy is just highly nuanced investment advice. Our deepest compassion is our best bargain hunting. Most of us would look with contempt on an emergency-room doctor who quintupled his hourly fee for patients on the brink of death. We would call him callously exploitive. We would ask, "Don't you have any *sympathy?*" And if he had read his Trivers, he would say, "Yes, I have lots of it. I'm just being honest about what my sympathy is." This might dampen our moral indignation.

Speaking of moral indignation: it, like sympathy, assumes a new cast in light of reciprocal altruism. Guarding against exploitation, Trivers notes, is important. Even in the simple world of Axelrod's computer, with its discrete, binary interactions, TIT FOR TAT had to punish creatures that abused it. In the real world, where people may, in the guise of friendship, run up sizable debts and then welch on them—or may engage in outright theft—exploitation should be discouraged even more emphatically. Hence, perhaps, the fury of our moral indignation, the visceral certainty that we've been treated *unfairly,* that the culprit deserves *punishment.* The intuitively obvious idea of just deserts, the very core of the human sense of justice, is, in this view, a by-product of evolution, a simple genetic stratagem.

What's puzzling at first is the intensity that righteous indignation reaches. It can start feuds that dwarf the alleged offense, sometimes causing the death of the indignant. Why would genes counsel us to take even a slight risk of death for something as intangible as "honor"? Trivers, in reply, noted that "small inequities repeated many times over a lifetime may exact a heavy toll," thus justifying a "strong show of aggression when the cheating tendency is discovered."[26]

A point he didn't make, but which has since been made, is that indignation is even more valuable when publicly observed. If word of your fierce honor gets around, so that a single, bloody fistfight deters scores of neighbors from cheating you—even slightly and occasionally—then the fight was worth the risk. And in a hunter-gatherer society, where almost all behavior is public, and gossip travels fast, the effective audience for a fistfight is all-encompassing. It is notable that, even in modern industrial societies, when males kill males they know, there is usually an audience.[27] This pattern seems perverse—why commit murder in front of witnesses?—except in terms of evolutionary psychology.

Trivers showed how complexly devious the real-life game of prisoner's dilemma could get, as feelings that evolved for one purpose were adapted to others. Thus, righteous indignation could become a pose that cheaters use—whether consciously or unconsciously—to escape suspicion ("How *dare* you impugn my integrity!"). And guilt, which may originally have had the simple role of prompting payment of overdue debts, could begin to serve a second function: prompting the preemptive confession of cheating that seems on the verge of discovery. (Ever notice how guilt does bear a certain correlation with the likelihood of getting caught?)

One hallmark of an elegant theory is its graceful comprehension of long-standing and otherwise puzzling data. In an experiment conducted in 1966, test subjects who believed they had broken an expensive machine were more inclined to volunteer for a painful experiment, but only if the damage had been discovered.[28] If guilt were what idealists assume it to be—a beacon for moral guidance—its intensity wouldn't depend on whether a misdeed had been uncovered. Likewise if guilt were what group selectionists believe it to be—an incentive for reparations that are good for the group. But if guilt is, as Trivers says, just a way of keeping everyone happy with your level of reciprocation, its intensity should depend not on your misdeeds but on who knows or may soon know about them.

The same logic helps explain everyday urban life. When we pass a homeless person, we may feel uncomfortable about failing to help. But what really gets the conscience twinging is making eye contact and still failing to help. We don't seem to mind not giving nearly so much as we mind being seen not giving. (And, as for why we should care about the opinion of someone we'll never encounter again: perhaps in our ancestral environment just about everyone encountered was someone we might well encounter again.)[29]

The demise of "good of the group" logic shouldn't be exaggerated or misconstrued. Reciprocal altruism is classically analyzed in one-on-one situations, and almost surely arose in that form. But the evolution of sacrifice may have grown more complex with time and fostered a sense of group obligation. Consider (not too literally) a "club-forming" gene. It gives you the capacity to think of two or three other people as parts of a unified team; in their presence, you target your altruism more diffusely, making sacrifices for the club as a whole. You might, for example, take a risk in the joint pursuit of wild game and (consciously or unconsciously) expect each of them to repay you on some future expedition. But rather than expect direct repayment, you expect them to sacrifice for "the group," as you did. The other club members expect this too, and people who fail to meet expectations may have their membership terminated, either gradually and implicitly or abruptly and explicitly.

A genetic infrastructure for clubbishness, being more complex than the infrastructure for one-on-one altruism, may sound less likely. But once the one-on-one variety is entrenched, the additional evolutionary steps aren't all that forbidding. So too for subsequent steps that might permit allegiance to even larger groups. Indeed, the growing success of a growing number of small groups within a hunter-gatherer village would be a Darwinian incentive to join larger ones, and get a leg up on the competition; genetic mutations that fostered such joining could flourish. Eventually, indeed, one can imagine a capacity for loyalty and sacrifice toward a group as large as the tribes that figured in Darwin's group-selectionist theory of the moral sentiments. Yet this scenario doesn't suffer from the complications of his scenario. It doesn't involve sacrifice for anyone who doesn't ultimately reciprocate.[30]

Actually, reciprocal altruism of the classic one-on-one variety can, by itself, yield seemingly collectivist behavior. In a species with language, one effective and almost effortless way to reward nice people and punish mean ones is to affect their reputations accordingly. Spreading the word that someone cheated you is potent retaliation, since it leads people to withhold altruism from that person for fear of getting burned. This may help explain the evolution of the "grievance"—not just the sense of having been wronged, but the urge to publicly articulate it. People spend lots of time sharing grievances, listening to grievances, deciding whether the grievances are just, and amending their attitudes toward the accused accordingly.

Perhaps Trivers, in explaining "moral indignation" as a fuel for retaliatory aggression, was getting ahead of the game. As Martin Daly and Margo Wilson have noted, if simple aggression is your goal, a sense of *moral* outrage isn't necessary; sheer hostility will do fine. Presumably it is because humans evolved amid bystanders—bystanders whose opinions mattered—that a moral dimension has emerged, that grievances crystallize.

Exactly *why* opinions of bystanders matter is another question. Bystanders may, as Daly and Wilson put it, be imposing "collective sanctions" as part of a "social contract" (or, at least, part of a "club contract"). Or they may, as I've just suggested, simply be shunning reputed offenders out of self-interest, creating de facto social sanctions. And they may do some of both. In any event, the airing of grievances can lead to widespread reactions that *function* as collective sanctions, and this has come to be a vital part of moral systems. Few evolutionary psychologists would quarrel with Daly and Wilson's basic view that "Morality is the device of an animal of exceptional cognitive complexity, pursuing its interests in an exceptionally complex social universe."[31]

Perhaps the most legitimately dispiriting thing about reciprocal altruism is that it is a misnomer. Whereas with kin selection the "goal" of our genes is to actually help another organism, with reciprocal altruism the goal is that the organism be left under the impression that we've helped; the impression alone is enough to bring the reciprocation. The second goal always entailed the first in Axelrod's computer, and in human society it often does. But when it doesn't— when we can look nice without really being so nice, or can be profitably mean without getting caught—don't be surprised if an ugly part of human nature surfaces. Hence secret betrayals of all gradations, from the everyday to the Shakespearean. And hence the general tendency of people to burnish their moral reputations; reputation is the object of the game for this "moral" animal. And hence hypocrisy; it seems to flow from two natural forces: the tendency toward grievance—to publicize the sins of others—and the tendency to obscure our own sins.

The evolution of George Williams's 1966 musings about reciprocal aid into a compelling body of explanation is one of the great feats of twentieth-century science. It involved ingenious and distinctly modern tools of analysis, and brought momentous results. Though the theory of reciprocal altruism isn't *proved* in the sense that theories of physics can be proved, it rightly commands much confidence within biology, and that confidence should grow as the connection of genes to the human brain becomes clearer in the coming decades. Though the theory isn't as arcane or as mind-bending as the theories of relativity or quantum mechanics, in the end it may alter the world-view of the human species more deeply and more problematically.

Chapter 10: **DARWIN'S CONSCIENCE**

Ultimately a highly complex sentiment, having its first origin in the social instincts, largely guided by the approbation of our fellow-men, ruled by reason, self-interest, and in later times by deep religious feelings, confirmed by instruction and habit, all combined, constitute our moral sense or conscience.

—*The Descent of Man* (1871)[1]

Darwin is sometimes thought of as an excessively decent man. Recall the assessment of one of his biographers, the psychiatrist John Bowlby. Bowlby found Darwin's conscience "overactive" and "overbearing." While admiring Darwin's lack of pretension and his "strong moral principles," Bowlby believed that "these qualities were unfortunately developed prematurely and to excessive degree," leaving him "prone to self-reproach" and to "periods of chronic anxiety and episodes of fairly severe depression."[2]

Self-reproach was indeed second nature to Darwin. He recalled, as a boy, "thinking that people were admiring me, in one instance for perseverance, and another for boldness in climbing a low tree," and at the same time feeling "that I was vain, and contempt of myself."[3] As he grew up, self-criticism became a kind of tick, a reflexive humility; an appreciable fraction of his voluminous correspondence consists of apologies for itself. "How shockingly untidy this letter is," he wrote as a teenager. "I find I am writing most precious nonsense," he wrote in his twenties. "I have written an unreasonably long & dull letter, so farewell," he wrote in his thirties.[4] And so on.

Nighttime was a feast for Darwin's doubts. Then, according to his son Francis, "anything which had vexed or troubled him in the day would haunt him." He might lie awake rehashing a conversation with a neighbor, worried that he had somehow caused offense. He might lie awake thinking about letters he hadn't yet answered. "He used to say that if he did not answer them, he had it on his conscience afterwards," Francis recalled.[5]

Darwin's moral sentiments covered much more than social obligations. Many years after the voyage of the *Beagle*, he was still plagued by the memory of slaves being tortured in Brazil. (Aboard the *Beagle* he had antagonized the captain by sarcastically probing his defense of slavery.) Even the suffering of animals Darwin found unbearable. Francis remembered him once returning from a walk "pale and faint from having seen a horse ill-used, and from the agitation of violently remonstrating with the man."[6] There is no denying Bowlby's point: Darwin's conscience was a very painful thing.

Then again, natural selection never promised us a rose garden. It doesn't "want" us to be happy. It "wants" us to be genetically prolific. And in Darwin's case it didn't do too badly. He had ten children, seven of whom survived to adulthood. So as we try to discern some of the finer features that natural selection has designed into the conscience, there is no glaring reason not to use Darwin's conscience as Exhibit A: an example of a basically sound adaptation. If it goaded him into doing things that amplified his genetic legacy, it may have been working as designed, even if the goading hurt.[7]

Of course, happiness is great. There's every reason to seek it. There's every reason for psychiatrists to try to instill it, and no reason for them to mold the kinds of people natural selection "wants." But therapists will be better equipped to make people happy once they understand what natural selection *does* "want," and how, with humans, it "tries" to get it. What burdensome mental appliances are we stuck with? How, if at all, can they be defused? And at what cost—to ourselves and to others? Understanding what is and isn't pathological from natural selection's point of view can help us confront things that are pathological from our point of view. One way to approach that understanding is to try and figure out when Darwin's conscience was and wasn't malfunctioning.

A SHAMELESS PLOY

One striking feature of the rewards and punishments dished out by the conscience is their lack of sensuality. The conscience doesn't make us feel bad the way hunger feels bad, or good the way sex feels good. It makes us feel as if we have done something that's wrong or something that's right. Guilty or not guilty. It is amazing that a process as amoral and crassly pragmatic as natural selection could design a mental organ that makes us feel as if we're in touch with higher truths. Truly a shameless ploy.

But effective—effective all over the world. Kin selection has ensured that people everywhere feel deeply guilty about, say, grievously harming or neglecting a brother or sister, a daughter or son, even a niece or nephew. And reciprocal altruism has extended the sense of obligation—selectively—beyond the circle of kin. Is there a single culture in which neglecting a friend is a guiltless and widely approved behavior? We would all be skeptical if some anthropologist claimed to have found one.

Reciprocal altruism may have left a more diffuse imprint on the conscience as well. Several decades ago, the psychologist Lawrence Kohlberg tried to construct a natural sequence of human moral development, ranging from the toddler's simple conception of "bad" (that which parents punish you for) to the detached weighing of abstract laws. The higher rungs of Kohlberg's ladder, the ones occupied by ethical philosophers (and, presumably, by Kohlberg), are far from species-typical. But progression through what he called "stage three" seems to be standard in diverse cultures.[8] That stage entails a desire to be known as "nice" and "good." Which is to say: a desire to be known as a reliable reciprocal altruist, a person with whom one could profitably associate. This impulse helps give consensual moral codes their tremendous power; we all want to do—or, more precisely, to be seen doing—what everyone says is good.

Beyond these sorts of basic and apparently universal dimensions

Honeypot ants of the genus *Myrmecocystus,* hanging from the roof of an underground nest, their abdomens swollen with liquid food. During dry spells, these living storage bins provide nourishment for their relatives. Darwin successfully explained the utter altruism of sterile insect castes such as the honeypot, but in very vague terms. Only with the theory of kin selection in the early 1960s did biologists see that Darwin's logic could be used to explain self-sacrifice in many species, including ours.

William Hamilton and Robert Trivers at Harvard University in 1977, more than a decade after Hamilton unveiled the theory of kin selection and a few years after Trivers published the theory of parent-offspring conflict and the Trivers-Willard hypothesis. Together, these ideas explain much about human family life, including sibling love, sibling rivalry, and the tendency of parents to favor some children over others.

On the opposite page are Darwin with son Willie ("Doddy"), Emma Darwin with son Leonard, and the Darwins' daughter Annie, whose death at age ten scarred her parents for life. Evolutionary psychologists have gathered evidence that the amount of grief parents feel over the loss of a child varies from case to case in accordance with Darwinian theory.

Above: After extracting blood from the foot of an unsuspecting chicken, this vampire bat may regurgitate some of it for a "friend" whose foraging has been less successful—and who will reciprocate in the future. *Below:* Two chimpanzee friends groom each other. The evolution of reciprocal altruism helps account for many human feelings, ranging from trust, sympathy, and gratitude to guilt, moral outrage, and even the sense of justice.

of moral sentiment, the contents of the conscience begin to vary. Not only do the particular norms enforced by collective praise and censure differ from culture to culture (another reminder of the huge variability human nature leaves room for); within any one culture, the strictness of obedience varies from person to person. Some people, like Darwin, have big and acute consciences and lie awake at night reflecting on their crimes. Some people don't.

Now, some aspects of Darwin's distinctively strong scruples presumably had to do with distinctive genes. Behavioral geneticists say the heritability of the cluster of traits they call "conscientiousness" is between .30 and .40.[9] That is: about one-third of differences among people (in a typical late-twentieth-century social environment, at least) can be traced to their different genes. But that still leaves two-thirds traceable to environment. In large part, the conscience seems to be an example of the genetically endowed knobs of human nature getting environmentally tuned to widely varied settings. Everyone feels guilt. But not everyone feels it acutely, as Darwin did, over everyday conversations. Everyone empathizes with human suffering at times, and at other times feels (if only briefly) that suffering is justified, that retribution is warranted. But the very fact that slaves were being brutally punished as Darwin visited Brazil suggests that not everyone shared his feelings about when empathy and retribution are, respectively, in order.

The questions are: Why has natural selection given us a fairly flexible conscience, rather than fixing its contents innately? And how has natural selection arranged for the conscience to be shaped? Why and how are the morality knobs of human nature tunable?

As for the "how" question: Darwin himself saw his moral tuning as beginning early, under the guidance of kin. That he could call himself, on balance, "a boy humane" he credited to the "instruction and example of my sisters. I doubt indeed whether humanity is a natural or innate quality." His plans to start an insect collection were complicated when, "on consulting my sister, I concluded that it was not right to kill insects for the sake of making a collection."[10]

Chief moralist was sister Caroline, nine years his elder, who functioned as surrogate mother after their mother's death in 1817, when Darwin was eight. Darwin recalls that Caroline was "too zealous in trying to improve me; for I clearly remember . . . saying to myself when about to enter a room where she was—'What will she blame me for now?' "[11]

Darwin's father was also a force to be reckoned with, a large, imposing, often austere man. His severity has spawned theories about the psychodynamics between father and son, and they have often not been flattering to the father. One Darwin biographer summarized a common profile of Robert Darwin: "his shape is that of a domestic bully and his effect on his son a continuing disaster of neurosis and disability."[12]

The emphasis placed by Darwin on the moral influence of kin has been affirmed by behavioral science. Parents and other authority figures, including older kin, serve as role models and as tutors, molding the conscience with praise and blame. This is basically the way Freud described the formation of the superego—which, in his scheme, encompasses the conscience—and he seems to have gotten it basically right. A child's peers also provide positive and negative feedback, encouraging conformity with playground norms.

It makes sense, of course, that kin should critically guide moral development. Because they share so many genes with the young child, they have a strong, though not unbounded, reason to give useful guidance. By the same token, the child has reason to follow. There is, as Robert Trivers noted, cause for skepticism on the part of children—cause, for example, to discount parental sermons about sharing equally with siblings. But in other realms—how to deal with friends, with strangers—the grounds for parental manipulation diminish, and thus the grounds for offspring obedience grow. In any event, it is clear that the voice of close kin carries a special resonance. Darwin says he reacted to sister Caroline's pedantic nagging by making himself "dogged so as not to care what she might say."[13] Whether he succeeded is another question. In his letters to Caroline from college, he apologizes for his penmanship, makes strained efforts to convince her of his religious piety, and generally evinces continued concern about what she might say.

The channels of paternal influence also seem to have been kept wide open by Darwin's brain. The young Darwin idolized his father and committed to lifelong memory both his sage advice and his cruelest rebuke—"You care for nothing but shooting, dogs, and rat-catching, and you will be a disgrace to yourself and all your family."[14] He devoutly wanted his father's approval, and worked hard to get it. "I think my father was a little unjust to me when I was young," he said, "but afterwards I am thankful to think I became a prime favourite with him." When Darwin made this remark to one of his daughters, it left her with "a vivid recollection of the expression of happy reverie that accompanied these words," as if "the remembrance left a deep sense of peace and gratitude."[15] The many people who share this sense of peace—and the many who instead suffer, well into adulthood, from a chafing sense of parental disapproval—attest to the power of the emotional equipment at work.

What about the "why" question? Why has natural selection made the conscience malleable? Granted, Darwin's kin were the natural providers of useful moral guidance; but what was useful about it? What is so valuable, from the genes' point of view, about the expansive guilt they infused in the young Darwin? And anyway, if a big conscience is so valuable, why don't the genes just hard-wire it into the brain?

The answer begins with the fact that reality is more complicated than Robert Axelrod's computer. In Axelrod's tournament, a bunch of electronic TIT FOR TAT organisms triumphed and then lived happily ever after in mutual cooperation. This exercise had value in showing how reciprocal altruism could evolve and thus suggesting why we all have the emotions that govern it. But of course, we don't use those emotions with the simple steadiness of TIT FOR TAT. People sometimes lie, cheat, or steal—and, unlike TIT FOR TAT, they may behave this way even toward people who have been nice to them. What's more: people sometimes prosper in this fashion. That we have this capacity for exploiting, and that it sometimes pays off, suggests that there have been times during evolution when being nice to nice people wasn't the genetically optimal strategy. We may all have the machinery of TIT FOR TAT, but we also have less admirable machinery. And the question we face is which machinery to use. Hence the adaptive value of a malleable conscience.

This, at least, was Trivers's suggestion in his 1971 paper on reciprocal altruism. He noted that the payoff from helping people—

and the payoff from cheating people—depends on the social environment in which we find ourselves. And environments change over time. So "one would expect selection to favor developmental plasticity of those traits regulating altruistic and cheating tendencies and responses to these tendencies in others." And thus, "the growing organism's sense of guilt" may "be educated, perhaps partly by kin, so as to permit those forms of cheating that local conditions make adaptive and to discourage those with more dangerous consequences."[16] In short: "moral guidance" is a euphemism. Parents are designed to steer kids toward "moral" behaviors only insofar as those behaviors are self-serving.

It's hard to specify the exact circumstances that, during evolution, made different moral strategies more or less valuable. There may have been recurrent changes in the size of villages, or the density of big, huntable game or menacing predators.[17] Any of these could affect the number and the value of cooperative endeavors locally available. And, besides, each person is born into a family that occupies a particular niche in the social ecology, and each person has particular social assets and liabilities. Some people can prosper without taking the risk of cheating, some people can't.

Whatever the reason natural selection first endowed our species with flexible reciprocally altruistic strategies, the advent of flexibility would further raise their value. Once the prevailing winds of cooperation are shifting—from generation to generation, from one village to an adjacent village, or from one family to the next—these shifts are a force to be reckoned with, and a flexible strategy is the way to reckon. As Axelrod showed, the value of a particular strategy depends utterly on neighborhood norms.

If Trivers is right, if the shaping of a young conscience amounts partly to instruction about profitable cheating (and profitable restraint from cheating), then you would expect young children to be good at learning to deceive. That seems to be, if anything, an understatement. Jean Piaget, in his 1932 study of moral development, wrote that "the tendency to tell lies is a natural tendency . . . spontaneous and universal."[18] Subsequent study has borne him out.[19]

Certainly Darwin seems to have been a natural-born liar—"much given to inventing deliberate falsehoods." For instance, "I once gathered much valuable fruit from my Father's trees and hid them in the shrubbery, and then ran in breathless haste to spread the news that I had discovered a hoard of stolen fruit." (As, in a sense, he had.) He rarely took a walk without claiming to have seen "a pheasant or some strange bird," whether or not this was true. And he once told a boy "that I could produce variously coloured Polyanthuses and Primroses by watering them with certain coloured fluids, which was of course a monstrous fable, and had never been tried by me."[20]

The idea here is that childhood lies are not just a phase of harmless delinquency we pass smoothly through, but the first in a series of test runs for self-serving dishonesty. Through positive reinforcement (for undetected and fruitful lies) and negative reinforcement (for lies that peers uncover, or through the reprimand of kin) we learn what we can and can't get away with, and what our kin do and don't consider judicious deceit.

That parents seldom lecture children on the virtues of lying doesn't mean they're not teaching them to lie. Children, it seems, will keep lying unless strongly discouraged. Not only are children whose parents often lie more likely than average to become chronic liars; so too are children who simply lack close parental supervision.[21] If parents refrain from discouraging the kinds of lies that have proven useful to them—and if they tell such lies in the presence of their children—they are giving an advanced course in lying.

One psychologist has written, "No doubt lying is exciting; that is, the manipulation itself, rather than the benefits that result from it, may spur children to lie."[22] This dichotomy is misleading. It is presumably *because* of the benefits of skillful lying that natural selection has made experimental lying exciting. Once again: natural selection does the "thinking"; we do the doing.

Darwin recalled making up stories for "the pure pleasure of exciting attention & surprise." On the one hand, "these lies, when not detected, I presume excited my attention [and] by having produced great effect on my mind, gave pleasure, like a tragedy."[23] On the other hand, at times they left him feeling shameful. He doesn't say why, but two possibilities spring to mind. One is that some of the lies were uncovered by suspicious children. Another is that lying got him chastised by older kin.

Either way, Darwin was getting feedback about the propitious-ness of lying in his particular social environment. And either way, the feedback had effect. By the time he reached adulthood, he was honest by any reasonable standard.

The transmission of moral instruction from old to young parallels the transmission of genetic instruction and is sometimes indis-tinguishable in its effects. In *Self-Help* Samuel Smiles wrote, "The characters of parents are thus constantly repeated in their chil-dren; and the acts of affection, discipline, industry, and self-control, which they daily exemplify, live and act when all else which they may have learned through the ear has long been forgotten. . . . Who can tell how much evil act has been stayed by the thought of some good parent, whose memory their children may not sully by the com-mission of an unworthy deed, or the indulgence of an impure thought?"[24]

This fidelity of moral transmission is plain in Darwin. When, in his autobiography, he extols his father—his generosity, his sympa-thy—he might just as well have been talking about himself. And Darwin would in turn work to endow his own children with solid reciprocal-altruism skills, ranging from moral probity to social nicety. To a son at school he wrote, "You must write to Mr. Wharton: you had better begin with 'My dear Sir.' . . . End by saying 'I thank you and Mrs. Wharton for all the kindness you have always done me. Believe me, Yours truly obliged.' "[25]

THE VICTORIAN CONSCIENCE

Natural selection had no way of anticipating Darwin's social envi-ronment. The human genetic program for custom-tailoring the con-science does not include an option labeled "Well-to-do Man in Victorian England." For this reason (among others) we shouldn't expect Darwin's early experience to have shaped his conscience in a wholly adaptive way. Still, some things that natural selection pre-sumably did "anticipate"—for example, that the sheer level of local cooperation would differ from environment to environment—are rel-evant to any time and place. It is worth seeing if Darwin's moral development left him reasonably well equipped to prosper.

The question of how Darwin's conscience paid off is really the question of how the Victorian conscience itself paid off. Darwin's moral compass was, after all, just a more acute version of the basic Victorian model. The Victorians are famous for their emphasis on "character," and many of them, if plopped down in our midst, would seem bizarrely earnest and conscientious, if less so than Darwin.

The essence of Victorian character was "truthfulness, integrity, and goodness," according to Samuel Smiles. "Integrity in word and deed is the backbone of character," he wrote in *Self-Help*, "and loyal adherence to veracity its most prominent characteristic."[26] Note the contrast with "personality," the amalgam of charm, pizzazz, and other social garnish that, we are told, has in the twentieth century largely replaced character as the measure of a human being. This change is sometimes noted with the wistful suggestion that the present century is one of moral relapse and rampant selfishness.[27] "Personality," after all, attaches so little value to honesty or honor and seems so plainly a vehicle for self-advancement.

The culture of personality does have a shallow feel, and it's easy to get nostalgic for the days when sheer glibness got a person less far in life. But that doesn't mean the reign of character was an era of pure integrity, unsullied by self-interest. If Trivers is right about why the conscience is malleable, then "character" may have been a self-serving thing.

The Victorians themselves made no bones about the uses of character. Samuel Smiles approvingly quoted a man of "sterling independence of principle and scrupulous adherence to truth" who noted that obedience to conscience is "the road to prosperity and wealth." Smiles himself observed that "character is power" ("in a much higher sense than that knowledge is power"). He quoted the stirring words of the statesman George Canning: "My road must be through Character to power; I will try no other course; and I am sanguine enough to believe that this course, though not perhaps the quickest, is the surest."[28]

If character was so conducive to advancement in those days, why was it more so then than now? This is no place for a Darwinian treatise on moral history, but one possible factor is obvious: most people in Victorian England lived in the rough equivalent of a small town. To be sure, urbanization was well under way, and thus the

era of anonymity was nearing. But, compared to today, neighborhoods, even urban ones, were stable. People tended to stay put, and
encounter the same small group of folks year after year. This is true
in spades of Darwin's hometown, the quaint village of Shrewsbury.
If Trivers is right—if the young conscience is molded, with the active
aid of kin, to fit the local social environment—then Shrewsbury
is the sort of place in which we might expect Darwin's scruples to
pay off.

There are at least two reasons integrity and honesty make particular sense in a small and steady social setting. One is that (as
everyone who has lived in a small town knows) there's no escaping
your past. In a section of *Self-Help* titled "Be What You Seem,"
Smiles wrote, "A man must really be what he seems or purposes to
be. . . . Men whose acts are at direct variance with their words,
command no respect, and what they say has but little weight." Smiles
relayed an anecdote about a man who says "I would give a thousand
pounds for your good name." "Why?" "Because I could make ten
thousand by it."[29] The Darwin described by young Emma Wedgwood—"the most open, transparent man I ever saw, and every word
expresses his real thoughts"—is a man well equipped to thrive in
Shrewsbury.[30]

Axelrod's computer world was a lot like Shrewsbury: the same
fairly small group of characters, day after day, all of whom remember
how you behaved on the last encounter. That, of course, is a central
reason why reciprocal altruism paid off inside the computer. If you
make the computer world even more like a small town, by allowing
its creatures to gossip about how scrupulous so-and-so is or isn't,
cooperative strategies flourish even faster. For then cheaters get away
with fewer swindles before people start shunning them.[31] (Axelrod's
computer has multiple uses. Once people have flexible moral equipment, cooperation can, over the generations, spread—or retreat—
without any changes in the gene pool. Thus the computer, in chron-
icling such waves, can model cultural change, as here, rather than
model genetic change, as in the last chapter.)

A second reason why being nice is so fruitful in a place like
Shrewsbury is that the people you're nice to will be around for a
long time. Even scattershot expenditures of social energy, such as the

diffuse bestowal of warm pleasantries, may be a sound investment. "Those little courtesies which form the small change of life, may separately appear of little intrinsic value, but they acquire their importance from repetition and accumulation," Smiles wrote. He observed that "benevolence is the preponderating element in all kinds of mutually beneficial and pleasant intercourse amongst human beings. 'Civility,' said Lady Montague, 'costs nothing and buys everything.' . . . 'Win hearts,' said Burleigh to Queen Elizabeth, 'and you have all men's hearts and purses.' "[22]

Actually, civility does cost something: a little bit of time and psychic energy. And these days it doesn't buy too much—at least, not unless laser-guided. Many, if not most, of the people we encounter each day don't know who we are and will never find out. Even our acquaintances may be fleeting. People move often, change jobs often. So a reputation for integrity matters less now, and sacrifices of all kinds—even for colleagues or neighbors—are less likely to be repaid far in the future. These days an upper-middle-class man who by example teaches his son to be slick and superficially sincere, to tell minor lies in profusion, to work harder on promise than delivery, may well be equipping him for success.

You can see this in Axelrod's computer. If you change the rules, and allow frequent migration into and out of the group, so that there are fewer chances to reap what you've sown, the power of TIT FOR TAT wanes visibly and the success of meaner strategies grows. (Here, again, we use the computer to model cultural, not genetic, evolution; the size of the average conscience is changing, but not because of underlying changes in the gene pool.)

In the computer, as in life, these trends are self-sustaining. When less cooperative strategies flourish, the amount of locally available cooperation declines, further devaluing cooperation, so that less cooperative strategies flourish all the more. It works in the other direction too: the more conscientious the Victorians got, the more conscientious it made sense to be. But when, for whatever reason, the pendulum finally reaches its apex and heads back down, it naturally picks up momentum.

To some extent, this analysis simply underscores the old truisms about the effects of urban anonymity: New Yorkers are rude, and

New York is full of pickpockets.[33] But that doesn't go far enough. The point here isn't just that people look around, see opportunities for cheating, and consciously seize them. Through a process they sense dimly if at all, a process that begins as soon as they learn to talk, the contours of their conscience get adjusted for them—by kin (who themselves may not grasp what's going on) and by other sources of environmental feedback. Cultural influence can be just as unconscious as genetic influence. This is not surprising, given how deeply intertwined the two are.

The same point applies to a sector whose ethos is the subject of much discussion these days: the poor, crime-ridden inner cities of America. Budding criminals needn't look around, appraise the situation, and rationally choose a life of crime. If this were the whole truth, then the standard solution to crime—"alter the incentive structure" by making sure crime doesn't pay—might work better. Darwinism suggests a more unsettling truth: from an early age, the conscience of many poor children, the very capacity for sympathy and guilt, is hemmed in by the environment, and as they grow up it settles somewhat firmly into this cramped form.

The source of this cramping presumably goes well beyond urban anonymity. Many people in the inner city face limited opportunities for "legitimate" cooperation with the wider world. And the males, risk-prone by virtue of their gender to begin with, don't have the long life expectancies that so many people take for granted. Martin Daly and Margo Wilson have argued that the "short time horizons" for which criminals are famous may be "an adaptive response to predictive information about one's prospects for longevity and eventual success."[34]

"Riches and rank have no necessary connection with genuine gentlemanly qualities," Samuel Smiles wrote. "The poor man may be a true gentleman,—in spirit and in daily life. He may be honest, truthful, upright, polite, temperate, courageous, self-respecting, and self-helping,—that is, be a true gentleman." For, "from the highest to the lowest, the richest to the poorest, to no rank or condition in life has nature denied her highest boon,—the great heart."[35] That's a nice thought, and it may hold true for the first few months of life.

But—under modern conditions, at least—it probably grows ever more false thereafter.

To some it will sound odd to hear Darwinians describing criminals as "victims of society" rather than as victims of faulty genes. But that's one difference between Darwinism at the turn of this century and at the turn of the last. Once you think of genes as programming behavioral *development,* and not just behavior, as molding the young mind to fit its context—then we all start to look like victims (or beneficiaries) of our environment, no less than of our genes.[36] Thus can a difference between two groups (socioeconomic, say, or even ethnic) be explained by evolution yet without reference to genetic differences.

There is, of course, no "urban underclass" notch in the developmental program that shapes the conscience, any more than there is a "Victorian" notch. (Indeed, the village of Shrewsbury is more like the kind of setting that natural selection "anticipated" than are today's large cities.) Still, the deftness with which urban opportunities for cheating are often exploited suggests that the ancestral environment did, at times, present opportunities for profitable crime.

One likely source of such opportunities would have been periodic contact with nearby villages. And the adaptation that would help seize those opportunities is precisely what we find in the human mind: a binary moral landscape, comprising an in-group that deserves consideration and an out-group that deserves exploitation.[37] Even urban gang members have people who can trust them. And even scrupulously polite Victorian men went to war, convinced of the justness of the death they dished out. Moral development is often a question not just of how strong the conscience will be, but of how long a reach it will have.

JUDGING THE VICTORIANS

How "moral" the Victorians really were is a subject of some contention. They are commonly accused of great hypocrisy. Well, as we've seen, a little hypocrisy is only natural in our species.[38] And, oddly, a lot of hypocrisy may be a sign of great morality. In a highly "moral" society—where daily life involves many acts of courtesy and altruism, where meanness and dishonesty are reliably punished by social sanction—a good moral reputation is vital and a bad one quite costly. This added weight of reputation is more incentive to do what people naturally do anyway: exaggerate their goodness. As Walter Houghton wrote in *The Victorian Frame of Mind,* "Although everyone at times pretends to be better than he is, even to himself, the Victorians were more given to this type of deception than we are. They lived in a period of much higher standards of conduct. . . ."[39]

Even if we accept that Victorian hypocrisy is a roundabout affirmation of Victorian morality, we might still ask whether *morality* is the right word. For most Victorians, after all, the prevailing ethos brought no net sacrifice. So many people were so diffusely considerate that everyone got a piece of the action. But that's no indictment of Victorian morality. That's the whole idea behind a robust morality: to encourage *informal* non-zero-sum exchanges, thus raising overall welfare; that is, to encourage non-zero-sum exchanges outside the realms of economic life and legal compulsion. One writer, lamenting "the rise of selfishness" and the passing of "Victorian America," has observed that, under the Victorian ethos, "the great central mass of Americans was living in a social system that was predictable, stable and basically decent. And it was so because—despite the hypocrisy— most people felt that they had duties and obligations to other people which came before their own gratification."[40] We may question the literal truth of that last sentence without doubting its drift. What sustained everyone's sense of duty was not self-abnegation, ultimately, but their implicit assent to a vast social contract, under which duties discharged to others would, however indirectly, be discharged to them in return. Still, the author is right: an immense amount of time and energy now spent on vigilance was not spent in those days.

One way to put the matter is to say that Victorian England was an admirable *society,* but not one composed of especially admirable *people.* They were only doing what we do—acting conscientiously, politely, and considerately to the extent that it pays. It just paid more in those days. And besides, their moral behavior, however laudable it was or wasn't, was more a heritage than a choice; the Victorian conscience got shaped in ways the Victorians never understood and were in some sense powerless to affect.

Here, then, is the verdict on Charles Darwin, rendered with the authority of everything we now know about genes: he was a product of his environment. If he was a good man, he was good in passive reflection of his society's goodness. And, anyway, much of his "goodness" paid off.

Still, Darwin does seem sometimes to have gone above and beyond the call of reciprocal altruism. While in South America, he planted gardens for the Fuegian Indians. And years later, while living in the village of Downe, he founded the Downe Friendly Society, which provided a savings plan for the local workers, as well as a "clubroom" (where their moral life would be improved by Skinnerian conditioning—swearing, fighting, and drunkenness were subject to a fine).[41]

Some Darwinians make a sport of reducing even this sort of niceness to self-interest. If they can't find a way that the Fuegian Indians might have reciprocated (and we don't know that they didn't), the next recourse is to talk about "reputation effects"; maybe the men of the *Beagle* would carry tales of Darwin's generosity back to England, where he would somehow be rewarded. But Darwin's moral sentiments were strong enough to strain such cynicism to its breaking point. When he heard that a local farmer had let some sheep starve to death, he personally gathered the evidence and brought the case to the magistrate.[42] The dead sheep were poorly positioned to repay him, and the farmer certainly wouldn't; and the "reputation effects" of behavior so fanatical might not have worked entirely to Darwin's advantage. Similarly, where was the payoff in losing sleep over the past suffering of South American slaves?

The simpler way to account for this sort of "excessively" moral behavior is to recall that human beings aren't "fitness maximizers" but rather "adaptation executers." The adaptation in question—the conscience—was *designed* to maximize fitness, to exploit the local environment in the name of genetic self-interest, but success in this endeavor is far from assured, especially in social settings alien to natural selection.

Thus the conscience can lead people to do things that aren't in their self-interest except in the sense of salving the conscience itself. Sympathy, obligation, and guilt, unless subjected to a veritable extermination campaign during youth, always have the potential to bring behaviors of which their "creator," natural selection, would not "approve."

We started this chapter with the working hypothesis that Darwin's conscience was a smoothly functioning adaptation. And in many ways it was. What's more, some of these ways are quite cheering: they show how some mental organs, though designed ultimately for self-interest, are at the same time designed to work harmoniously with other people's mental organs, and in the process may yield a large amount of social welfare. Still, in some ways Darwin's conscience did not function adaptively. This too is cause for cheer.

Part Three: **SOCIAL STRIFE**

Chapter 11: **DARWIN'S DELAY**

My health has improved a good deal, since I have been in the country, & I believe to a stranger's eyes, I should look quite a strong man, but I find I am not up to any exertion, & I am constantly tiring myself by very trifling things. . . . [I]t has been a bitter mortification for me, to digest the conclusion, that the "race is for the strong"—& that I shall probably do little more, but must be content to admire the strides others make in Science— So it must be. . . .

—Letter to Charles Lyell (1841)[1]

After discovering natural selection in 1838, Darwin spent the next two decades not telling the world about it. He didn't start writing a book on his theory until 1855, and that book he never really finished. Only in 1858, when he learned that another naturalist had arrived at the same theory, did he decide to produce what he called an "abstract"—*The Origin of Species,* published in 1859.

But Darwin didn't spend the 1840s idly. Though slowed by frequent illness—violent shivering and vomiting attacks, gastric pain and epic flatulence, faintness, heart palpitations—he was prolific.[2] During the first eight years of his marriage, he published scientific papers, finished editing the five volumes of *The Zoology of the Voyage of H.M.S. Beagle,* and wrote three books based on the voyage: *The Structure and Distribution of Coral Reefs* (1842), *Geological Obser-*

vations on the Volcanic Islands (1844), and *Geological Observations on South America* (1846).

On October 1, 1846, Darwin made this entry in his personal journal: "Finished last proof of my Geolog. Obser. on S. America; This volume, including Paper in Geolog. Journal on the Falkland Islands took me 18 & ½ months: the M.S., however, was not so perfect as in the case of Volcanic Islands. So that my Geology has taken me 4 & ½ years: now it is 10 years since my return to England. How much time lost by illness!"[3]

This is vintage Darwin in several respects. There is the grim resignation with which, as his illness wore on, he often trudged through his work; though he had on this day finished a grand trilogy (at least one volume of which is still considered a classic), he doesn't sound as if he's poised to crack open a bottle of champagne. There is his never-ending self-criticism; he can't savor the project's end for even a day before turning to its imperfections. There is his sharp awareness of the passage of time, and his obsession with using it well.

You might think that this moment was an auspicious one for Darwin finally to start moving with some briskness toward his appointment with destiny. Certainly one great spur to productivity—a sense of mortality—was now honed to a keen edge. In 1844 he had given Emma a 230-page sketch of the theory of natural selection, along with written instructions to publish it—and "take trouble in promoting it"—in the event of his death. The very fact that the Darwins had now moved out of London, to the rural village of Downe, was testament to his physical decline. There he was to be insulated from the distractions and disequilibriums of city life, draw warmth from his growing family, and, under a tightly structured regimen of work, recreation, and rest, try to extract from his constitution a few good hours of output each day—seven days a week—so long as he could stay alive. This was the environment he had built for himself by the time he finished his books on the geology of South America. In a letter to Captain FitzRoy written that same day (October 1, 1846), Darwin reported: "My life goes on like Clockwork, and I am fixed on the spot where I shall end it."[4]

Given all this—a secure workplace, the faint sound of the grim reaper's footsteps, and completion, at last, of all scholarly obligations from the *Beagle* expedition—given all this, what cause could there possibly be to further postpone the writing of Darwin's book on natural selection?

In a word: barnacles. Darwin's long involvement with barnacles began innocently enough, with curiosity about a species found along the coast of Chile. But one species led to another, and before long his house was world barnacle headquarters, replete with specimens solicited from collectors by mail. For so long did the study of barnacles figure in Darwin's life, and so centrally, that one of Darwin's young sons, upon visiting a neighbor's home, asked, "Where does he do his barnacles?"[5] By the end of 1854—eight years after Darwin predicted that his barnacle work would take him a few months, maybe a year—he had published two books on living species of barnacles and two on barnacle fossils, and established an enduring reputation within this realm. His books are consulted to this day by biologists studying the subclass Cirripedia of the subphylum Crustacea (that is to say, barnacles).

Now, there is nothing wrong with being a leading barnacle authority. But some people are capable of greater things. Why Darwin took so long to realize his greatness has been the subject of much reflection. The most common theory is the most obvious: writing a book that affronts the religious beliefs of virtually everyone in your part of the world—including many colleagues and your wife—is a task not to be approached without circumspection.

The task had already been approached by a few people, and the result was never unalloyed praise. Darwin's grandfather Erasmus, a noted naturalist and poet, had himself advanced a theory of evolution in 1794 in the book *Zoonomia*. He had wanted the book to be published posthumously but finally changed his mind, after some twenty years, saying, "I am now too old and hardened to fear a little abuse," which is what he got.[6] Jean-Baptiste de Lamarck's grand exposition of a similar evolutionary scheme appeared in 1809, the year of Darwin's birth, and was denounced as immoral. And in 1844, a book called *Vestiges of the Natural History of Creation* appeared, outlining a theory of evolution and making a commotion. Its author, a Scottish publisher named Robert Chambers, chose to keep his name secret, perhaps wisely. The book was called, among other things, "a foul and filthy thing, whose touch is taint, whose breath is contamination."[7]

And none of these heretical theories was quite as godless as Darwin's. Chambers had a "Divine Governor" guiding evolution. Erasmus Darwin, being a deist, said that God had wound up the great clock of evolution and let it tick. And though Lamarck was denounced by Chambers as being "disrespectful of Providence,"[8] Lamarckian evolution, as compared to Darwinian, was downright spiritual; it featured an inexorable tendency toward greater organic complexity and more highly conscious life. Imagine, if these men were due a severe scolding, what was in store for Darwin, whose theory involved no Divine Governor, no clock-winder (though Darwin pointedly left open the possibility of one), and no inherent progressive tendency—nothing but the slow accretion of fortuitous change.[9]

There's no doubt that Darwin was, from early on, worried about public reaction. Even before his belief in evolution had crystallized into the theory of natural selection, he weighed rhetorical tactics that might blunt criticism. In the spring of 1838, he wrote in his notebook, "Mention persecution of early Astronomers."[10] In later years, Darwin's fear of censure is evident in his correspondence. The letter in which he confessed his heresy to his friend Joseph Hooker features one of the most defensive passages he ever produced—no small achievement. "I am almost convinced (quite contrary to opinion I started with) that species are not (it is like confessing a murder) immutable," he wrote in 1844. "Heaven forfend me from Lamarck nonsense of a 'tendency to progression' 'adaptations from the slow willing of animals' &c,—but the conclusions I am led to are not widely different from his—though the means of change are wholly so—I think I have found out (here's presumption!) the simple way by which species become exquisitely adapted to various ends.— You will now groan, & think to yourself 'on what a man have I been wasting my time in writing to.' I sh'd, five years ago, have thought so."[11]

SICK AND TIRED

The theory that Darwin was slowed by a hostile social climate assumes many forms, ranging from baroque to simple, and these depict his delay in various ways, ranging from pathological to wise.

In the more elaborate versions of the theory, Darwin's illness—which was never clearly diagnosed and remains a mystery—figures as a psychosomatic procrastination device. Darwin was feeling heart palpitations in September 1837, a couple of months after opening his first evolution notebook, and his reports of illness are fairly frequent as those notebooks unfold toward the theory of natural selection.[12]

It has been suggested that Emma, who held her religion dear, and was pained by her husband's evolutionism, heightened the tension between his science and his social environs; and that, by so devotedly nursing him, she made his unwellness easier than was healthy. A letter to Charles just before their marriage contains a passage to that effect: "[N]othing *could* make me so happy as to feel that I could be of any use or comfort to my own dear Charles when he is not well. If you knew how I long to be with you when you are not well! . . . So don't be ill any more my dear Charley till I can be with you to nurse you. . . ."[13] These sentences may represent the high-water mark of Emma's premarital ardor.

Not all theories linking Darwin's illness to his ideas imply a subconscious plot to conceal them. Darwin may have simply had what is known today as an emotionally induced illness. Anxiety about social rejection is, after all, an ultimately physiological thing, as Darwin would have been the first to point out. It takes a physiological toll.[14]

Some people accept that Darwin had a bona fide disease, probably contracted in South America (perhaps Chagas' disease or chronic fatigue syndrome), but say he used barnacles to subconsciously forestall the day of reckoning. Certainly, as Darwin entered his barnacle phase, insisting it would be brief, he had some misgivings about what lay beyond. He wrote to Hooker in 1846: "I am going to begin some papers on the lower marine animals, which will last me some months, perhaps a year, & then I shall begin looking over my ten-year-long accumulation of notes on species & varieties which, with writing, I daresay I shall stand infinitely low in the opinion of all sound naturalists—so this is my prospect for the future."[15] That's the sort of attitude that could lead to an eight-year barnacle research project.

Some observers, including some of Darwin's contemporaries, have said the barnacles did him a great service.[16] They immersed him fully in the details of taxonomy (good experience for someone who purports to have a theory explaining how all valid taxonomies came to be) and gave him an entire subclass of animals to examine in light of natural selection.

Besides, there were things other than taxonomy that he hadn't yet mastered—which leads to the simplest of all theories about his delay. The fact is that in 1846—and in 1856, and, really, in 1859, when the *Origin* was published—Darwin had not fully figured out natural selection. And it is only logical, before unveiling a theory that will get you defamed and hated, to try to get it into good shape.

One of the puzzles about natural selection that faced Darwin was the puzzle of extreme selflessness, of insect sterility. Not until 1857 did he solve it, with his precursor of the theory of kin selection.[17]

Another of the puzzles Darwin never solved.[18] This is the problem of heredity itself. A great virtue of Darwin's theory is that it doesn't depend, as Lamarck's did, on the inheritance of acquired traits; for natural selection to work, it isn't necessary that a giraffe's stretching for higher leaves affect the neck length of its offspring. But Darwinian evolution does depend on *some* form of change in the range of inherited traits; natural selection needs an ever-changing menu to "choose" from. Today any good high school biology student can tell you how the menu keeps changing—through sexual recombination and genetic mutation. But neither of these mechanisms made obvious sense before people knew about genes. For Darwin to have talked about "random mutations" when asked how the pool of traits changes would have been like saying, "It just does—trust me."[19]

It is possible to assess Darwin's delay from the standpoint of evolutionary psychology. This view doesn't yield a whole new theory about the delay, but it does help drain the episode of some mystery.

It can be best appreciated after the evolutionary roots of his ambitions and fears have become clear. For now, let us leave the story in 1854, when the last of the barnacle books was published and the time had come for Darwin to muster his full reserves of enthusiasm for the coming culmination of his life's work. He wrote to Hooker: "How awfully flat I shall feel, if, when I get my notes together on species, &c. &c., the whole thing explodes like an empty puff-ball."[20]

Chapter 12: **SOCIAL STATUS**

Seeing how ancient these expressions are, it is no wonder that they are so difficult to conceal.—a man insulted may forgive his enemy & not wish to strike him, but he will find it far more difficult to look tranquil.—He may despise a man & say nothing, but without a most distinct will, he will find it hard to keep his lip from stiffening over his canine teeth.—He may feel satisfied with himself, & though dreading to say so, his step will grow erect & stiff like that of turkey.

—M Notebook (1838)[1]

Among the things Charles Darwin found troubling about the Fuegian Indians was their apparent lack of social inequality. "At present," he wrote in 1839, "even a piece of cloth is torn into shreds and distributed; and no one individual becomes richer than another." Such "perfect equality," he feared, would "for a long time retard their civilization." Darwin noted, by way of example, that "the inhabitants of Otaheite, who, when first discovered, were governed by hereditary kings, had arrived at a far higher grade than another branch of the same people, the New Zealanders—who although benefited by being compelled to turn their attention to agriculture, were republicans in the most absolute sense." The upshot: "In tierra del Fuego, until some chief shall arise with power sufficient to secure any acquired advantages, such as the domesticated animals or other valuable presents, it seems scarcely possible that the political state of the country can be improved."

Then Darwin added, "On the other hand, it is difficult to understand how a chief can arise till there is property of some sort by which he might manifest and still increase his authority."[2]

Had Darwin mulled this afterthought a little longer, he might have begun to wonder whether the Fuegians were, in fact, a people of "perfect equality." Naturally, to an affluent Englishman, reared amid servants, a society never far from starvation will seem starkly egalitarian. There will be no opulent displays of status, no gross disparities. But social hierarchy can assume many forms, and in every human society it seems to find one.

This pattern has been slow to come to light. One reason is that lots of twentieth-century anthropologists have, like Darwin, come from highly stratified societies, and been struck, sometimes charmed, by the relative classlessness of hunter-gatherer peoples. Anthropologists have been burdened, also, by a hopeful belief in the almost infinite malleability of the human mind, a belief fostered especially by Franz Boas and his famous students, Ruth Benedict and Margaret Mead. The Boasian bias against human nature was in some ways laudable—a well-meant reaction against crude political extensions of Darwinism that had countenanced poverty and various other social ills as "natural." But a well-meant bias is still a bias. Boas, Benedict, and Mead left out large parts of the story of humanity.[3] And among those parts are the deeply human hunger for status and the seemingly universal presence of hierarchy.

More recently, anthropologists of a Darwinian bent have started looking closely for social hierarchy. They have found it in even the least likely places.

The Ache, a hunter-gatherer people in South America, seem at first to possess an idyllic equality. Their meat goes into a communal pool, so the best hunters routinely aid their less fortunate neighbors. But during the 1980s, anthropologists took a closer look and found that the best hunters, though generous with meat, hoard a resource more fundamental. They have more extramarital affairs and more illegitimate children than lesser hunters. And their offspring have a better chance of surviving, apparently because they get special treatment.[4] Being known as a good hunter, in other words, is an informal rank that carries clout with men and women alike.

The Aka pygmies of central Africa also appear at first glance to be lacking in hierarchy, as they have no "headman," no ultimate political leader. But they do have a man called a *kombeti* who subtly but powerfully influences big group decisions (and who often earns that rank through his hunting prowess). And it turns out that the *kombeti* gets the lion's share of the food, the wives, and the offspring.[5]

And so it goes. As more and more societies are reevaluated in the unflattering light of Darwinian anthropology, it becomes doubtful that any truly egalitarian human society has ever existed. Some societies don't have sociologists, and thus may not have the *concept* of status, but they do have status. They have people of high status and low status, and everyone knows who is who. In 1945 the anthropologist George Peter Murdock, swimming against the prevailing Boasian current, published an essay called "The Common Denominator of Cultures," in which he ventured that "status differentiation" (along with gift giving, property rights, marriage, and dozens of other things) was a human universal.[6] The closer we look, the righter he seems.

In one sense, the ubiquity of hierarchy is a Darwinian puzzle. Why do the losers keep playing the game? Why is it in the genetic interest of the low men on the totem pole to treat their betters with deference? Why lend your energy to a system that leaves you with less than your neighbors?

One can imagine reasons. Maybe hierarchy makes the whole group so cohesive that most or all members benefit, even if they benefit unequally—exactly the fate that Darwin hoped would someday befall the Fuegians. In other words, maybe hierarchies serve "the good of the group" and are thus favored by "group selection." This theory was embraced by the popular writer Robert Ardrey, a prominent member of the generation of group selectionists whose decline marked the rise of the new Darwinian paradigm. If people weren't inherently capable of submission, Ardrey wrote, then "organized society would be impossible and we should have only anarchy."[7]

Well, maybe so. But judging by the large number of essentially asocial species, natural selection doesn't seem to share Ardrey's concern for social order. It is perfectly willing to let organisms pursue inclusive fitness amid anarchy. Besides, if you start thinking carefully about this group-selectionist scenario, problems arise. Granted, when two tribes meet in combat, or compete for the same resource, the more hierarchical and cohesive may win. But how did it get hierarchical and cohesive in the first place? How would genes counseling submission, and thus lowering fitness, manage to gain a foothold amid the everyday competition among genes within the society? Wouldn't they tend to be banished from the gene pool before they had a chance to demonstrate their goodness for the group? These are the questions group selection theories—such as Darwin's theory of the moral sentiments—often face and often fail to surmount.

The most widely accepted Darwinian explanation for hierarchy is simple, straightforward, and nicely compatible with observed reality. It is only with this theory in hand—only after taking a clear look at human social status, uncolored by morality and politics—that we can get back to the moral and political questions. In exactly what senses *is* social inequality inherent in human nature? Is inequality indeed, as Darwin suggested, a prerequisite for economic or political advancement? Are some people "born to serve" and others "born to lead"?

THE MODERN THEORY OF STATUS HIERARCHIES

Throw a bunch of hens together, and, after a time of turmoil, including much combat, things will settle down. Disputes (over food, say) will now be brief and decisive, as one hen simply pecks the other, bringing quick deferral. The deferrals form a pattern. There is a simple, linear hierarchy, and every hen knows its place. A pecks B with impunity, B pecks C, and so on. The Norwegian biologist Thorleif Schjelderup-Ebbe noticed this pattern in the 1920s and gave it the name "pecking order." (Schjelderup-Ebbe also observed, in a frenzy of politically loaded overextrapolation: "Despotism is the basic idea of the world, indissolubly bound up with all life and existence. . . . There is nothing that does not have a despot."[8] No wonder anthropologists shied away from evolutionary accounts of social hierarchy for so long.)

The order of the pecking is not arbitrary. B had a marked tendency to defeat C in early conflicts, and A tended to prevail over B. So it isn't, after all, such a great challenge to explain the emerging social hierarchy as merely the sum of individual self-interest. Each hen is deferring to hens that will probably win anyway, saving itself the costs of battle.

If you've spent much time with chickens, you may doubt their ability to process a thought as complex as "Chicken A will beat me anyway, so why bother to fight?" Your doubt is well placed. Pecking orders are yet another case where the "thinking" has been done by natural selection, and so needn't be done by the organism. The organism must be able to tell its neighbors apart, and to feel a healthy fear of the ones that have brutalized it, but it needn't grasp the logic behind the fear. Any genes endowing a chicken with this selective fear, reducing the time spent in futile and costly combat, should flourish.

Once such genes pervade the population, hierarchy is part of the social architecture. The society may look, indeed, as if designed by someone who valued order over liberty. But that doesn't mean it was. As George Williams put it in *Adaptation and Natural Selection*, "The dominance-subordination hierarchy shown by wolves and a wide variety of vertebrates and arthropods is not a functional organization. It is the statistical consequence of a compromise made by each individual in its competition for food, mates, and other resources. Each compromise is adaptive, but not the statistical summation."[9]

This isn't the only conceivable explanation of hierarchy that skirts the pitfalls of group selectionism. Another is based on John Maynard Smith's concept of an evolutionarily stable state—more specifically, on his "hawk-dove" analysis of a hypothetical bird species. Imagine dominance and submission as two genetically based strategies, the success of each depending on their relative frequency. Being a dominant (for example, walking around intimidating submissives into giving you half their food) is fine so long as there are lots of submissives around. But as the strategy spreads, it grows less fruitful: there are fewer and fewer submissives to exploit, and meanwhile dominants encounter one another more and more, engaging in costly combat. That's why the submissive strategy can thrive; a submissive animal must often surrender some of its food, but it avoids the fighting that takes an increasingly large toll on dominants. The population should in theory equilibrate, with a fixed ratio of dominants to submissives. And, as with all evolutionarily stable states (recall the bluegill sunfish from chapter three), this equilibrium ratio is the point at which each strategy enjoys the same reproductive success.[10]

There are species that this explanation seems to fit. Among Harris sparrows, darker birds are aggressive and dominant, and lighter ones more passive and submissive. Maynard Smith has found indirect evidence that the two strategies are equally conducive to fitness—the hallmark of an evolutionarily stable state.[11] But when we move to the human species—and, indeed, when we move to other hierarchical species—this explanation for social hierarchy encounters problems. Prominent among them is the number of findings—in the Ache, the Aka, many other human societies, and many other species—that low status brings low reproductive success.[12] This is not the hallmark of an evolutionarily stable mix of strategies. It is the hallmark of low-status animals trying to make the best of a bad situation.

For decades, while many anthropologists have downplayed social hierarchy, psychologists and sociologists have studied its dynamics, watching the facility with which members of our species sort themselves out. Put a group of children together, and before long they fall into distinct grades. The ones at the top are best liked, most frequently imitated, and, when they try to wield influence, best obeyed.[13] The rudiments of this tendency are seen among children only a year old.[14] At first, status equals toughness—high-ranking children are the ones that don't back down—and indeed, for males, toughness matters well through adolescence. But as early as kindergarten, some children ascend the hierarchy via skill in cooperation.[15] Other talents—intellectual, artistic—also carry weight, especially as we grow older.

Many scholars have studied these patterns without bringing a Darwinian slant to their work, but it's hard not to suspect an innate underpinning for such robotic patterns of learning. Besides, status hierarchies run in our family. They emerge with great clarity and complexity in our nearest relatives, the chimps and bonobos, and are found also, if in simpler form, in gorillas, our next closest kin, and in many other primates.[16] If you took a zoologist from another planet, showed him our family tree, and pointed out that the three species nearest our limb were inherently hierarchical, he would probably guess that we are too. If you then told him that hierarchy is indeed found in every human society where people have looked closely for it, and among children too young to talk, he might well consider the case closed.

There is more evidence. Some of the ways people signify their status, and the status of others, seem to hold steady across cultures. Darwin himself, after widely questioning missionaries and other world travelers, concluded that "scorn, disdain, contempt, and disgust are expressed in many different ways, by movements of the features, and by various gestures; and that these are the same throughout the world." He also noted that "a proud man exhibits his sense of superiority over others by holding his head and body erect."[17] A century later, studies would show that posture becomes straighter immediately after social triumph—as, say, when a student gets a high test score.[18] And the ethologist Irenaüs Eibl-Eibesfeldt would find that children in diverse cultures, after losing a fight, lower their heads in self-abasement.[19] These universals of expression have reflections within. People in all cultures feel pride upon social success, embarrassment, even shame, upon failure, and, at times, anxiety pending these outcomes.[20]

Nonhuman primates send some of the same status signals as people. Dominant male chimps—and dominant primates generally—strut proudly and expansively. And after two chimpanzees fight over status, the loser crouches abjectly. This sort of bowing is thereafter repeated to peacefully express submission.

STATUS, SELF-ESTEEM, AND BIOCHEMISTRY

Beneath the behavioral parallels between human and nonhuman primates lie biochemical parallels. In vervet monkey societies, dominant males have more of the neurotransmitter serotonin than do their subordinates. And one study found that in college fraternities, officers, on average, have more serotonin than do their less powerful fraternity brothers.[21]

This is a good opportunity to extinguish a once-flourishing misconception that, though in decline, has yet to die its richly deserved death. It is *not* the case that all behavior under "hormonal control," or some other "biological control," is "genetically determined." Yes, there is a correlation between serotonin (a hormone, like all neurotransmitters) and social status. But no, that doesn't mean that a given person's social status was "in the genes," preordained at birth. If you check the serotonin levels of a fraternity president well before his political ascent, or of an alpha vervet monkey well before his, you may find them unexceptional.[22] Serotonin level, though a "biological" thing, is largely a product of the social environment. It isn't nature's way of destining people at birth for leadership; it's nature's way of equipping them for leadership once they've gotten there (and, some evidence suggests, of encouraging them to make a bid for leadership at a politically opportune moment).[23] You too can have a high serotonin level, if you can get elected president of a college fraternity.

Certainly genetic differences matter. Some people's genes dispose them to be unusually ambitious, or clever, or athletic, or artistic, or various other things—including unusually rich in serotonin. But these traits depend, for their flowering, on the environment (and sometimes on each other), and their eventual translation into status can rest heavily on chance. No one is born to lead, and no one is born to follow. And to the extent that some people are born with a leg up in the race (as they surely are), that birthright probably lies at least as much in cultural as in genetic advantage. In any event, there are good Darwinian reasons to believe that *everyone* is born with the capacity for high serotonin—with the equipment to function as a high-status primate given a social setting conducive to their ascent. The whole point of the human brain is behavioral flexibility, and it would be very unlike natural selection, given that flexibility, to deny anyone a chance at the genetic payoffs of high status, should the opportunity arise.

What does serotonin do? The effect of neurotransmitters is so subtle, and so dependent on chemical context, that simple generalizations are risky. But often, at least, serotonin seems to relax people, make them more gregarious, more socially assertive, much as a glass of wine does. In fact, one of alcohol's effects is to release serotonin. As a slight and useful oversimplification, you might say that serotonin raises self-esteem; it makes you behave in ways befitting an esteemed primate. Extremely low levels of serotonin can accompany not just low self-esteem, but severe depression, and may precede suicide. Antidepressants such as Prozac boost serotonin.[24]

So far this book has said little about neurotransmitters like serotonin, or about biochemistry in general. That is partly because the biochemical links among genes, brain, and behavior are largely unfathomed. It is also because the elegant logic of evolutionary analysis often lets us figure out the role of genes without worrying about the nuts and bolts of their influence. But, of course, there always *are* nuts and bolts. Whenever we talk about the influence of genes (or environment) on behavior, thought, or emotion, we are talking about a biochemical chain of influence.

As these chains become clearer, they can give form to inchoate data, and help graft the data onto a Darwinian framework. Psychologists found several decades ago that artificially lowering self-esteem (by giving false reports about scores on a personality test) made people more likely to cheat in a subsequent game of cards. A more recent study finds that people with lower serotonin levels are more likely to commit impulsive crimes.[25] Maybe both of these findings, translated into evolutionary terms, are saying the same thing: that "cheating" is an adaptive response, triggered when people are shunted to the bottom of the heap and thus find it hard to get resources legitimately. Maybe there's some truth to that ostensibly simplistic refrain about inner-city crime—that it grows out of "low self-esteem," as poor children are reminded, via TV and movies, that they're nowhere near the top of the roost. Again we see how Darwinism, often caricatured as genetically determinist and right-wing, can mesh with the sort of environmental determinism favored on the left.

We also see another way to test group-selectionist theories. If the acceptance of low status had evolved mainly as an ingredient of group success, success that then trickles down and benefits even the lowly, you wouldn't expect low-status animals to spend their time subverting the group's order.[26]

Confirming a link between serotonin and status in nonhuman primates is a messy task, and no one has tried it with our first cousins, the chimpanzees. But the smart money says the link is there. Indeed, so striking are the parallels between the human and the chimpanzee pursuit of status, and so closely related are we to chimps, that there may well be many biochemical mechanisms—and corresponding mental or emotional states—that we share with chimps by common descent. Chimpanzee striving is worth taking a look at.

Of the lavish attention that chimpanzees pay to status, much is merely ritual: greetings humbly offered to a social superior. Chimps often bow down and may literally kiss their master's feet.[27] (The foot kissing seems to be a cultural quirk, not found in all chimp colonies.) But in the case of males, at least, the rankings so peacefully acknowledged are set by struggle. If you see a chimp that regularly inspires great homage, he has won some pivotal fights.

The stakes are very real. Resources are allotted in rough accordance with status, and the alpha male tends to get the lion's share. In particular, the alpha jealously guards desirable females during estrus, their conspicuous phase of fertility.

Once this status ladder exists, and the higher rungs bring reproductive payoffs, genes that help a chimp climb it at acceptable cost will spread. The genes may work by instilling drives that, in humans, get labeled "ambition" or "competitiveness"; or by instilling feelings such as "shame" (along with an aversion to it and a tendency to feel it after conspicuous failure); or "pride" (along with an attraction to it and a tendency to feel it after doing impressive things). But whatever the exact feelings, if they raise fitness, they will become part of the species' psychology.

Male chimps seem more dramatically in the thrall of these sorts of forces than female chimps; they work harder for status. For that reason, male hierarchies are unstable. There seems always to be some young Turk mounting a challenge to the alpha male, and alpha males spend a lot of time spotting these threats and trying to head them off. Females settle into a hierarchy with less conflict (seniority often counts for a lot), and are thereafter less preoccupied with their status. In fact, the female hierarchy is so subdued that it takes an experienced eye to discern it, whereas spotting a pompous, imperious alpha male is something a schoolchild can do. Female social coalitions—friend-ships—often last a lifetime, whereas male coalitions shift with stra-tegic utility.[28]

MEN, WOMEN, AND STATUS

Some of this has a familiar ring. Human males, too, have a reputation for being ambitious, egotistical, and opportunistic. The linguist Deb-orah Tannen, author of *You Just Don't Understand*, has observed that for men, unlike women, conversation is "primarily a means to preserve independence and negotiate and maintain status in a hier-archical social order."[29] Many people have argued, especially during the second half of this century, that this difference is wholly cultural, and Tannen, in her book, accepts this view. It is almost surely wrong. The evolutionary dynamics behind the male chimpanzee's fevered pursuit of status are well understood, and they have been at work during human evolution.

These dynamics are the same ones that explain the male and female approaches to sex: the huge reproductive potential of a male, the limited potential of a female, and the resulting disparity in repro-ductive success among males. At one extreme, a low-status male may have zero offspring—a fact that, via natural selection, could readily come to imply an energetic aversion to low status. At the other extreme, alpha status can mean fostering dozens of offspring by nu-merous mothers—a fact that, via natural selection, could embed in males a boundless lust for power. For females, the reproductive stakes of the status game are lower. A female chimp in ovulation, regardless of her status, faces no shortage of suitors. She is not fundamentally in sexual competition with other females.

Of course, females in *our* species *do* compete for mates—for mates with the most parental investment to offer. But there's no evidence that, during evolution, social status was a primary tool in that competition. Besides, the evolutionary pressure behind male competition for sex seems to have been stronger than the pressure behind female competition for investment. The reason, again, is that potential differences in fitness are so much greater among males than among females.

The *Guinness Book of World Records* vividly makes the point. The most prolific human parent in world history is credited with 888 children—about 860 more than a woman could dream of having, unless she had a knack for multiple births. His name and title were, respectively, Moulay Ismail the Bloodthirsty, the Sharifian emperor of Morocco.[30] It's a little chilling to think that the genes of a man nicknamed "Bloodthirsty" found their way into nearly 1,000 off-spring. But that's the way natural selection works: the most chilling genes often win. Of course, it's not certain that Moulay Ismail's bloodthirstiness lay in distinctive genes; maybe he just had a rough childhood. Still, you get the point: sometimes genes *are* responsible for a male's inordinate drive for power, and so long as that power translates into viable offspring, those genes thrive.[31]

Shortly after the *Beagle*'s voyage, Darwin wrote to his cousin Fox that his work was being "favourably received by the great guns, & this gives me much confidence, & I hope not a very great deal of vanity; though I confess I feel too often like a peacock admiring his tail."[32] At that point, before natural selection dawned on him, and long before sexual selection did, Darwin could not have known how apt the comparison was. But later he would see that, indeed, the man-sized ego was produced by the same forces that created the peacock's tail: sexual competition among males. "Woman seems to differ from man in mental disposition, chiefly in her greater tenderness and less selfishness," he wrote in *The Descent of Man*. "Man is the rival of other men; he delights in competition, and this leads to ambition which passes too easily into selfishness. These latter qualities seem to be his natural and unfortunate birthright."[33]

Darwin also saw that this birthright wasn't just a vestige of our ape days, but a product of forces at work long after our species became human. "The strongest and most vigorous men,—those who could best defend and hunt for their families, and during later times the chiefs or head-men,—those who were provided with the best weap-ons and who possessed the most property, such as a larger number of dogs or other animals, would have succeeded in rearing a greater average number of offspring, than would the weaker, poorer and lower members of the same tribes. There can, also, be no doubt that such men would generally have been able to select the more attractive women. At present the chiefs of nearly every tribe throughout the world succeed in obtaining more than one wife."[34] Indeed, studies of the Ache, the Aka, the Aztecs, the Inca, the ancient Egyptians, and many other cultures suggest that, until the common use of contraception, male power translated into lots of offspring. And even now that contraception has broken this link, a link remains between status and the amount of sex a man has.[35]

Certainly male competitiveness has a cultural as well as genetic basis. Though male toddlers, generally speaking, are naturally more assertive than female toddlers, they're also given guns and signed up for Little League. Then again, this treatment may itself lie partly in the genes. Parents may be programmed to mold their children into optimal reproductive machines (or, strictly speaking, into machines that would have been optimally reproductive in the environment of our evolution). Margaret Mead once made an observation about primitive societies that probably applies in some measure to all societies: "[T]he small girl learns that she is a female and that if she simply waits, she will some day be a mother. The small boy learns that he is a male and that if he is successful in manly deeds some day he will be a man, and will be able to show how manly he is."[36] (The relative strength of these messages may depend on how much Darwinian sense they make locally. There is evidence that in polygynous societies, where high-status males are astronomically prolific, parents nurture their sons' competitiveness with special care.)[37]

None of this is to say that males have a monopoly on ambition. For female primates—ape or human—status can bring benefits, such as more food or favored treatment of offspring; accordingly, they do seek status with *some* enthusiasm. Female chimps routinely dominate young adolescent males and, given a vacuum in the male power structure, can even reach for great political heights. When colonies of captive chimps contain no adult males, a female may assume alpha status and then defend her rank ably after male rivals show up. And bonobos—our other evolutionary first cousins—evince even more female lust for power. In several small captive populations, females are the unquestioned leaders. Even in the wild, the more formidable females can prevail over the lowliest adult males.[38]

So as we look at status battles among chimps, the lessons will apply—in part, at least—to females. We'll focus on battles fought by males, because males battle in such high style. But the mental forces fueling these battles, if they reside in humans, probably reside in women as well as in men, albeit in smaller doses.

Both chimp and human hierarchies are subtler than chicken hierarchies. Which animal defers to which may change from day to day—not just because the hierarchies get reshuffled (which they do) but because dominance can depend on context; which primate gets its way can depend on which other primates are around. The reason is that chimps and humans have something chickens don't: reciprocal altruism. Living in a society with reciprocal altruism means having friends. And friends help each other during social conflicts.

This may seem obvious. What, after all, are friends for? But it really is remarkable. The evolutionary mixture that generated it—of reciprocal altruism and status hierarchy—is exceedingly rare in the annals of animal life.

The catalyst for the compound is the fact that, once hierarchies exist, status is a resource.[39] If status expands your access to food or sex, then it makes sense to seek status in the abstract, just as it makes sense to seek money even though you can't eat it. So an exchange of status-enhancing assistance between two animals is not different in kind from an exchange of food: so long as the exchange is non-zero-sum, natural selection will encourage it, given the opportunity. Indeed, after looking closely at chimp and human society, one might suspect that, from natural selection's point of view, status assistance is the main purpose of friendship.

The evolutionary fusion of hierarchy and reciprocal altruism accounts for a good part of the average human life. Many, if not most, of our swings in mood, our fateful commitments, our changes of heart about people, institutions, even ideas, are governed by mental organs that this fusion wrought. It has done much to form the texture of everyday existence.

It has also formed much of the *structure* of existence. Life within and among corporations, within and among national governments, within and among universities—it is all governed by these same mental organs. Both reciprocal altruism and status hierarchies evolved as an aid to the survival of individual genes, yet together they're holding up the world.

You can see the foundation in the daily life of chimpanzees. Look at the structure of their society, then imagine a huge growth in their intelligence—in memory, cunning, long-range planning, language—and suddenly you can picture whole buildings full of well-dressed chimps: office buildings, capitol buildings, campus buildings, all functioning much as they do now, for better or worse.

CHIMPANZEE POLITICS

Status for chimps, like status for people, depends on more than ambition and raw strength. True, an alpha chimp's ascent almost always involves beating up the incumbent alpha at least once. And the new alpha may thereafter make a habit of daunting his predecessor, and all other subjects; he runs through the colony, pounding the ground, heading toward a series of apes that, by ducking, acknowledge his supremacy—and he may slap one or two of them anyway for good measure. Still, it often takes strategic savvy, as well, to reach dominance and hang on to it.

The most famous example of cleverly sought status comes courtesy of Mike, one of the chimpanzees studied by Jane Goodall in Africa. Mike, though not a hulking specimen, discovered that by running toward more manly chimps while propelling empty kerosene cans loudly in their direction, he could earn their respect. Goodall writes: "Sometimes Mike repeated this performance as many as four times in succession, waiting until his rivals had started to groom once more before again charging toward them. When he eventually stopped (often in the precise spot where the other males had been sitting), they sometimes returned and with submissive gestures began to groom Mike. . . . Mike made determined efforts to secure other human artifacts to enhance his displays—chairs, tables, boxes, tripods, anything that was available. Eventually we managed to secure all such items."[40]

Mike's particular genius is not especially typical and may not be of utmost relevance to human evolution. Among chimpanzees, the most common use of wit in the quest for status has to do not with technological wizardry but with social savvy: the manipulation of reciprocally altruistic allegiances to personal advantage. Machiavellianism.

After all, chimpanzees, like human beings, seldom lead alone. To sit atop a heap of apes, some of whom are ambitious young males, is precarious, so alphas tend to arrange a regular source of support. The support may lie mainly in a single strong lieutenant that helps the alpha fend off challengers and in return is granted favors, such as sexual access to ovulating females. Or the support may lie in a close relationship with the dominant female; she will come to the alpha's defense, and perhaps in return get preferential treatment for her and her offspring. The support may be more complex and diffuse as well.

The best illustration of the fluidity of chimpanzee power, and the attendant emotional and cognitive complexity of chimps, is the primatologist Frans de Waal's gripping, almost soap-operatic, account of life among chimps housed on a two-acre island in a zoo in the Dutch town of Arnhem. Some people find de Waal's book—indeed its very title, *Chimpanzee Politics*—problematic. They think he too easily attributes to chimps an almost human nature. But no one can deny that this book is unique in its minutely detailed account of life among apes. I'll tell the story as de Waal tells it, complete with his engrossingly anthropomorphic tone, and we'll deal with problems of interpretation afterward.

Yeroen, a leading character in the drama, knew well the precariousness of power. While occupying the alpha position, he relied on the allegiance of various females, most notably Mama, a highly influential ape who occupied the dominant female slot throughout de Waal's narrative. It was to the females that Yeroen turned for help when challenged by the younger, stronger Luit.

Luit's challenge escalated relentlessly. First it was sexual intercourse with a female around ovulation, blatantly performed within sight of the jealous and possessive (like any alpha) Yeroen; next came a series of aggressive "displays," or threats, aimed at Yeroen; and finally a physical assault: Luit descended on Yeroen from a tree, struck him, and ran away. This is not the sort of treatment to which alpha males are accustomed. Yeroen started screaming.

He then ran over to a group of chimps, mostly females, embraced each of them, and, having thus consolidated his strategic ties, led them toward Luit. Yeroen and company cornered Luit, who then dissolved into a temper tantrum. He had lost the first battle.

Yeroen seems to have sensed in advance that this challenge was in the works. De Waal's records showed that during the weeks before Luit's first overt defiance, Yeroen had more than doubled the amount of time he spent in friendly contact with adult females. Politicians do most of their baby kissing around election time.

Alas for poor Yeroen, his victory was fleeting. Luit set about to undermine the governing coalition. For weeks on end he punished Yeroen's supporters. When he saw a female grooming Yeroen, he would approach the two and threaten or actually assault the female, sometimes jumping up and down on her. Yet later, Luit might be seen grooming the same female, or playing with her children—so long as she wasn't with Yeroen. The females got the message.

Perhaps if Yeroen had defended his allies better, he could have hung on to alpha status. But this option was rendered dicey by an alliance between Luit and a young male named Nikkie. Nikkie would accompany Luit during his persecution of females, sometimes giving them a hard slap of his own. Their partnership was a natural: Nikkie, just emerging from adolescence, was struggling to establish dominance over all females—a rite of passage for young male chimps—and the affiliation with Luit made this simple. Eventually, after some hesitation, Luit gave Nikkie the added incentive of special sexual privileges.

Having isolated Yeroen, Luit could ascend to alpha rank. The transition came through several hostile encounters, though it wasn't sealed until Yeroen finally mustered the humility to greet Luit submissively.

Luit proved a wise and mature leader. Under his rule, life was orderly and just. If two chimps were fighting, he would step between them with calm authority, ending hostilities without fear or favor. And when he did side with one combatant, it was almost always the one who was losing. This pattern of support for the downtrodden—

populism, we call it—had also been employed by Yeroen. It seemed to impress the females especially; being less caught up in the pursuit of status than males, they seemed to place a premium on social stability. Luit could now count on their support.

In the long run, however, populism would not be enough. Luit still faced, on the one hand, Yeroen's persistent fondness for power (and perhaps some lingering enmity, though the two had lavishly reconciled, with much mutual grooming, after Yeroen conceded defeat); and, on the other hand, the conspicuous ambition of Nikkie. Luit must have found the latter the more threatening, for he sought alliance with Yeroen, thus freezing Nikkie out of the leadership circle. But Yeroen, seemingly aware of his pivotal place in the balance of power, proved a coy ally, and played the two off against each other. Finally, he shifted his weight toward Nikkie and, in league with him, toppled Luit. Alpha status went to Nikkie, but Yeroen continued to play his cards so deftly that for the next year he, not Nikkie, led all males in sexual activity. De Waal considered Nikkie a "figurehead" and Yeroen the power behind the throne.

The story has a morbid epilogue. After de Waal's book was published, Nikkie and Yeroen had a falling out. But their sense of common purpose was revived after Luit resumed alpha status. One night during a brutal fight, they wounded Luit mortally—even going so far, in a gratuitous bit of Darwinian symbolism, as to rip out his testicles. De Waal had little doubt about which of the two suspected killers deserved more blame. "Nikkie, ten years younger, seemed only a pawn in Yeroen's games," he later observed. "I found myself fighting this moral judgment, but to this day I cannot look at Yeroen without seeing a murderer."[41]

WHAT IS IT LIKE TO BE A CHIMP?

That's the story of the Arnhem chimps, told as if they were people. Does de Waal deserve to be condemned for anthropomorphizing? Ironically, even a jury of evolutionary psychologists might vote for a conviction—on one count of the indictment, at least.

De Waal suspects that, just before Luit's bid for alpha status, when Yeroen started spending more time with the females, he had "already sensed that Luit's attitude was changing and he knew that his position was threatened."[42] Yeroen probably did "sense" a change of attitude, and this may well account for his sudden interest in the politically pivotal females. But must we assume, with de Waal, that Yeroen "knew" about—consciously anticipated—the coming challenge and rationally took measures to head it off? Why couldn't Luit's growing assertiveness simply have inspired pangs of insecurity that pulled Yeroen into closer touch with his friends?

Certainly genes encouraging an unconsciously rational response to threats can fare well in natural selection. When a baby chimp or a baby human, sighting a spooky-looking animal, retreats to its mother, the response is logical, but the youngster presumably isn't conscious of the logic. Similarly, when I suggested earlier that Darwin's recurrent illness may have periodically replenished his affection for Emma, I didn't mean he consciously reappraised her value in view of his poor health (though he may have). Threats of various kinds seem to nourish our affection for the people who help us face threats—kin and friends.

The point is that too readily imputing strategic brilliance to chimps may obscure a basic theme of evolutionary psychology: everyday human behavior is often a product of subterranean forces—rational forces, perhaps, but not consciously rational. Thus, de Waal may be creating a misleading dichotomy when he speaks of Yeroen's and Luit's "policy reversals, rational decisions and opportunism" and then asserts that "there is no room in this policy for sympathy and antipathy."[43] What look like policies may be products of sympathy and antipathy; the ultimate policy maker is natural selection, and it calibrates these feelings to execute its policies.

With that verdict rendered, our jury of evolutionary psychologists would probably go on to acquit de Waal of many other counts of anthropomorphism. For often what he imputes to chimps is not human calculation but human feelings. During the early, inconclusive phase of Luit's challenge to Yeroen, the two periodically fought. And a fight (among chimps and many other primates, including us) is typically followed, sooner or later, by rituals of reconciliation. De Waal notes how reluctant each chimp was to start the rapprochement and ascribes their hesitation to a "sense of honor."[44]

He gingerly puts that phrase in quotation marks, but they may not be needed. In chimp society, as in human society, a peace overture can carry intimations of submission; and submission during a leadership struggle carries real Darwinian costs, as it may bring secondary or still lower status. So a genetically based aversion to such submissions (up to a point, at least) makes evolutionary sense. In our species, we call this aversion a sense of honor, or pride. Is there any reason we shouldn't use the same terms when talking about chimps? As de Waal has noted, given the close kinship of the two species, to assume a deep mental commonality is good parsimonious science: a single hypothesis that plausibly accounts for two separate phenomena.

Wives have been known to say of their husbands, "He can *never* bring himself to admit he's wrong," or "He's *never* the first to apologize," or "He *hates* to ask for directions." Men seem loath to concede the superiority of another human being, even in such trivial realms as municipal geography. The reason, perhaps, is that during human evolution males who too readily sought reconciliation after a fight, or otherwise needlessly submitted to others, saw their status drop, and with it their inclusive fitness. Presumably females did too; women, like men, are reluctant to apologize or admit they're wrong. But if folk wisdom can be trusted, the average woman is less reluctant than the average man. And that shouldn't surprise us, as the fitness of our female ancestors depended less on such reluctance than did that of our male ancestors.

De Waal also speaks of "respect." When Luit's dominance was finally undeniable, and Yeroen faintly sought rapprochement, Luit ignored him until he heard some "respectful grunts," unambiguous signs of submission.[45] A beta chimp may well feel toward the alpha much the way a losing prizefighter feels toward an opponent he says he now "respects." And at moments of utter ape dominance, when the vanquished crouches in abject submission, *awe* may be an apt word.

Jane Goodall, like de Waal, saw "respect" in the apes she came to know, though she used that word somewhat differently. Recalling the apprenticeship of a young chimp, Goblin, under the alpha male Figan, she writes that "Goblin was very respectful of his 'hero,' followed him around, watched what he did, and often groomed him."[46] Everyone who has been through adolescence and has had a role model can imagine how Goblin felt. In fact, some might suggest that *reverent* is a better word than *respectful*.

All of this may sound facile—a grand leap from surface parallels between us and apes to the depths of primate psychology. And maybe it *will* turn out to have been facile; maybe the uncanny resemblance between chimp and human life isn't grounded in a common evolutionary origin or a common biochemistry. Still, if we are not going to explain such things—respect, reverence, awe, honor, stubborn pride, contempt, disdain, ambition, and so on—as natural selection's way of equipping us for life in a status hierarchy, how, then, are we to explain them? Why are they found in cultures everywhere? Is there an alternative theory? If so, does it explain, as well, why pride and ambition, for example, seem to reach greater heights in men, on average, than in women? Modern Darwinism has an explanation for all of this, and it's simple: natural selection in a context of status hierarchy.

MIGHT AND RIGHT

One of de Waal's alleged anthropomorphisms puts flesh on a skeletal speculation made by Robert Trivers in his 1971 paper on reciprocal altruism. De Waal believes conduct among chimps may be "governed by the same sense of moral rightness and justice as it is among humans." This thought was provoked by a female chimp named Puist, who "had supported Luit in chasing Nikkie. When Nikkie later displayed at Puist she turned to Luit and held out her hand to him in search of support. Luit, however, did nothing to protect her against Nikkie's attack. Immediately Puist turned on Luit, barking furiously, chased him across the enclosure and even hit him."[47] It doesn't take great imagination to see in such fury the heated indignation with which you might chastise a friend who had deserted you in time of need.

The deepest source of this "sense of fairness" is, as Trivers noted, reciprocal altruism. No status hierarchy need be involved. Indeed, what de Waal calls two of the basic rules of chimpanzee conduct— "One good turn deserves another," and "An eye for an eye, a tooth for a tooth"—amount to a description of TIT FOR TAT, which evolved in the absence of status.

Still, it is competition for social status—and the attendant phenomenon of social alliance, of collective enmity—that has given these deeply held philosophical intuitions much of their weight. Human coalitions competing for status often feature a vague sense of moral entitlement, a sense that the other coalition *deserves* to lose. The fact that our species evolved amid both reciprocal altruism and social hierarchy may underlie not just personal grudges and reprisals, but race riots and world wars.

That war may in this sense be "natural" doesn't mean it's good, of course; or even that it's inevitable. And much the same can be said of social hierarchy. That natural selection has opted for social inequality in our species certainly doesn't make inequality right; and it makes it inevitable in only a limited sense. Namely: when groups of people—especially males—spend much time together, some sort of hierarchy, if implicit and subtle, is pretty sure to appear. Whether we know it or not, we tend naturally to rank one another, and we signify the ranking through patterns of attention, agreement, and deference—whom we pay attention to, whom we agree with, whose jokes we laugh at, whose suggestions we take.[48] But social inequality in the larger sense—gross disparities in wealth and privilege across a whole nation—is another matter. That is a product of government policy, or lack of policy.

Of course, public policy, in the end, must comply with human nature. If people are basically selfish—and they are—then asking them to work hard yet earn no more than their unproductive neighbor is asking more than they'll readily give. But we already know that; communism has failed. We also know that mildly redistributive taxation does not snuff out the will to work. Between these two extremes is a large menu of policies. Each has its cost, but the cost is a product of plain old human selfishness—something that's not exactly news— and not of the human hunger for status per se.

Indeed, the hunger for status may actually lower the costs of redistribution. Humans, it seems, tend to compare themselves to those very near them in the status hierarchy—to those just above them, in particular.[49] This makes evolutionary sense as a ladder-climbing technique, but that's not the point. The point is that if the government takes a thousand dollars more from everyone in your middle-class neighborhood, you're in about the same position relative to your neighbors as you were before. So if keeping up with the Joneses is what drives you, your work incentive shouldn't be dampened as it would if calibrated in absolute monetary terms.

The modern view of social hierarchy also deals a heavy blow to one of the cruder philosophical excuses for inequality. As I've tried to stress, there is no reason to derive our values from natural selection's "values," no reason to deem "good" what natural selection has "deemed" expedient. Still, some people do. They say that hierarchy is nature's way of keeping the group strong, so inequality can be justified in the name of the greater good. Since it now looks as if nature *didn't* invent human hierarchies for the good of the group, that piece of logic is twice as flawed as it used to be.

The crowning (alleged) anthropomorphism in de Waal's book is its title, *Chimpanzee Politics.* If politics is, as political scientists say, the process by which resources are divvied up, then chimps demonstrate, in de Waal's view, that the origins of human politics long predate humanity. In fact, he sees not just a political process, but "even a democratic structure" at work in the Arnhem chimp colony.[50] Alpha males have trouble ruling without the consent of the governed.

Nikkie, for example, lacked Luit's common touch and never became as popular as either Luit or Yeroen were during their tenures. The females were especially sparing in their submissive greetings, and when Nikkie was needlessly violent, they would pursue him en masse. On one occasion he was chased up a tree by the entire colony. There he sat, alone, surrounded, and screaming—the dominant male, dominated. Maybe this wasn't modern representative democracy, but it wasn't a very smooth dictatorship either. (There's no telling how long Nikkie would have stayed trapped had Mama, the troop's chief conciliator, not climbed the tree, given him a kiss, and led him back down, after which he humbly sought mass forgiveness.)[51]

Here's a useful exercise: when watching a politician speak on TV, turn down the volume. Notice the gestures. Note their similarity to the gestures politicians everywhere in the world use—exhortation, indignation, and so on. Then turn up the volume. Listen to what the politician is saying. Here's a virtual guarantee: he (or, more rarely, she) is saying things that appeal to the group of voters most likely to get him into power or keep him there. The interests of the governed—or of some crucial slice of the governed—governs what human politicians say, just as it governs what chimpanzee politicians do. In both cases, the politician's ultimate aim (whether he knows it or not) is status. And in both cases we may see a certain flexibility as to what the politician is willing to do, or say, to get that status and keep it. Even the most stirring oratory can boil down to convenient coalition. In turning up the volume, you've capsulized several million years of evolution.

THE ZUNI WAY

For all the suggestive parallels between ape and human striving, the differences remain large. Human status often has relatively little to do with raw power. It's true that overt physical dominance is often a key to social hierarchy among boys. But, especially in adults, the status story is much more complex, and in some cultures its plainly political aspects have been quite subdued. Here is one scholar's description of life among the Navajo: "No one who actively seeks power is to be trusted. Leaders arise out of example and emulation. If someone is successful at growing corn, he is emulated and to that extent is a leader. If someone knows many verses to a curing chant, he is respected for that accomplishment and his status as a 'singer' is considerable. Politicking, handshaking . . . have no place in traditional Navajo society."[52]

This isn't to say that Navajos don't seek power—only that they seek it subtly. Nor is it to say that status is severed from the goal of reproductive advantage. The expert corn grower and the expert singer probably make for attractive mates. And it's easy to guess why; one has a knack for providing material resources and both show signs of intelligence. Still, these two Navajo didn't gain their reproductive advantage by physically intimidating or otherwise controlling people; they simply found their calling and excelled.

The range of things that can bring status in different cultures and subcultures is astonishing. Making beads, making music, delivering sermons, delivering babies, inventing drugs, inventing tales, collecting coins, collecting scalps. Yet the mental machinery driving these various activities is fundamentally the same. Human beings are designed to assess their social environment, and, having figured out what impresses people, do it; or, having found what people disfavor, avoid it. They're pretty open-minded about what "it" is. The main thing is that they be able to succeed at it; people everywhere want to feel pride, not shame; to inspire respect, not disdain.

This tendency of humankind's psychic unity to hide behind behavioral diversity is what enabled the Boasian anthropologists to minimize human nature. Ruth Benedict wrote in 1934, "We must accept all the implications of our human inheritance, one of the most important of which is the small scope of biologically transmitted behaviour, and the enormous role of the cultural process of the transmission of tradition."[53] Strictly speaking, she was right. Once you get past stereotyped acts such as walking, eating, and suckling, "behaviors" don't get transmitted biologically. Mental organs do, and they're usually limber enough to yield lots of different behaviors, depending on circumstance.

It is easy to see how the mental machinery of status seeking, in particular, eluded Benedict's emphasis. She studied the Zuni, who, like the nearby Navajo, play down competition and overt political striving. She wrote, "The ideal man in Zuni is a person of dignity and affability who has never tried to lead. . . . Any conflict, even though all right is on his side, is held against him. . . . The highest praise . . . runs: 'He is a nice polite man.' "[54] Note the subtext. There is an "ideal man," and anyone who approaches the ideal gets "praise," while anyone who falls short has his failure "held against him." In other words: the Zuni confer status on those who don't seek status too fiercely, and deny status to those who do. The very strength of the status-seeking machinery is what keeps Zuni status hierarchies subtle. (Also, as we've seen, the social infrastructure of reciprocal altruism tends in all cultures to exert some pressure toward friendliness, as well as generosity and honesty. Zuni culture may have harnessed this pressure with unusual efficiency, reinforcing the natural link between niceness and status.)

You can look at life among the Zuni as a tribute either to the power of culture or to the suppleness of mental adaptations. It is both, but let's ponder the latter: mental organs, it seems, are so flexible that they can participate in a virtual rebellion against the Darwinian logic behind them. Though the status-seeking machinery has long energized fistfights and macho politicking, it can also be used to suppress both. In a monastery, serenity and asceticism can be sources of status. In some strata of Victorian England, a nearly ludicrous amount of gentility and humility could help earn status (rather like among the Zuni, perhaps).

In other words, what we call cultural "values" are expedients to social success.[55] People adopt them because other people admire them. By controlling a child's social environment, by selectively dishing out respect and scorn, we can program his values as if he were a robot. Some people find this troubling. Well, that just goes to show that you can't please everybody. During the sociobiology controversy of the 1970s, a major source of outrage was the fear that, if the sociobiologists were right, people couldn't be programmed as B. F. Skinner and other behaviorists had promised.

The new paradigm does have room for Skinnerian conditioning, complete with positive and negative reinforcement. To be sure, some drives and emotions—say, lust and jealousy—may never be wholly erasable. Still, the great moral diversity among cultures—that is, diversity in the tolerated behavioral *expressions* of, say, lust and jealousy—suggests much leeway in the values department. Such is the power of social approval and disapproval.

The big question is: How deeply can the patterns of approval and disapproval themselves be shaped? Or to put it another way: How flexible is society about what it will find pleasing?

Here, no doubt, lie some pretty firm tendencies. Social assets that mattered consistently during evolution may stubbornly continue to carry weight. Big strong men and beautiful women may always have a head start in status competition. Stupidity may never provoke widespread admiration. The command of resources—that is, money—will tend to hold a certain appeal. Still, resistance is possible. There *are* cultures and subcultures that try to put less emphasis on the material and more emphasis on the spiritual. And their success is sometimes impressive, if less than total. And, moreover, there is no reason to believe that any of them have reached the limits of biological potential.

Even our own culture, for all its materialistic excess, starts to seem almost admirable when you look at some of the alternatives. Among the Yanomamo of South America, one route to status for a young man is to kill lots of men in neighboring villages.[56] If, in the process, he can participate in the abduction and gang rape of women from that village, so much the better. If his wife tries to leave him for another man, he can feel free to, say, cut off her ears. At the risk of sounding morally nonrelativistic, we've come a long way.

In some modern urban neighborhoods, values have lately grown closer to those of the Yanomamo. Young men who kill get respect— at least within the circle of young men whose opinions they care about. This is evidence that the worst parts of human nature are always near the surface, ready to rise when cultural restraint weakens. We are not blank slates, as some behaviorists once imagined. We are organisms whose more egregious tendencies can be greatly, if arduously, subdued. And a primary reason for this tenuous optimism is the abject flexibility with which status is sought. We will do almost anything for respect, including not act like animals.

Chapter 13: **DECEPTION AND SELF-DECEPTION**

What wretched doings come from the ardor of fame; the love of truth alone would never make one man attack another bitterly.
—Letter to J. D. Hooker (1848)[1]

Natural selection's disdain for the principle of truth in advertising is widely evident. Some female fireflies in the genus *Photuris* mimic the mating flash of females in the genus *Photinus* and then, having attracted a *Photinus* male, eat him. Some orchids look quite like female wasps, the better to lure male wasps that then unwittingly spread pollen. Some harmless snakes have evolved the coloration of poisonous snakes, gaining undeserved respect. Some butterfly pupa bear an uncanny resemblance to a snake's head—fake scales, fake eyes—and, if bothered, start rattling around menacingly.[2] In short: organisms may present themselves as whatever it is in their genetic interest to seem like.

People appear to be no exception. In the late 1950s and early sixties, the (non-Darwinian) social scientist Erving Goffman made a stir with a book called *The Presentation of Self in Everyday Life,* which stressed how much time we all spend on stage, playing to one audience or another, striving for effect. But there is a difference between us and many other performers in the animal kingdom. Whereas the female *Photuris* is, presumably, under no illusion as to its true identity, human beings have a way of getting taken in by their acts. Sometimes, Goffman marveled, a person is "sincerely convinced that the impression of reality which he stages is the real reality."[3]

What modern Darwinism brings to Goffman's observation is, among other things, a theory about the function of the confusion: we deceive ourselves in order to deceive others better. This hypothesis was tossed out during the mid-1970s by both Richard Alexander and Robert Trivers. In his foreword to Richard Dawkins's *The Selfish Gene*, Trivers noted Dawkins's emphasis on the role of deception in animal life and added, in a much-cited passage, that if indeed "deceit is fundamental to animal communication, then there must be strong selection to spot deception and this ought, in turn, to select for a degree of self-deception, rendering some facts and motives unconscious so as not to betray—by the subtle signs of self-knowledge—the deception being practised." Thus, Trivers ventured, "the conventional view that natural selection favors nervous systems which produce ever more accurate images of the world must be a very naive view of mental evolution."[4]

It should come as no surprise that the study of self-deception makes for murky science.[5] "Awareness" is a region with ill-defined and porous borders. The truth, or certain aspects of it, may float in and out of awareness, or hover on the periphery, present yet not distinct. And even assuming we could confirm that someone is wholly unaware of information relevant to some situation, whether this constitutes self-deception is another question altogether. Is the information somewhere in the mind, blocked from consciousness by a censor designed for that function? Or did the person just fail to take note of the information in the first place? If so, is that selective perception itself a result of specific evolutionary design for self-deception? Or a more general reflection of the fact that the mind can hold only so much information (and the conscious mind even less)? Such difficulties of analysis are one reason the science Trivers envisioned two decades ago—a rigorous study of self-deception, which might finally yield a clear picture of the unconscious mind—has not arrived.

Still, the intervening years have tended to validate the drift of Dawkins's and Trivers's and Alexander's worldview: our accurate depiction of reality—to others, and, sometimes, to ourselves—is not high on natural selection's list of priorities. The new paradigm helps us map the terrain of human deception and self-deception, if at a low level of resolution.

We've already explored one realm of deception: sex. Men and women may mislead each other—and even, in the process, themselves—about the likely endurance of their commitment or about their likely fidelity. There are two other large realms in which the presentation of self, and the perception of others, has great Darwinian consequence: reciprocal altruism and social hierarchy. Here, as with sex, honesty can be a major blunder. In fact, reciprocal altruism and social hierarchy may together be responsible for most of the dishonesty in our species—which, in turn, accounts for a good part of the dishonesty in the animal kingdom. We are far from the only dishonest species, but we are surely the *most* dishonest, if only because we do the most talking.

LEAVING A GOOD IMPRESSION

People don't seek status per se. They don't chart out their desired ascent and pursue it as methodically as a field general prosecutes a war. Well, okay, some do. Maybe all of us do sometimes. But the quest for status is also built more finely into the psyche. People in all cultures, whether they fully realize it or not, want to wow their neighbors, to rise in local esteem.

The thirst for approval appears early in life. Darwin had crystalline memories of impressing people with his tree-climbing skills: "My supposed admirer was old Peter Hailes the bricklayer, & the tree the Mountain Ash on the lawn."[6] The other side of this coin is an early and continuing aversion to disdain or ridicule. Darwin wrote that his oldest son, at age two and a half, became "extremely sensitive to ridicule, and was so suspicious that he often thought people who were laughing and talking together were laughing at him."[7]

Darwin's son may have been abnormal in this regard, but that's beside the point. (Though it's interesting to note how many psycho-pathologies, including paranoia, may simply be evolutionarily in-grained tendencies turned up a notch too high.)[8] The point is that if he was abnormal, he was abnormal in degree and not in kind. For all of us, avoiding ridicule is, from an early age, little short of an obsession. Recall Darwin's remarks about "the burning sense of shame which most of us have felt even after the interval of years, when calling to mind some accidental breach of a trifling though fixed rule of etiquette."[9] Such a hair-trigger mechanism suggests large stakes. Indeed: just as high public esteem can bring great genetic rewards, very low public esteem can be genetically calamitous. In numerous nonhuman primate communities—and in not a few human ones—extremely unpopular individuals are pushed to the margins of the society and even beyond, where survival and reproduction become perilous.[10] For that matter, a drop in status at any rung on the ladder carries costs. Whatever your place in society, leaving the sort of impression that exerts upward pressure on it is often worth the trouble (in Darwinian terms), even if the effect is slight.

Whether the impression is accurate is, by itself, irrelevant. When a chimp threatens a rival, or responds to a threat from one (or from a predator), its hair stands on end, making it seem larger than life. Vestiges of this illusion can be seen in people whose hair stands on end when they're frightened. But as a rule, humans do their self-inflating verbally. Darwin, in speculating about when in evolution the regard for public opinion became so strong, noted that "the rudest savages" show such a regard "by preserving the trophies of their prowess" and "by their habit of excessive boasting."[11]

In Victorian England, boasting was frowned on, and Darwin was an expert on how not to do it. Many modern cultures share this taste, and in them "excessive boasting" is merely a phase through which children pass.[12] But what is the next phase? A lifetime of more meas-ured boasting. Darwin himself was good at this. In his autobiography, he noted that his books "have been translated into many languages, and passed through several editions in foreign countries. I have heard it said that the success of a work abroad is the best test of its enduring value. I doubt whether this is at all trustworthy; but judged by this standard my name ought to last for a few years."[13] Well, if he really doubted that this standard is trustworthy, why judge by it?

Presumably, how much blatant boasting you do depends on the credible means of self-advertisement in your social environment (and was probably calibrated by feedback from kin and peers early on). But if you don't feel even *some* urge to disseminate news of your triumphs, however subtly, and *some* reluctance to talk widely about your failures, you aren't functioning as designed.

Does such self-advertisement often involve deception? Not in the grossest sense. To tell huge lies about ourselves, and believe them, would be dangerous. Lies can be found out, and they force us to spend time and energy remembering which lies we've told to whom. Samuel Butler, himself a Victorian evolutionist (and the man who noted that a hen is just an egg's way of making another egg) observed that "the best liar is he who makes the smallest amount of lying go the longest way."[14] Indeed. There are kinds of lies that, being slight, or hard to discredit, are hard to get tangled up in, and these are the sorts of lies we should expect people to tell. Among fishermen, the notorious and heartfelt embellishment of "the one that got away" has become a staple source of humor.

Such distortion may initially be conscious, or, at least, half-conscious. But if it goes unchallenged, the vague awareness of exaggeration can subside upon successive retellings. Cognitive psychologists have shown how the details of a story, even if false, embed themselves in the original memory with repetition.[15]

It goes without saying that the fish got away through no fault of the fisherman's. The assignment of blame and of credit, an area where objective truth is elusive, offers rich terrain for self-inflation. The tendency to attribute our successes to skill, and our failures to circumstance—luck, enemies, Satan—has been demonstrated in the laboratory and, anyway, is obvious.[16] In games where chance plays a role, we tend to chalk up our losses to the luck of the draw and our victories to cleverness.

And we don't just *say* this; we believe it. Darwin was an enthusiastic backgammon player and, not surprisingly, he often won when playing against his children. One of his daughters recalls that "we kept a list of the doublets thrown by each, as I was convinced that he threw better than myself."[17] This conviction is familiar to losing backgammon players everywhere. It helps preserve our belief in our competence and thus helps us convince others of it. It also provides a steady source of income for backgammon hustlers.

Self-aggrandizement always comes at the expense of others. To say that you lost a game through luck is to say that your opponent won through luck. And even leaving aside games and other openly competitive endeavors, to toot your own horn is to mute other horns, for status is a relative thing. Your gain is someone else's loss.

And vice versa: someone else's loss is your gain. This is where the unconscious pursuit of status can turn nasty. In a small group (a group, say, the size of a hunter-gatherer village), a person has a broad interest in deflating the reputations of others, especially others of the same sex and similar age, with whom there exists a natural rivalry. And again, the best way to convince people of something, including their neighbors' shortcomings, is to believe what you're saying. One would therefore expect, in a hierarchical species endowed with language, that the organisms would often play up their own feats, downplay the feats of others, and do both things with conviction. Indeed, in the social psychology laboratory, people not only tend to attribute success to skill and failure to circumstance; they tend to reverse the pattern when evaluating others.[18] Luck is the thing that makes you fail and other people succeed; ability works the other way around.

Often the derogation of others hovers at a barely detectable level, and it may disappear if they are kin or friends. But expect it to reach high volume when two people are vying for something that there's only one of—a particular woman, a particular man, a particular professional distinction.[19] One reviewer who savaged *The Origin of Species* was Richard Owen, an eminent zoologist and paleontologist who had his own ideas about how species might change. After the review came out, Darwin noted that "the Londoners say he is mad with envy because my book has been talked about."[20] Had Owen self-servingly convinced himself (and hence others) that a rival's work was inferior? Or had Darwin self-servingly convinced himself (and hence others) that a man who threatened his status was driven by selfish motives? Probably one or the other, and possibly both.

The keen sensitivity with which people detect the flaws of their rivals is one of nature's wonders. It takes a Herculean effort to control this tendency consciously, and the effort must be repeated on a regular basis. Some people can summon enough restraint not to *talk* about their rivals' worthlessness; they may even utter some Victorian boilerplate about a "worthy opponent." But to rein in the perception itself—the unending, unconscious, all-embracing search for signs of unworthiness—is truly a job for a Buddhist monk. Honesty of evaluation is simply beyond the reach of most mortals.

SELF-DEFLATION

If self-advertisement is so deeply ingrained in people, why are there self-deprecators? One answer is that self-deprecation is without cost when everyone knows better, and can actually have some benefit; a reputation for humility boosts the credibility of subtle boasting. (Witness Darwin.) Another answer is that the genetic program for mental development is very complex and unfolds in a world full of uncertainty (and a world quite unlike the ancestral environment); don't expect all human behavior to serve genetic interest. The third answer is the most interesting: social hierarchy has, via natural selection, had some ironic effects on the human mind. There are times when it makes good evolutionary sense to have a genuinely low opinion of yourself and to share that opinion with others.

The whole origin of status, remember, lies in the fact that some neighbors—some of a chicken's fellow chickens, say—are too formidable to challenge profitably. Genes that build brains that tell the animal which neighbors are worth challenging, and which aren't, flourish. How exactly do the brains convey this message? Not by sending little "Challenge" or "Don't Challenge" subtitles across the eyeball. Presumably, the message travels via feeling; animals feel either up to the challenge or not up to it. And animals at the very bottom of the hierarchy—animals that get pummeled by all comers—will get the latter feeling chronically. You could call it low self-esteem.

In fact, you could say that low self-esteem evolved as a way to reconcile people to subordinate status when reconciliation is in their genetic interest.

Don't expect people with low self-esteem to hide it. It may be in their genetic interest not only to accept low status, but, in at least some circumstances, to convey their acceptance of it—to behave submissively so that they aren't erroneously perceived as a threat and treated as such.[21]

There's nothing necessarily self-deceptive about low self-esteem. Indeed, any feeling designed to keep people from aspiring to more than they can attain should, in theory, bear at least rough correspondence to reality. But not always. If one function of low self-esteem is to keep high-status people satisfied with your deference, then its level, strictly speaking, should depend on how much deference it takes to do that; you may, in the presence of someone powerful, feel a deeper humility—about your intelligence, for example—than an objective observer would see as warranted. The anthropologist John Hartung, who in 1988 raised the possibility of self-deceptively lowering self-esteem—"deceiving down," he called it—has come up with another kind of example. Women, he suggested, may sometimes falsely subordinate themselves to men. If, say, household income depends partly on the husband having high self-esteem at the workplace, a woman may find herself unwittingly "building her husband's self-confidence by providing a standard of lower competence."[22]

An ingenious experiment has shown how deeply the truth about ourselves can be buried. When people hear a recorded voice, their galvanic skin response (GSR) rises, and it rises even more if the voice they hear is their own. Surprisingly, when people are asked whether the voice is theirs, they are, on average, right less often than is their GSR. What's intriguing is the pattern of error. After self-esteem is lowered, by making subjects "fail" on some contrived task, they tend to deny that the voice is theirs even though their GSR shows that at some level they "know" the truth. When self-esteem is raised, they start claiming other voices as their own, although again their GSR shows that somewhere within, the information is tallied correctly. Robert Trivers, reviewing this experiment, wrote, "it is as if we expand ourselves . . . when succeeding and shrink our presentation of self when failing, yet we are largely unconscious of this process."[23]

Feeling bad about yourself is good for things other than sending people self-serving signals. To begin with, there is the function, mentioned above, of burning shame: a wrist-slapping for social blunders, a way of discouraging the repeat of status-reducing behaviors. Also, as the evolutionary psychiatrist Randolph Nesse has stressed, mood can efficiently focus energy.[24] People of all statuses may get lethargic and glum when social, sexual, or professional prospects look dim, and then grow optimistic and energetic when opportunities arise. It's as if they had been resting up for a big match. And if no opportunities arise, and lethargy passes into mild depression, this mood may goad them into a fruitful shift of course—changing careers, jettisoning ungrateful friends, abandoning the pursuit of an elusive mate.

Darwin offers a good example of the manifold utility of bad feelings. In July of 1857, two years before publishing *The Origin of Species,* he wrote to his friend Joseph Hooker, "I have been making some calculations about varieties &c. & talking yesterday with Lubbock, he has pointed out to me the grossest blunder which I have made in principle, & which entails 2 or 3 weeks lost work." This left Darwin feeling even less inclined than usual to stress his worth. "I am the most miserable, bemuddled, stupid Dog in all England," he wrote, "& am ready to cry at [sic] vexation at my blindness & presumption."[25]

Count the ways this glumness might be valuable. One: as a *self-esteem deflator.* Darwin had suffered a social humiliation. In a face-to-face encounter, he was shown to be gravely confused on an issue within his supposed expertise. Perhaps some long-term slippage in self-esteem was in order; perhaps he should tone down the ambition of his scholarship, so as not to be perceived as a threat to England's great intellectual stars, who would in the end outshine him anyway.

Two: as a *negative reinforcement.* Lingering pain from this incident may have served to discourage Darwin from repeating behaviors (the confused analysis, in this case) that lead to humiliation. Perhaps he'd be more careful next time.

And three: as a *course changer.* If this gloom had persisted, even verging on depression, it might have altered Darwin's behavior more radically, diverting his energy into wholly new channels. "It is enough to make me tear up all my M.S. & give up in despair," he wrote the same day to Lubbock, thanking him for the correction and apologizing for being so "muddled."[26] As we know, Darwin didn't tear up the manuscript. But if he had encountered a string of setbacks of this magnitude, he might well have abandoned the project. And this would probably have been a good thing for his long-term social status, if he indeed had been too consistently muddled to write an impressive book on the origin of species.

These three explanations for Darwin's gloom aren't mutually exclusive. Natural selection is a frugal and resourceful process, making multiple use of existing chemicals, and of the feelings those chemicals carry. This is one reason simple statements about the function of any neurotransmitter, such as serotonin, or any one mood, such as gloom, are tricky. But it is also a reason that a Darwinian doesn't feel stymied when something like a low (or high) opinion of oneself turns out to have several equally plausible purposes. They may all be genuine.

Where does truth belong on the spectrum of self-esteem? If one month, following a string of professional and social successes, you're fairly brimming with serotonin and feel enduringly competent, likable, and attractive, and the next month, after a few setbacks, and some serotonin slippage, you feel enduringly worthless, you can't have been right both times. Which time were you wrong? Is serotonin truth serum or a mind-numbing narcotic?

Maybe neither. When you're feeling either very good or very bad about yourself, it probably means that a large body of evidence is being hidden from view. The most truthful times come between the extremes.

Anyway, maybe "truth" is best left out of this altogether. Whether you're a "good" or a "worthless" person is a question whose objective meaning is, at best, elusive. And even when "truth" can be clearly defined, it is a concept to which natural selection is indifferent. To be sure, if an accurate portrayal of reality, to oneself or to others, can help spread one's genes, then accuracy of perception or communication may evolve. And often this will be the case (when, say, you remember where food is stored, and share the data with offspring or siblings). But when accurate reporting and genetic interest do thus intersect, that's just a happy coincidence. Truth and honesty are never favored by natural selection in and of themselves. Natural selection neither "prefers" honesty nor "prefers" dishonesty. It just doesn't care.[27]

STRONG YET SENSITIVE

Reciprocal altruism brings its own agenda to the presentation of self, and thus to the deception of self. Whereas status hierarchies place a premium on our seeming competent, attractive, strong, smart, etcetera, reciprocal altruism puts its accent on niceness, integrity, fairness. These are the things that make us seem like worthy reciprocal altruists. They make people want to strike up relationships with us. Puffing up our reputations as decent and generous folks can't hurt, and it often helps.

Richard Alexander, in particular, has stressed the evolutionary importance of moral self-advertisement. In *The Biology of Moral Systems* he writes that "modern society is filled with myths" about our goodness: "that scientists are humble and devoted truth-seekers; that doctors dedicate their lives to alleviation of suffering; that teachers dedicate their lives to their students; that we are all basically law-abiding, kind, altruistic souls who place everyone's interests before our own."[28]

There's no reason moral self-inflation has to involve self-deception. But there's little doubt that it can. The unconscious convolutions by which we convince ourselves of our goodness were seen in the laboratory before the theory of reciprocal altruism was around to explain them. In various experiments, subjects have been told to behave cruelly toward someone, to say mean things to him or even deliver what they thought were electric shocks. Afterwards, the subjects tended to derogate their victim, as if to convince themselves that he deserved his mistreatment—although they knew he wasn't being punished for any wrongdoing and, aside from that, knew only what you can learn about a person by briefly mistreating him in a laboratory setting. But when subjects delivered "shocks" to someone after being told he would get to retaliate by shocking them later, they tended *not* to derogate him.[29] It is as if the mind were programmed with a simple rule: so long as accounts are settled, no special rationalization is in order; the symmetry of exchange is sufficient defense of your behavior. But if you cheat or abuse another person who doesn't cheat or abuse you, you should concoct reasons why he deserved it. Either way, you'll be prepared to defend your behavior if challenged; either way, you'll be prepared to fight with indignation any allegations that you're a bad person, a person unworthy of trust.

Our repertoire of moral excuses is large. Psychologists have found that people justify their failure to help others by minimizing, variously, the person's plight ("That's not an assault, it's a lover's quarrel"), their own responsibility for the plight, and their own competence to help.[30]

It's always hard to be sure that people really believe such excuses. But a famous series of experiments shows (in a quite different context) how oblivious the conscious mind can be to its real motivation, and how busily it sets about justifying the products of that motivation.

The experiments were conducted on "split-brain" patients—people who have had the link between left and right hemispheres cut to stop severe epileptic seizures. The surgery has surprisingly little effect on everyday behavior, but under contrived conditions, strange things can happen. If the word *nut* is flashed onto the left half of the visual field (which is processed by the right hemisphere), but not onto the right half (processed by the left hemisphere), the subject reports no conscious awareness of the signal; the information never enters the left hemisphere, which in most people controls language and seems to dominate consciousness. Meanwhile, though, the subject's left hand—controlled by the right hemisphere—will, if allowed to rummage through a box of objects, seize on a nut. The subject reports no awareness of this fact unless allowed to see what his left hand is up to.[31]

When it comes time for the subject to justify his behavior, the left brain passes from professed ignorance into unknowing dishonesty. One example: the command *walk* is sent to a man's right brain, and he complies. When asked where he's going, his left brain, not privy to the real reason, comes up with another one: he's going to get a soda, he says, convinced. Another example: a nude image is flashed to the right brain of a woman, who then lets loose an embarrassed laugh. Asked what's so funny, she gives an answer that's less racy than the truth.[32]

Michael Gazzaniga, who conducted some of the split-brain experiments, has said that language is merely the "press agent" for other parts of the mind; it justifies whatever acts they induce, convincing the world that the actor is a reasonable, rational, upstanding person.[33] It may be that the realm of consciousness itself is in large part such a press agent—the place where our unconsciously written press releases are infused with the conviction that gives them force. Consciousness cloaks the cold and self-serving logic of the genes in a variety of innocent guises. The Darwinian anthropologist Jerome Barkow has written, "It is possible to argue that the primary evolutionary function of the self is to be the organ of impression management (rather than, as our folk psychology would have it, a decision-maker)."[34]

One could go further and suggest that the folk psychology itself is built into our genes. In other words, not only is the feeling that we are "consciously" in control of our behavior an illusion (as is suggested by other neurological experiments as well); it is a purposeful illusion, designed by natural selection to lend conviction to our claims. For centuries people have approached the philosophical debate over free will with the vague but powerful intuition that free will does exist; *we* (the conscious we) are in charge of our behavior. It is not beyond the pale to suggest that this nontrivial chunk of intellectual history can be ascribed fairly directly to natural selection—that one of the most hallowed of all philosophical positions is essentially an adaptation.

DUBIOUS ACCOUNTING

The warping effect of reciprocal altruism goes beyond a general belief in our own uprightness. It can also be seen in our skewed social accounting systems. Central to reciprocal altruism is the monitoring of exchanges—the record of whom you owe, who owes you, and how much is owed. From the gene's point of view, monitoring the two sides of the record with equal diligence would be foolish. If you end up getting slightly more than you give, so much the better. But if you give more than you get by even the smallest increment, that's an increment of loss.

That people keep closer track of what they're owed than of what they owe is hardly a news flash from the frontiers of behavioral science. It has been so obvious for so long that a century and a half ago it served as the unspoken basis for a little joke Darwin relayed to his sister Caroline. In a letter from the *Beagle*, he wrote of a man who "in one of Lord Byrons [sic] letters is said to be so altered after an illness that his *oldest Creditors* would not know him."[35] Darwin himself amassed some debts in college, and one biographer reports that he "felt rather badly about these debts and, when mentioning his extravagances in after years, seems to have scaled them down by a half."[36]

Darwin selectively remembered debts of an intellectual sort as well. At a young age, he had read his grandfather Erasmus's writings on evolution. They include a sentence that strikingly anticipates sexual selection, the variant of natural selection that has made males so combative: "The final cause of this contest amongst the males seems to be, that the strongest and most active animal should propagate the species, which should thence become improved." Yet when Darwin included, in the third edition of the *Origin*, a prefatory outline of intellectual precursors, he dismissed his grandfather in a footnote as a pre-Lamarckian harbinger of Lamarck's confusion. And in his *Autobiography*, Darwin spoke disparagingly of Erasmus's *Zoonomia*, the book that, judging by the above quotation, may well have planted in Darwin's mind the seed not only of evolutionism, but of the theory of natural selection. It is a safe bet that Darwin's ever-vigilant conscience wouldn't have let him *consciously* give such short shrift to his own grandfather.[37]

Darwin was not generally remiss in giving intellectual credit. He was selectively remiss. As one biographer wrote, "generous though Darwin always was to those whose empirical observations he found useful, he barely acknowledged those whose ideas had influenced him."[38] What a useful pattern. Darwin lavished credit on scores of minor-league researchers, while diminishing the few predecessors who might have been even remote contenders for his crown; he thus incurred the debt of many young, rising scientists, while risking the offense mainly of the old and the dead. All in all, a fairly sound formula for high status. (Of course, the formula itself—"don't credit people who foreshadowed your theory"—isn't written in the genes.

Sigmund Freud's model of the unconscious mind is being amended by modern Darwinians. Their approach often begins with the simple observation that animals may present themselves as whatever it is in their genetic interest to seem like. Examples include this leaf-mimicking katydid and this harmless snake, which looks like a poisonous coral snake. Human animals, unlike insects and snakes, may best execute deception (about their intentions, for example) by first deceiving themselves. This is one, though not the only, Darwinian explanation for unconscious motivation.

Luit and Nikkie (*above, right*) formed a coalition to dethrone alpha male Yeroen (*above, left*). *Left:* Luit sometimes jumped up and down on females to punish them for having socialized with Yeroen.

Charles Lyell (*above*) and
Joseph Hooker (*left*) were
part of Darwin's coalition,
and maneuvered to elevate
his status at the expense
of Alfred Russel Wallace
(*below*). Their tactics were
more subtle than Luit's.

Darwin around 1855, four years before *The Origin of Species* appeared and three years before he found that Alfred Russel Wallace had separately discovered the theory of natural selection. He wrote to Charles Lyell in 1858, "So all my originality, whatever it may amount to, will be smashed. . . ." His ensuing struggle to avoid that fate seems to have involved convenient self-deception.

But there could well be a built-in tendency to refrain from bestowing status-enhancing benefits on people whose status threatens your own.)

The egocentric bias in accounting ranges from the epic to the minor. Wars routinely feature a deep and sure sense of grievance on both sides, a weighty belief in the enemy's guilt. And next-door neighbors, even good friends, can bring comparable conviction to their differing historical records. This fact can get lost in some strata of modern society, where a gloss of cordiality covers everyday life. But there is every reason to believe that through history and prehistory, reciprocal altruism has carried an everyday tension, an implicit or explicit haggling. Bronislaw Malinowski observed that the Trobriand Islanders seemed absorbed in the giving of gifts and were "inclined to boast of their own gifts, with which they are entirely satisfied, while disputing the value and even quarrelling over what they themselves receive."[39]

Was there ever a culture in which people didn't regularly disagree—over goods in a market, over salaries at work, over zoning variances, over whose child was wronged by whose child? The resulting arguments can have real consequence. They're seldom, by themselves, life or death matters, but they affect material well-being, and during human evolution a small slice of material well-being has at times been the difference between life and death, between attracting a mate and not attracting one, between three surviving offspring and two. So there is reason to suspect an innate basis for biased social accounting. The bias appears to be universal, and seems intuitively to be a corollary of the theory of reciprocal altruism.

Still, once you look at the situation with something other than intuition, things grow less clear. In Axelrod's computer, the key to TIT FOR TAT's success was that it didn't try to get the better of its neighbors; it was always willing to settle for an exactly equal exchange. Creatures that weren't this easily satisfied—creatures that tried to "cheat," to get more than they gave—went extinct. If evolution thus punishes the greedy, why do humans seem unconsciously compelled to give a bit less than they get?

The first step toward an answer is to see that getting more than you give isn't the same as "cheating."[40] Axelrod's computer conflated the two by making life binary: either you cooperate, or you don't; you're nice or you're a cheater. Real life is more finely graded. So bountiful are the benefits of non-zero-sumness that slightly uneven exchanges can make sense for both people. If you do forty-nine favors for your friend and get fifty-one in return, the friendship is probably still worth your friend's while. You haven't really "cheated" him. You've gotten the better of him, yes, but not so much that he should prefer no deal at all to the deal he got.

So it's possible, in theory, to be a little more stingy than TIT FOR TAT without really cheating, and thus without triggering painful retaliation. This sort of stinginess, as ingrained by natural selection, might well assume the form of shady accounting—a deep sense of justice slightly slanted toward the self.

Why would it be so important that the bias be unconscious? A clue may lie in a book called *The Strategy of Conflict* by the economist and game theorist Thomas Schelling. In a chapter called "An Essay on Bargaining"—which isn't about evolution, but could apply to it—Schelling noted an irony: in a non-zero-sum game, "the power to constrain an adversary may depend on the power to bind oneself." The classic example is the non-zero-sum game of "chicken." Two cars head toward each other. The first driver to swerve loses the game, along with some stature among his adolescent peers. On the other hand, if neither driver bails out, both lose in a bigger way. What to do? Schelling suggests tossing your steering wheel out the window in full view of the other driver. Once convinced that you're irrevocably committed to your course, he will, if rational, do the swerving himself.

The same logic holds in more common situations, like buying a car. There is a range of prices within which a deal makes sense for both buyer and seller. Within that range, though, interests diverge: the buyer prefers the low end, the seller the high end. The path to success, says Schelling, is essentially the same as in the game of chicken: be the first to convince the other party of your rigidness. If the dealer believes you're walking away for good, he'll cave in. But if the dealer stages a preemptive strike, and says "I absolutely cannot accept less than x," and appears to be someone whose pride wouldn't let him swallow those words, then he wins. The key, said Schelling, is to make a "voluntary but irreversible sacrifice of freedom of choice"—and to be the first to do it.

For our purposes, take out the word *voluntary*. The underlying logic may be excluded from consciousness to make the sacrifice seem truly "irreversible." Not when we're on a used-car lot, maybe. Car salesmen, like game theorists, actually think about the dynamics of bargaining, and the savvier car buyers do too. Still, everyday haggling—over fender benders, salaries, disputed territory—often begins with an actual belief, on each side, in its own rightness. And such a belief, a quickly reached and hotly articulated sense of what we deserve, is a quick route to the preemptive strikes Schelling recommends. Visceral rigidity is the most convincing kind.

Still, puzzles remain. Utter rigidity could be self-defeating. As "shady accounting" genes spread through the population, shady accountants would more and more often run into each other. With each insisting on the better half of the deal, both would fail to strike any deal. Besides, in real life, the rigidity wouldn't know where to set in, because it's often hard to say what deals the other party will accept. A car buyer doesn't know how much the car actually cost the dealer or how much other buyers are offering. And in less structured situations—swapping favors with someone, say—these calculations are even dimmer, because things are less quantifiable. Thus has it been throughout evolution: hard to fathom precisely the range of deals that are in the interest of the other party. If you begin the bargaining by insisting irreversibly on a deal outside of that range, you're left without a deal.

The ideal strategy, perhaps, is a pseudorigidity, a flexible firmness. You begin the discourse with an emphatic statement of what you deserve. Yet you should retreat—up to a point, at least—in the face of evidence as to the other person's firmness. And what sort of evidence might that be? Well, evidence. If people can explain the reasons behind their conviction, and the reasons seem credible (and sound heartfelt), then some retreat is in order. If they talk about how much they've done for you in the past, and it's true, you have to concede the point. Of course, to the extent that you can muster countervailing evidence, with countervailing conviction, you should. And so it goes.

What we've just described are the dynamics of human discourse. People do argue in precisely this fashion. (In fact, that's what the word *argue* means.) Yet they're often oblivious to what they're doing, and to why they're doing it. They simply find themselves constantly in touch with all the evidence supporting their position, and often having to be reminded of all the evidence against it. Darwin wrote in his autobiography of a habit he called a "golden rule": to immediately write down any observation that seemed inconsistent with his theories—"for I had found by experience that such facts and thoughts were far more apt to escape from the memory than favourable ones."[41]

The reason the generic human arguing style feels so effortless is that, by the time the arguing starts, the work has already been done. Robert Trivers has written about the periodic disputes—contract renegotiations, you might call them—that are often part of a close relationship, whether a friendship or a marriage. The argument, he notes, "may appear to burst forth spontaneously, with little or no preview, yet as it rolls along, two whole landscapes of information appear to lie already organized, waiting only for the lightning of anger to show themselves."[42]

The proposition here is that the human brain is, in large part, a machine for winning arguments, a machine for convincing others that its owner is in the right—and thus a machine for convincing its owner of the same thing. The brain is like a good lawyer: given any set of interests to defend, it sets about convincing the world of their moral and logical worth, regardless of whether they in fact have any of either. Like a lawyer, the human brain wants victory, not truth; and, like a lawyer, it is sometimes more admirable for skill than for virtue.

Long before Trivers wrote about the selfish uses of self-deception, social scientists had gathered supporting data. In one experiment, people with strongly held positions on a social issue were exposed to four arguments, two pro and two con. On each side of the issue, the arguments were of two sorts: (a) quite plausible, and (b) implausible to the point of absurdity. People tended to remember the plausible arguments that supported their views and the implausible arguments that didn't, the net effect being to drive home the correctness of their position and the silliness of the alternative.[43]

One might think that, being rational creatures, we would eventually grow suspicious of our uncannily long string of rectitude, our unerring knack for being on the right side of any dispute over credit, or money, or manners, or anything else. Nope. Time and again—whether arguing over a place in line, a promotion we never got, or which car hit which—we are shocked at the blindness of people who dare suggest that our outrage isn't warranted.

FRIENDSHIP AND COLLECTIVE DISHONESTY

In all the psychological literature that predates and supports the modern Darwinian view of deception, one word stands out for its crisp economy: *beneffectance*. It was invented in 1980 by the psychologist Anthony Greenwald to describe the tendency of people to present themselves as being both beneficial and effective. The two halves of this compound coinage embody the legacies, respectively, of reciprocal altruism and status hierarchies.[44]

This distinction is a bit oversimplified. In real life, the mandates of reciprocal altruism and status—to seem beneficial and effective—can merge. In one experiment, when people who had been part of a team effort were asked about their role in it, they tended to answer expansively if first told that the effort was a success. If told it had failed, they left more room for the influence of a teammate.[45] This hoarding of credit and sharing of blame makes both kinds of evolutionary sense. It makes a person seem beneficial, having helped others in the group achieve success, and thus deserving future repayment; it also makes that person seem effective, deserving high status.

One of the most famous triumphs for Darwin's supporters came in 1860, when Thomas Huxley, a.k.a. "Darwin's bulldog," took on Bishop Samuel Wilberforce during a debate on *The Origin of Species*. Wilberforce sarcastically asked on which side of his family Huxley was descended from an ape, and Huxley replied that he would rather have an ape as an ancestor than a man "possessed of great means and influence and yet who employs these faculties and that influence for

the mere purpose of introducing ridicule into a grave scientific discussion." At least, that's how Huxley told the story to Darwin—
and Huxley's account is the one that made it into the history books.
But Darwin's close friend Joseph Hooker was also present, and he
remembered things differently. He told Darwin that Huxley "could
not throw his voice over so large an assembly, nor command the
audience; & he did not allude to *Sam's* [Bishop Wilberforce's] weak
points nor put the matter in a form or way that carried the audience."

Fortunately, reported Hooker, he himself had taken on Wilberforce: "I smacked him amid rounds of applause" and went on to
show "that he could never have read your book" and that he "was
absolutely ignorant" of biology. Wilberforce "had not one word to
say in reply & the meeting *was dissolved forthwith* leaving you master
of the field after 4 hours battle." Since the encounter, said Hooker,
"I have been congratulated & thanked by the blackest coats & whitest
stocks in Oxford." Huxley, meanwhile, reported being "the most
popular man in Oxford for full four & twenty hours afterwards."[46]
Both Huxley and Hooker were telling stories that would do two
things: raise their stature in Darwin's eyes, and leave him indebted
to them.

Reciprocal altruism and status intersect in a second way. A common exception to our tendency to deflate the contributions of others
comes when those others have high status. If we have a friend who
is, say, mildly famous, we cherish even his meager gifts, forgive his
minor offenses, and make extra sure not to let him down. In one
sense this is a welcome corrective to egocentrism; our balance sheets
are perhaps more honest for high-status people than for others. But
the coin has two sides. These high-status people, meanwhile, are
viewing us with even greater distortion than usual, as our side of the
ledger is discounted steeply to reflect our lowliness.

Still, we seem to consider the relationship worthwhile. A high-
status friend may, in time of need, wield decisive influence on our
behalf, often at little cost. Just as an alpha male ape can protect an
ally by looking askance at the would-be attacker, a highly placed
sponsor can, with a two-minute phone call, make a world of difference for an upstart.

Seen in this light, social hierarchy and reciprocal altruism not only intersect but merge into a single dimension. Status is simply another kind of asset that people bring to the bargaining table. Or, more precisely: it is an asset that leverages other assets; it means that at little cost a person can do big favors.

Status can also be one of the favors. When we ask friends for help, we are often asking not only that they use their status, but that they raise ours in the process. Among the chimps of Arnhem, the swapping of status support was sometimes simple; chimp A helps chimp B fend off a challenger and maintain its status; chimp B later returns the favor. Among people, status support is less tangible. Except in barrooms, junior-high schoolyards, and other venues of high testosterone, the support consists of information, not muscle. Backing a friend means verbally defending him when his interests are in dispute—and, more generally, saying good, status-raising things about him. Whether these things are true doesn't especially matter. They're just the things friends are supposed to say. Friends engage in mutual inflation. Being a person's true friend means endorsing the untruths he holds dearest.

Whether this bias toward a friend's interests is deeply unconscious is a matter for research that hasn't yet been done. A purely positive answer would clash with the treachery that has been known to infest friendships. Still, it may be that the hallmark of the strongest, longest friendships is the depth of the shared bias; the best friends are the ones who see each other least clearly. Anyway, however conscious or unconscious the lies, one effect of friendship is to take individual nodes of self-serving dishonesty and link them up into webs of collective dishonesty. Self-love becomes a mutual-admiration society.

And enmity becomes two mutual-detestation societies. If your true friend has a true enemy, you're supposed to adopt that enemy as your own; that's how you support your friend's status. By the same token, that enemy—and that enemy's friends—are expected to dislike not just your friend, but you. This isn't a rigid pattern, but it's a tendency. To maintain close friendship with two avowed enemies is to be in a position whose awkwardness is viscerally felt.

The malevolent conspiracy between reciprocal altruism and status hierarchies runs one level deeper. For enmity itself is a cocreation of the conspirators. On the one hand, enmity grows out of rivalry, the mutual and incompatible pursuit of status. On the other hand, it is the flip side of reciprocal altruism. Being a successful reciprocal altruist, as Trivers noted, means being an enforcer—keeping track of those who take your aid but don't return it, and either withholding future aid or actively punishing them.

Once again, all of this enmity may be expressed not plainly and physically, as among chimps, but verbally. When people are our enemies, or when they support our enemies, or fail to support us after we've supported them, the standard response is to convincingly say bad things about them. And, again, the best way to convincingly say such things is to believe them—believe that the person is incompetent or stupid or, best of all, *bad*, morally deficient, a menace to society. In *The Expression of the Emotions in Man and Animals*, Darwin captured the morally charged nature of enmity: "[F]ew individuals . . . can long reflect about a hated person, without feeling and exhibiting signs of indignation or rage."[47]

Darwin's own assessments of people sometimes had a flavor of retaliation. While at Cambridge, he met a man named Leonard Jenyns, a gentleman entomologist who, like Darwin, collected beetles. It seemed possible that, notwithstanding a natural rivalry between them, the two men could become friends and allies. Indeed, Darwin made the overture, giving Jenyns "a good many insects" for which, Darwin reported, Jenyns seemed "very grateful." But when the time came to reciprocate, Jenyns "refused me a specimen of the Necroph. sepultor . . . although he has 7 or 8 specimens." In relaying this news to his cousin, Darwin commented not only on Jenyns's selfishness, but on his "weak mind." Eighteen months later, though, Darwin considered Jenyns "an excellent naturalist." This revised opinion may be related to Jenyns having bestowed on Darwin, in the meanwhile, a "magnificent present of Diptera."[48]

When grudges are expanded into networks, as friends form coalitions to support each other's status, the result is vast webs of self-deception and, potentially, of violence. Here is a sentence from the *New York Times:* "In a week's time, both sides have constructed deeply emotional stories explaining their roles, one-sided accounts that are offered with impassioned conviction, although in many respects they do not stand up, in either case, under careful scrutiny."[49]

The sentence refers to an incident in which Israeli soldiers shot Palestinian civilians, and each side clearly saw that the other had started the trouble. But the sentence could be applied with equal accuracy to all kinds of clashes, big and little, through the centuries. By itself this sentence tells a large part of human history.

The mental machinery that drives modern wars—patriotic fervor, mass self-righteousness, contagious rage—has often been traced by evolutionists to eons of conflict among tribes or bands. Certainly such large-scale aggression has surfaced repeatedly during the life of our species. And no doubt warriors have often gotten Darwinian rewards through the rape or abduction of enemy women.[50] Still, even if the psychology of war has indeed been shaped by out-and-out wars, they may well have been of secondary importance.[51] Feelings of enmity, of grievance, of righteous indignation—of *collective* enmity and grievance and righteous indignation—probably have their deepest roots in ancient conflicts *within* bands of humans and prehumans. In particular: in conflicts among coalitions of males for status.

INTEREST GROUPS

The tendency of friends to dislike each other's enemies needn't be merely an exchange of favors. Often it's simple redundancy. One of the strongest bonds two friends can have—the great starter and sustainer of friendships—is a common enemy. (Two people playing a game of prisoner's dilemma will play more cooperatively in the presence of someone they both dislike.)[52]

This strategic convenience is often obscured in modern society. Friendships may rest not on common enemies but on common interests: hobbies, tastes in movies or sports. Affinities emerge from shared passions of the most innocent sort. But this reaction presumably evolved in a context in which shared passions tended to be less innocent: a context of frankly political opinions about who should lead a tribe, say, or how meat should be divided. In other words, the affinity of common interest may have evolved as a way to cement fruitful political alliances, and only later attached itself to matters of little consequence. This, at any rate, would help explain the absurd gravity surrounding disputes over seemingly trivial matters. Why is it that a smooth dinner party can turn suddenly awkward over a disagreement about the merits of John Huston's movies?

And, moreover, "matters of little consequence" often turn out, on close examination, to involve real stakes. Take two Darwinianly minded social scientists, for example. Their binding interest is "purely intellectual"—a fascination with the evolutionary roots of human behavior. But this is also a common political interest. Both scholars are tired of being ignored or attacked by the academic establishment, tired of the dogma of cultural determinism, tired of its stubborn prevalence within so many anthropology and sociology departments. Both scholars want to be published in the most esteemed journals. They want tenure at the best universities. They want power and status. They want to depose the ruling regime.

Of course, if they do depose the ruling regime, and thus become famous and write best-selling books, there may be no Darwinian payoff. They may not convert their status into sex, and if they do they may use contraception. But in the environment in which we evolved—indeed, until the last few hundred years—status got converted into Darwinian currency more efficiently. This fact seems to have deeply affected the texture of intellectual discourse, especially among men.

We'll explore an example of this effect in the next chapter, in describing the particular intellectual discourse that made Darwin famous. For now, let's simply note Darwin's delight, in 1846, at discovering common scientific interests with Joseph Hooker, who more than a decade later would join with Darwin in the scientific battle of the century and devote much energy to the elevation of Darwin's social status. "[W]hat a good thing is community of tastes," Darwin wrote to Hooker. "I feel as if I had known you for fifty years. . . ."[53]

Chapter 14: **DARWIN'S TRIUMPH**

I am got most deeply interested in my subject; though I wish I could set less value on the bauble fame, either present or post-humous, than I do, but not I think, to any extreme degree; yet, if I know myself, I would work just as hard, though with less gusto, if I knew that my Book would be published for ever anonymously.

—Letter to W. D. Fox (1857)[1]

Darwin was one of our finest specimens. He did superbly what human beings are designed to do: manipulate social information to personal advantage. The information in question was the prevailing account of how human beings, and all organisms, came to exist; Darwin reshaped it in a way that radically raised his social status. When he died in 1882, his greatness was acclaimed in newspapers around the world, and he was buried in Westminster Abbey, not far from the body of Isaac Newton.[2] Alpha-male territory.

And to top it all off: he was a good guy. The *Times* of London observed, "Great as he was, wide as was the reach of his intelligence, what endeared him to his many friends, what charmed all those who were brought into even momentary contact with him, was the beauty of his character."[3] Darwin's legendary lack of pretense persisted until the very end, when it slipped beyond his control. The local coffin maker recalled: "I made his coffin just the way he wanted it, all rough, just as it left the bench, no polish, no nothin'." But then, after the sudden decision to bury him at Westminster Abbey, "my coffin wasn't wanted and they sent it back. This other one you could see to shave in."[4]

This is the basic, oft-noted paradox of Charles Darwin. He became world famous yet seemed to lack the traits that typically fuel epic social ascents. He appears, as one biographer put it, "an unlikely survivor in the immortality stakes, having most of the decent qualities that deter a man from fighting with tooth and claw."[5]

The paradox can't be resolved simply by noting that Darwin authored the correct theory of how people came to exist, for he wasn't alone in doing this. Alfred Russel Wallace, who arrived at natural selection independently, began circulating a written description of it before Darwin had gone public. The two men's versions of the theory were formally unveiled on the same day, in the same forum. But today Darwin is Darwin, and Wallace is an asterisk. Why did Darwin triumph?

In chapter ten we partly reconciled Darwin's decency with his fame, noting that he lived in a society in which doing good was typically a prerequisite for doing well. Moral reputation meant much, and just about everything you did caught up with your reputation.

But the story is more complex than that. A closer look at Darwin's long and winding road to fame calls into question some common assessments of him—that, for example, he had little ambition and not a shred of Machiavellianism, that his commitment to truth was unadulterated by the thirst for fame. Viewed through the new paradigm, Darwin looks a bit less like a saint and a bit more like a male primate.

SOCIAL CLIMBING

From early on, Darwin exhibited a common ingredient of social success: ambition. He competed with rivals for status, and longed for the esteem it brings. "My success . . . has been very good amongst the water beettles [sic]," he wrote to a cousin from Cambridge. "I think I beat Jenyns in Colymbetes." When his insect collecting got him cited in *Illustrations of British Insects*, he wrote,

"You will see my name in Stephens' last number. I am glad of it if it is merely to spite Mr. Jenyns."[6]

The notion of Darwin as a typical young male, bent on conquest, seems at odds with standard appraisals. The Darwin described by John Bowlby—"nagging self-contempt," a "tendency to disparage his own contributions," an "ever-present fear of criticism, both from himself and from others," an "exaggerated respect for authority and the opinions of others"—doesn't sound like an alpha male in the making.[7] But remember: often in chimpanzee societies, and almost always in human societies, the social scale can't be ascended alone; a common first step is to forge a bond with a primate of higher status, and this involves an act of submission, a profession of inferiority. One biographer has described Darwin's purported pathology in especially suggestive terms: "some flaw of self-confidence, some absence of certainty, that made him emphasize his shortcomings when dealing with those in authority."[8]

In his autobiography, Darwin recalled the "glow of pride" he felt when, as a teenager, he heard that an eminent scholar, after chatting with him, had said, "There is something in that young man that interests me." The compliment, Darwin said, "must have been chiefly due to his perceiving that I listened with much interest to everything which he said, for I was as ignorant as a pig about his subjects of history, politicks and moral philosophy."[9] Here, as usual, Darwin is too humble by half, but he is probably right to suggest that his humility itself had played a role. (Darwin goes on to note: "To hear of praise from an eminent person, though no doubt apt or certain to excite vanity, is, I think, good for a young man, as it helps to keep him in the right course."[10] Yes: upward.)

To call Darwin's humility tactically sound isn't to call it disingenuous. The tendency of people to view the next rung on the social ladder with respect is most effective when they're thoroughly in its thrall, and not conscious of its purpose: we feel genuinely in awe of people before whom, it so happens, we might profitably grovel. Thomas Carlyle, one of Darwin's contemporaries (and acquaintances), was probably right to say that hero worship is an essential part of human nature. And it is probably no coincidence that hero worship grows powerful at the time of life when people begin their social competition in earnest. "Adolescence," one psychiatrist has observed, "is a time of a renewed search for ideals. . . . [T]he adolescent is seeking a model, a perfect person to emulate. It's much like the moment in infancy before they realized their parents' imperfections."[11]

Yes, the awe for role models feels much like the early awe for a parent—and may spring from the same neurochemistry. But its function is not only to encourage instructive emulation; it also helps write the implicit contract between senior and junior partners in a coalition. The latter, lacking the social status that counts heavily in reciprocal altruism, will compensate for this shortcoming through deference.

While Darwin was at Cambridge, his most extreme deference was reserved for the professor (and reverend) John Stevens Henslow. Darwin had heard from his older brother that Henslow was "a man who knew every branch of science, and I was accordingly prepared to reverence him."[12] After striking up an acquaintance, Darwin reported that "he is quite the most perfect man I ever met with."[13]

Darwin became known at Cambridge as "the man who walks with Henslow." Their relationship was like the millions of other such relationships in the history of our species. Darwin benefited from Henslow's example and counsel and drew on his social connections, and repaid him with, among other things, subservience, arriving early for Henslow's lectures to help set up equipment.[14] One is reminded of Jane Goodall's description of Goblin's social ascent: he was "respectful" of his mentor Figan, followed him around, watched what he did, and often groomed him."[15]

After earning Figan's acceptance and absorbing his wisdom, Goblin turned on him, displacing him as alpha male. But Goblin may have felt truly reverent until the moment when greater detachment was in order. And so it is with us: our gauging of people's worth—their professional caliber, their moral fiber, whatever—reflects partly the place they occupy in our social universe at the time. We are selectively blinded to those qualities that it would be inconvenient to acknowledge.

Darwin's worship of Henslow isn't the best example of this blindness, as Henslow was a widely admired man. But consider the captain of the *Beagle*, Robert FitzRoy. When Darwin met FitzRoy for the interview that would decide whether he sailed with the *Beagle*, the situation was simple: here was a man of high status whose approval might eventually elevate Darwin's own status markedly. So it's not surprising that Darwin seems to have come prepared to "reverence" FitzRoy. After the meeting, he wrote to his sister Susan: "[I]t is no use attempting to praise him as much as I feel inclined to do, for you would not believe me. . . ." He wrote in his diary that FitzRoy was "as perfect as nature can make him." To Henslow (who was the rung on Darwin's ladder that had led to the *Beagle*) he wrote, "Cap. FitzRoy is every thing that is delightful. . . ."[16]

Years later, Darwin would describe FitzRoy as a man who "has the most consummate skill in looking at everything & every body in a perverted manner." But then, years later he could afford to. Now was no time to be scanning FitzRoy for flaws, or probing beneath the civil façade commonly mustered for first encounters. Now was a time for deference and amity, and their deployment proved a success. On the evening Darwin was writing his letters, FitzRoy was writing to a naval officer—"I like what I see and hear of him, much"—and requesting that Darwin be named the ship's naturalist. Darwin, in one of the calmer passages in his letter to Susan, had written, "I hope I am judging reasonably, & not through prejudice about Cap. Fitz."[17] He was doing both; he was rationally pursuing long-term self-interest with the aid of short-term prejudice.

Toward the end of the *Beagle*'s voyage, Darwin got his strongest early taste of professional esteem. He was (aptly enough) on Ascension Island when he got a letter from Susan relaying the interest aroused by his scientific observations, which had been read before the Geological Society of London. Most notably, Adam Sedgwick, the eminent Cambridge geologist, had said that some day Darwin would "have a great name among the Naturalists of Europe." It's not yet clear exactly which neurotransmitters are unleashed by status-raising news such as this (serotonin, we've seen, is one candidate), but Darwin described their effect clearly: "After reading this letter I clambered over the mountains of Ascension with a bounding step and made the volcanic rocks resound under my geological hammer!"

In reply Darwin affirmed to Susan that he would now live by the creed "that a man who dares to waste one hour of time, has not discovered the value of life."[18]

Elevations of status may bring a reevaluation of one's social constellation. The relative positions of the stars have changed. People who used to be central are now peripheral; the focus must be shifted toward brighter bodies that once seemed beyond reach. Darwin was not the sort of person to perform this maneuver crudely; he never forgot the little people. Still, there are hints of a shifting social calculus while he was aboard the *Beagle*. His older cousin, William Fox, had introduced him to entomology (and to Henslow); at Cambridge Darwin had profited much from their ongoing exchange of insect lore and specimens. During that correspondence, while seeking guidance and data from Fox, Darwin had assumed his customary stance of abject submission. "I should not send this very shamefully stupid letter," he wrote, "only I am very anxious to get some *crumbs* of information about yourself & the insects." He sometimes reminded Fox of "how long I have been hoping in vain to receive a letter from my old master" and enjoined him to "remember I am your pupil. . . ."[19]

It is thus poignant when, six years later, as Darwin's researches aboard the *Beagle* signal his rise in stature, Fox senses a new asymmetry in their friendship. Suddenly it is he who apologizes for the "dullness" of his letter, he who stresses that "You are never a whole Day absent from my thoughts," he who begs for mail. "It is now so long since I saw your handwriting that I cannot tell you the pleasure it would give me. I feel however that your time is valuable & mine worth nothing, which makes a vast difference."[20] This shifting balance of affection is a regular feature of friendships amid sharp changes in status, as the reciprocal-altruism contract is silently renegotiated. Such renegotiations may have been less common in the ancestral environment, where, to judge by hunter-gatherer societies, status hierarchies were less fluid after early adulthood than they are now.[21]

LOVING LYELL

During the voyage, Henslow, Darwin's mentor, remained his main link to British science. The geological reports that had so impressed Sedgwick were extracts from letters to Henslow, which he had dutifully publicized. It was to Henslow that Darwin wrote near the voyage's end, asking him to lay the groundwork for membership in the Geological Society. And throughout, Darwin's letters left no doubt about his continuing allegiance to "my President & Master." Upon arriving in Shrewsbury after the *Beagle* docked, he wrote: "My dear Henslow, I do long to see you; you have been the kindest friend to me, that ever Man possessed."[22]

But Henslow's days as main mentor were numbered. On the *Beagle,* Darwin had (at Henslow's suggestion) read *Principles of Geology,* by Charles Lyell. Therein, Lyell championed the much-disputed theory, advanced earlier by James Hutton, that geological formations are mainly the product of gradual, ongoing wear and tear, as opposed to catastrophic events, such as floods. (The catastrophist version of natural history had found favor with the clergy, since it seemed to suggest divine interventions.) Darwin's work on the *Beagle*—his evidence, for example, that the coast of Chile had been rising imperceptibly since 1822—tended to support the gradualist view, and he soon was calling himself a "zealous disciple" of Lyell.[23]

As John Bowlby notes, it's not surprising that Lyell should become Darwin's chief counselor and role model; "their partnership in advocating the same geological principles gave them a common cause that was lacking in Darwin's relationship with Henslow."[24] Common causes, as we've seen, are a frequent sealer of friendships, apparently for Darwinian reasons. Once Darwin had endorsed Lyell's view of geology, both men's status would rise or fall with its fortunes.

Still, the bond of reciprocal altruism between Lyell and Darwin was more than mere "common cause." Each man brought his own assets to the table. Darwin brought mountains of fresh evidence for the views to which Lyell's reputation was inseverably attached. Lyell, in addition to providing a sturdy theoretical rack on which Darwin could array his researches, brought the guidance and social sponsorship for which mentors are known. Within weeks of the *Beagle*'s return, Lyell was inviting Darwin to dinner, counseling him on the wise use of time, and assuring him that, as soon as a spot opened in the elite Athenaeum Club, he could fill it.[25] Darwin, Lyell told a colleague, would make "a glorious addition to my society of geologists. . . ."[26]

Though Darwin could at times be a detached and cynical student of human motivation, he seems to have been numb to the pragmatic nature of Lyell's interest. "Amongst the great scientific men, no one has been nearly so friendly & kind, as Lyell," he wrote to Fox a month after his return. "You cannot imagine how good-naturedly he entered into all my plans."[27] What a nice man!

It is time for yet another reminder that self-serving behavior needn't involve conscious calculation. In the 1950s, social psychologists showed that we tend to like people we find we can influence. And we tend to like them even more if they have high status.[28] It isn't necessary that we think, "If I can influence him, he could be useful, so I should nourish this friendship," or "His compliance will be especially useful if he has high status." Once again, natural selection seems to have done the "thinking."

Of course, people may supplement this "thinking" with their own thinking. There must have been *some* awareness within both Lyell and Darwin of the other man's utility. But they surely also felt, at the same time, a substratum of solid and innocent-feeling amity. It probably *was*, as Darwin wrote to Lyell, "the *greatest* pleasure to me to write or talk Geolog. with you." And Darwin was no doubt sincerely overwhelmed by the "*most* goodnatured manner" in which Lyell gave him guidance, "almost without being asked."[29]

Darwin was probably equally sincere when, several decades later, he complained that Lyell had been "very fond of society, especially of eminent men, and of persons high in rank; and this over-estimation of a man's position in the world, seemed to me his chief foible."[30] But this was after Darwin, now world famous, had acquired some, shall we say, perspective. Earlier Darwin had been too dazzled by Lyell's own position in the world to pay much mind to his flaws.

DARWIN'S DELAY REVISITED

We've seen how Darwin spent the two decades after his return to England: discovering natural selection and then doing a series of things other than disclose it. We have also seen several theories about this delay. The Darwinian slant on Darwin's delay isn't really an alternative to existing theories so much as a backdrop for them. To begin with, evolutionary psychology silhouettes the two forces that wrenched Darwin, one attracting him toward publication, the other repelling him.

First is the inherent love of esteem, a love that Darwin had his share of. One route to esteem is to author a revolutionary theory.

But what if the theory fails to revolutionize? What if it's roundly dismissed—dismissed, indeed, as a threat to the very fabric of society? In that event (the sort of event Darwin was the type to dwell on) our evolutionary history weighs against publication. There's hardly been a genetic payoff, over the ages, for loudly espousing deeply unpopular views, especially when they antagonize the powers that be.

The human bent for saying things that please people was clear long before its evolutionary basis was. In a famous experiment from the 1950s, a surprisingly large number of people were willing to profess incorrect opinions—patently, obviously incorrect opinions— about the relative length of two lines if placed in a room with other people who professed them.[31] Psychologists also found decades ago that they can strengthen or weaken a person's tendency to offer opinions by adjusting the rate at which a listener agrees.[32] Another fifties-era experiment showed that a person's recollections vary according to the audience he is to share them with: show him a list of the pros and cons of raising teachers' salaries, and which ones make a lasting impression depends on whether he expects to address a teachers' or a taxpayers' group. The authors of this experiment wrote, "It is likely that a good deal of a person's mental activity consists, in whole or part, of imagined communication to audiences imagined or real, and that this may have a considerable effect on what he remembers and believes at any one point in time. . . ."[33] This jibes with a Darwinian angle on the human mind. Language evolved as a way of manipulating people to your advantage (your advantage in this case being popularity with an audience that holds firm opinions); cognition, the wellspring of language, is warped accordingly.

In light of all this, the question of Darwin's delay becomes less of a question. Darwin's famous bent for self-doubt in the face of disagreement (especially, it is said, disagreement from authority figures) is quintessentially human—unusual in degree, maybe, but not in kind. It isn't remarkable that he spent many years studying barnacles rather than unveil a theory widely considered heretical—heretical in a sense that is hard to grasp today, when the word *heresy* is almost always used with irony. Nor is it remarkable that Darwin, in the many years of the *Origin*'s gestation, often felt anxious or even mildly depressed; natural selection "wants" us to feel uneasy when pondering actions that augur a massive loss of public esteem.

What's amazing, in a way, is that Darwin could be steadfast in his *belief* in evolution, given the pervasive hostility toward the idea. Leading the assault on *Vestiges*, Robert Chambers's 1844 evolutionist tract, had been Adam Sedgwick, the Cambridge geologist (and reverend) whose praise, relayed to Darwin at Ascension Island, had so thrilled him. Sedgwick's review of the Chambers book was candid about its own agenda. "The world cannot bear to be turned upside down; and we are ready to wage an internecine war with any violation of our modest principles and social manners."[34] Not encouraging.

What was Darwin to do? The standard view is that he vacillated, like a laboratory rat eyeing food whose procurement will bring a shock. But there's also a minority view: during his celebrated barnacle detour, while failing to publish his theory about evolution, he was busy paving the way for its eventual reception. The strategy can be seen as three-pronged.

First, Darwin strengthened his argument. While immersed in barnacles, he continued to gather evidence for his theory, partly through the postal interrogation of far-flung experts on flora and fauna. One reason for the *Origin*'s ultimate success was Darwin's meticulous anticipation of, and preemptive response to, criticism. Two years before the book's publication, he correctly wrote, "[I] think I go as far as almost anyone in seeing the grave difficulties against my doctrine."[35]

This thoroughness grew out of self-doubt—out of Darwin's legendary humility and grave fear of criticism. Frank Sulloway, an authority on both Freud and Darwin, has made this point by comparing the two men: "Although both were revolutionary personalities, Darwin was unusually concerned about personal error and was modest to a fault. He also erected a new scientific theory that has successfully stood the test of time. Freud, in contrast, was tremendously ambitious and highly self-confident—a self-styled 'conquistador' of science. Yet he developed an approach to human nature that was largely a collection of nineteenth-century psychobiological fantasies masquerading as real science."[36]

In reviewing John Bowlby's biography of Darwin, Sulloway made the point Bowlby failed to make. "[I]t seems reasonable to argue that a moderate degree of lowered self-esteem, which in Darwin was coupled with dogged persistence and unflagging industry, is actually a valuable attribute in science by helping to prevent an overestimation of one's own theories. Constant self-doubt, then, is a methodological hallmark of good science, even if it is not especially congenial to good psychological health."[37]

The question naturally arises as to whether such useful self-doubt, however painful, might be part of the human mental repertoire, preserved by natural selection because of its success, in some circumstances, at propelling social ascent. And the question grows only more intriguing in light of Darwin's father's role in forging his son's self-doubt. Bowlby asks: Was Charles "the disgrace to his family his father had so angrily predicted, or had he perhaps made good? . . . Throughout his scientific career, unbelievably fruitful and distinguished though it would be, Charles's ever-present fear of criticism, both from himself and from others, and never satisfied craving for reassurance, seep through." Bowlby also notes that "a submissive and placatory attitude towards his father became second nature to Charles" and suggests that his father is at least partly to blame for Charles's "exaggerated" respect for authority and his "tendency to disparage his own contributions."[38]

The speculation is irresistible: perhaps the elder Darwin, in implanting this lifelong source of discomfort, was functioning as designed. Parents may be programmed—whether they know it or not—to adjust their children's psyches, even if painfully, in ways that promise to raise social status. For that matter, the younger Darwin, in absorbing the painful adjustment, may have been functioning as designed. We are built to be effective animals, not happy ones.[39] (Of course, we're designed to *pursue* happiness; and the attainment of Darwinian goals—sex, status, and so on—often brings happiness, at least for a while. Still, the frequent *absence* of happiness is what keeps us pursuing it, and thus makes us productive. Darwin's heightened fear of criticism kept him almost chronically distanced from serenity, and therefore kept him busy trying to reach it.)

Thus, Bowlby may be right about all the painful paternal influence on Darwin's character yet wrong to make it sound so pathological. Of course, even things that aren't pathological in the strict sense may be regrettable, and valid targets of psychiatric intervention. But presumably psychiatrists can more ably intervene once they get clear on what sorts of pain are and aren't "natural."

The second prong of Darwin's three-pronged strategy was to beef up his credentials. It's a commonplace of social psychology that credibility grows with prestige.[40] Forced to believe either a college professor or a grade-school teacher on some question of biology, we usually choose the professor. In one sense, this is a valid choice, as the professor is more likely to be right. In another sense, this is just another arbitrary by-product of evolution—a reflexive regard for status.

Either way, an air of mastery is a handy thing when you're trying to change minds. Hence barnacles: even aside from what Darwin *learned* from the barnacles, he knew that the sheer weight of his four volumes on the subclass Cirripedia would lend prestige to his theory of natural selection.

That, at least, is the suggestion of one biographer, Peter Brent: "[P]erhaps . . . Darwin was not training himself with the Cirripede, he was *qualifying* himself."[41] Brent cites an exchange between Darwin and Joseph Hooker. In 1845, Hooker had offhandedly professed doubts about the grand pronouncements of a French naturalist who "does not know what it is to be a specific Naturalist himself." Darwin, characteristically, took the remark to reflect on his own "presumption in accumulating facts & speculating on the subject of variation, without having worked out my due share of species."[42] A year later Darwin went to work on barnacles.

Brent may be right. Several years after the *Origin* was published, Darwin advised a young botanist, "*let theory guide your observations,* but till your reputation is well established, be sparing in publishing theory. It makes persons doubt your observations."[43]

The third prong of Darwin's strategy was to marshal potent social forces—to meld a coalition that included men of stature, men of rhetorical power, and men who fit both descriptions. There was Lyell, who would bring Darwin's first paper on natural selection before the Linnean Society of London, lending it his authority[44] (though Lyell was then an agnostic on natural selection); Thomas Huxley, who would famously confront Bishop Wilberforce in the Oxford evolution debate; Hooker, who would less famously confront Wilberforce and would join Lyell in unveiling Darwin's theory; and Asa Gray, the Harvard botanist who, through his writings in the *Atlantic Monthly,* would become Darwin's chief publicist in America. One by one, Darwin let these men in on his theory.

Was Darwin's assembly of troops really so calculated? Certainly Darwin was aware, by the time the *Origin* was published, that the battle for truth is fought by people, not just ideas. "[W]e are now a good and compact body of really good men, and mostly not old men," he assured one supporter only days after publication. "In the long-run we shall conquer." Three weeks after the *Origin*'s publication, he wrote his young friend John Lubbock, whom he had sent a copy, and asked, "Have you finished it? If so, pray tell me whether you are with me on the *general* issue, or against me." He assured Lubbock in a postscript that "I have got—I wish and hope I might say that *we* have got—a fair number of excellent men on our side of the question. . . ."[45] Translation: If you act now, you can be part of a winning coalition of male primates.

Darwin's pleas for Charles Lyell's full support—almost pathetic in their persistence—are similarly pragmatic. Darwin sees that it is the prestige of his allies, not just their number, that will shape public opinion. September 11, 1859: "Remember that your verdict will probably have more influence than my book in deciding whether such views as I hold will be admitted or rejected at present. . . ." September 20: "[A]s I regard your verdict as far more important in my own eyes, and I believe in the eyes of the world than of any other dozen men, I am naturally very anxious about it."[46]

Lyell's long delay in granting unequivocal support would bring Darwin to the point of bitterness. He wrote to Hooker in 1863: "I am deeply disappointed (I do not mean personally) to find that his timidity prevents him giving any judgment. . . . And the best of the joke is that he thinks he has acted with the courage of a martyr of old."[47] But in terms of reciprocal altruism, Darwin was asking for too much. Lyell was by then sixty-five years old, with an ample intellectual legacy that wouldn't much benefit from his endorsing another man's theory and which could suffer appreciably from identification with a radical doctrine that later proved false. Besides, Lyell had opposed evolutionism in its Lamarckian guise, and thus might be viewed as backtracking. So Darwin's theory wasn't "common cause" for the two men, as Lyell's had been two decades earlier, when Darwin needed a display case for his freshly gathered data. And Lyell, having repaid Darwin's support in various ways, had little if any debt outstanding. Darwin seems to have suffered here from a quaintly pre-Darwinian conception of what friendship is. Or, perhaps, he was under the sway of an egocentric accounting system.

That Darwin was urgently recruiting allies as of 1859 does not, of course, prove that he had for years been plotting strategy. The origin of his alliance with Hooker seems ingenuous enough. Their bond matured during the 1840s as a friendship of the classic variety—based on common interests and common values and consecrated by affection.[48] As it became clear that one of those common interests was openness to the possibility of evolution, Darwin's affection can only have deepened. But we needn't assume that Darwin then envisioned Hooker becoming an avid defender of his theory. The affection inspired by common interests is natural selection's *implicit* recognition of the political usefulness of friends.

Much the same can be said of the way Darwin warmed to Hooker's sterling character. ("One can see at once that he is honourable to the back-bone.")[49] Yes, Hooker's trustworthiness would prove essential; Darwin used him as a confidential sounding board long before natural selection entered public discourse. But no, that doesn't mean Darwin was from the beginning calibrating the value of Hooker's trustworthiness. Natural selection has given us an affinity for people who will be reliable reciprocal-altruism partners. In all cultures, trust joins common interest as the sine qua non of friendship.

Darwin's very compulsion to *have* a confidant—and, as he comes closer to making the theory public, to have additional confidants in Lyell, Gray, Huxley, and others—can be viewed as the product of evolutionary, and not just conscious, calculation. "I do not think I am brave enough to have stood being odious without support," he wrote days after the *Origin* was published.[50] Who would have been? You would have to be just about literally not human to launch a massive attack on the status quo without first seeking social support. In fact, you would almost have to be non-hominoid.

Imagine how many times since our ape days social challenges have hinged on the challenger's success in forging a sturdy coalition. Imagine how many times the challengers have suffered from acting too soon, or from being too open in their machinations. And imagine the ample reproductive stakes. Is it any wonder that mutinies of all kinds, in all cultures, begin with whispers? That even an untutored six-year-old schoolboy feels intuitively the wisdom of discreetly eliciting opinions about the local bully before mounting a challenge? When Darwin confided his theory in a select few, employing his trademark defensiveness (to Asa Gray: "I know that this will make you despise me"),[51] he was probably driven as much by emotion as by reason.

THE PROBLEM OF WALLACE

The greatest crisis of Darwin's career began in 1858. While trudging along on his epic manuscript, he found he had waited too long. Alfred Russel Wallace had now discovered the theory of natural selection—two decades after Darwin did—and stood poised to preempt him. In response, Darwin fiercely pursued his self-interest, but he pursued it so smoothly, and shrouded it in so much moral angst, that, ever since, observers have been calling the episode yet another example of his superhuman decency.

Wallace was a young British naturalist who, like the young Darwin, had set sail for foreign lands to study life. Darwin had known for some time that Wallace was interested in the origin and distribution of species. In fact, the two men had corresponded about the matter, with Darwin noting that he already had a "distinct & tangible idea" on the subject and claiming that "it is really *impossible* to explain my views in the compass of a letter." But Darwin continued to resist any impulse to publish a short paper outlining his theory. "I rather hate the idea of writing for priority," he had written to Lyell, who had urged him to get his views on the record. "[Y]et I certainly should be vexed if any one were to publish my doctrines before me."[52]

The vexation hit on June 18, 1858, when the mail brought a letter from Wallace. Darwin opened it and found a precise sketch of Wallace's theory of evolution, whose likeness to his own theory was stunning. "Even his terms now stand as heads of my chapters," he observed.[53]

The panic that must have struck Darwin that day is a tribute to natural selection's resourcefulness. The biochemical essence of the panic probably goes back to our reptilian days. Yet it was triggered not by its primordial trigger—threat to life and limb—but rather by a threat to status, a concern more characteristic of our primate days. What's more, the threat wasn't of the physical sort common among our primate relatives. Instead it came as an abstraction: words, sentences—symbols whose comprehension depended on brain tissue acquired only within the past few million years. Thus does evolution take ancient raw materials and continually adapt them to current needs.

Presumably Darwin did not pause to reflect on the natural beauty of his panic. He sent Wallace's paper to Lyell—whose opinion of it Wallace had asked Darwin to solicit—and sought advice. Actually, "sought" is a little strong; I'm reading between the lines. Darwin proposed a pious course of action and left it for Lyell to propose a less pious one. "Please return me the MS., which he does not say he wishes me to publish, but I shall, of course, at once write and offer to send to any journal. So all my originality, whatever it may amount to, will be smashed, though my book, if it will ever have any value, will not be deteriorated; as all the labour consists in the application of the theory."[54]

Lyell's reply—which, oddly, has not survived, even though Darwin saved correspondence religiously—seems to have succeeded in checking Darwin's piety. Darwin wrote back: "There is nothing in Wallace's sketch which is not written out much fuller in my sketch, copied out in 1844, and read by Hooker some dozen years ago. About a year ago I sent a short sketch, of which I have a copy, of my views . . . to Asa Gray, so that I could most truly say and prove that I take nothing from Wallace."

Then Darwin gets into an epic wrestling match with his conscience, in full view of Lyell. At the risk of sounding cynical, I include in brackets the letter's subtext, as I interpret it: "I should be extremely glad now to publish a sketch of my general views in about a dozen pages or so; but I cannot persuade myself that I can do so honourably. [Maybe you can persuade me.] Wallace says nothing about publication, and I enclose his letter. But as I had not intended to publish any sketch, can I do so honourably, because Wallace has sent me an outline of his doctrine? [Say yes. Say yes.] . . . Do you not think his having sent me this sketch ties my hands? [Say no. Say no.] . . . I would send Wallace a copy of my letter to Asa Gray, to show him that I had not stolen his doctrine. But I cannot tell whether to publish now would not be base and paltry. [Say nonbase and nonpaltry.]" In a postscript added the next day, Darwin washed his hands of the affair, appointing Lyell arbitrator: "I have always thought you would make a first-rate Lord Chancellor; and I now appeal to you as a Lord Chancellor."[55]

Darwin's anguish was deepened by events at home. His daughter Etty had diphtheria, and his mentally retarded baby, Charles Waring, had just contracted scarlet fever, from which he would soon die.

Lyell consulted with Hooker, whom Darwin had also alerted to the crisis, and the two men decided to treat Darwin's and Wallace's theories as equals. They would introduce Wallace's paper at the next meeting of the Linnean Society, along with the sketch Darwin had sent to Asa Gray and parts of the 1844 draft he had given Emma, and all of this would then be published together. (Darwin had sent

Gray the 1,200-word sketch only a few months after telling Wallace it would be "*impossible*" to sketch the theory in a letter. Whether he wanted to produce unimpeachable evidence of his priority, after sensing Wallace gaining on him, will never be known.) Since Wallace was then in the Malay Archipelago, and the next meeting of the society was imminent, Lyell and Hooker decided to proceed without consulting him. Darwin let them.

When Wallace learned what had happened, he was in a position much like Darwin's during the *Beagle*'s voyage, when the thrilling word of Sedgwick's endorsement arrived. Wallace was a young naturalist, eager to make a name for himself, isolated from professional feedback, still not sure if he had much to give science. Suddenly he found that his work was being read by great men before a great scientific society. He wrote proudly to his mother that, "I sent Mr. Darwin an essay on a subject on which he is now writing a great work. He showed it to Dr. Hooker and Sir Charles Lyell, who thought so highly of it that they immediately read it before the Linnean Society. This assures me the acquaintance and assistance of these eminent men on my return home."[56]

DARWIN'S BIGGEST MORAL BLEMISH?

This ranks as one of the most poignant passages in the history of science. Wallace had just been taken to the cleaners. His name, though given equal billing with Darwin's, was now sure to be eclipsed by it. After all, it wasn't news that some young upstart had declared himself an evolutionist and proposed an evolutionary mechanism; it *was* news that the well-known and respected Charles Darwin had done so. And any lingering doubt about whose name should be attached to the theory would be erased by Darwin's book, which he would now finally produce with due speed. Lest the relative status of the two men escape anyone's attention, Hooker and Lyell, in introducing the papers to the Linnean Society, had noted that, "while the scientific world is waiting for the appearance of Mr. Darwin's complete work, some of the leading results of his labours, as well as those of his able correspondent, should together be laid before the public."[57] "Able correspondent" isn't a phrase likely to wind up at the top of a marquee.

Now, it may be that Darwin's having put the pieces together so many years before Wallace makes Wallace's eventual obscurity just.[58] But the fact is that as of June 1858, Wallace, unlike Darwin, had written a paper on natural selection that he was ready to publish, even if he didn't ask Darwin to publish it. If Wallace had sent his paper to a journal instead of to Darwin—indeed, if he had sent it almost *anywhere* instead of to Darwin—he might be remembered today as the first man to posit the theory of evolution by natural selection.[59] Darwin's great book, technically speaking, would have been an extension and popularization of another scientist's idea. Whose name the theory would then have carried will forever be an open question.

However just Darwin's worldwide fame, it seems hard to argue that when given the toughest moral test of his life, he passed with flying colors. Consider the options confronting him, Lyell, and Hooker. They could publish only Wallace's version of the theory. They could write Wallace and offer to thus publish his version, as Darwin had originally suggested—without, perhaps, even mentioning Darwin's version. They could write Wallace and explain the situation, suggesting joint publication. Or they could do what they did. Since Wallace might, for all they knew, have resisted joint publication, the option they pursued was the only one which ensured that natural selection would go down in history as Darwin's theory. And that option entailed publishing Wallace's paper without his expressed permission—an act whose propriety someone with Darwin's king-size scruples might normally question.

Remarkably, observers have time and again depicted this ploy as some sort of testament to human morality. Julian Huxley, Thomas Huxley's grandson, called the outcome "a monument to the natural generosity of both the great biologists."[60] Loren Eiseley called it an example of "that mutual nobility of behavior so justly celebrated in the annals of science."[61] They're both half right. Wallace, ever gracious, would long insist—correctly, but still generously and nobly—that Darwin's length and depth of thought about evolution had earned him the title of premier evolutionist. Wallace even titled a book of his *Darwinism.*

Wallace defended the theory of natural selection for the rest of his life, but he crucially narrowed its scope. He began to doubt that the theory could account for the full powers of the human mind; people seemed smarter than they really *had* to be to survive. He concluded that although man's body was built by natural selection, his mental capacities were divinely implanted. It may be too cynical (even by Darwinian standards) to suggest that this revision would have been less likely had the theory of natural selection been called "Wallacism." At any rate, the man whose name *was* synonymous with the theory mourned the weakening of Wallace's faith. "I hope you have not murdered too completely your own and my child," Darwin wrote to him.[62] (This from the man who, after mentioning Wallace in the introduction of the *Origin*, referred to natural selection in subsequent chapters as "my theory.")

The common idea that Darwin behaved like a perfect gentleman throughout the Wallace episode rests partly on the myth that he had some option other than those outlined above—that he could have rushed his theory to press without so much as mentioning Wallace. But unless Wallace was even more saintly than he seems to have been, this would have brought a scandal that left Darwin's name tainted, even to the point of endangering its connection to his theory. In other words: this option was not an option. The biographer who admiringly observes that Darwin "hated losing his priority, but he hated even more the chance of being suspected of ungentlemanly or nonsporting conduct"[63] is creating a distinction where none existed; to have been thought unsporting would have threatened his priority. When Darwin wrote to Lyell, on the day he received Wallace's sketch, "I would far rather burn my whole book, than that he or any other man should think that I had behaved in a paltry spirit," he wasn't being conscientious so much as savvy.[64] Or rather: he was being conscientious, which, especially in his social environment, was the same as being savvy. Savviness is the function of the conscience.

The other source of retrospective naïveté about Darwin's behavior is his brilliant decision to place the matter in Lyell's and Hooker's hands. "In despair, he abdicated," as one biographer obligingly puts it.[65] Darwin would forever use this "abdication" as moral camouflage. After Wallace signaled his approval of the affair, Darwin wrote to him: "Though I had absolutely nothing whatever to do in leading

Lyell and Hooker to what they thought a fair course of action, yet I naturally could not but feel anxious to hear what your impression would be. . . ."[66] Well, if he wasn't sure Wallace would approve, why didn't he bother to check? Couldn't Darwin, having gone two decades without publishing his theory, have waited a few months longer? Wallace had asked that his paper be sent to Lyell, but he hadn't asked that Lyell determine its fate.

For Darwin to say he exerted no influence "whatever" on Hooker and Lyell strains the facts and, anyway, is irrelevant; these were two of his closest friends. Surely Darwin wouldn't have felt he could appoint his brother Erasmus as a disinterested judge. Yet, there is every reason to believe that evolution, in embedding friendship in the human species, has resourcefully used many of the impulses of affection, devotion, and loyalty that it first used to bind kin.

Darwin didn't know this, of course, but surely he knew that friends tend toward partiality—that the whole idea of a friend is someone who at least partly shares your self-serving biases. For him to depict Lyell as impartial—"a Lord Chancellor"—is remarkable. And it only appears more so in light of Darwin's later appeals to their friendship, when he virtually asks Lyell to endorse the theory of natural selection as a personal favor.

POSTGAME ANALYSIS

Enough moral outrage. Who am I to judge? I've done things worse than this, Darwin's biggest single crime. In fact, my ability to muster all this righteous indignation, and assume a stance of moral superiority, is a tribute to the selective blindness with which evolution has endowed us all. Now, I'll try to transcend biology and summon enough detachment for a brisk appraisal of the salient Darwinian features of the Wallace episode.

Note, first of all, the exquisite pliability of Darwin's values. As a rule he was gravely disdainful of academic territoriality; for scientists to guard against rivals who might steal their thunder was, he believed, "unworthy of searchers after truth."[67] And though he was too perceptive and honest to deny that fame had a tempting effect on him, he generally held the effect to be minor. He claimed that even without it he would work just as hard on his species book.[68] Yet when his turf was threatened, he took steps to defend it—which included producing the *Origin* at a rather stepped-up pace once there was doubt as to whose name would become synonymous with evolutionism. Darwin saw the contradiction. Weeks after the Wallace episode, he wrote to Hooker that, as far as priority goes, he had always "fancied that I had a grand enough soul not to care; but I found myself mistaken and punished."[69]

As the crisis receded into the past, though, Darwin's old pieties resurfaced. He claimed in his autobiography that he "cared very little whether men attributed most originality to me or Wallace."[70] Anyone who has read Darwin's distraught letters to Lyell and Hooker will have to marvel at the power of Darwin's self-deception.

The Wallace episode highlights a basic division within the conscience, the line between kin selection and reciprocal altruism. When we feel guilty about having harmed or cheated a sibling, it is, generally, because natural selection "wants" us to be nice to siblings, since they share so many of our genes. When we feel guilty about having harmed or cheated a friend, or a casual acquaintance, it is because natural selection "wants" us to *look like* we're being nice; the perception of altruism, not the altruism itself, is what will bring the reciprocation. So the aim of the conscience, in dealings with nonkin, is to cultivate a reputation for generosity and decency, whatever the reality.[71] Of course, gaining and holding this reputation will often entail actual generosity and decency. But sometimes it won't.

In this light we see Darwin's conscience working in top form. It made him generally reliable in his bestowal of generosity and decency—in a social environment so intimate that actual generosity and decency were essential to maintaining a good moral reputation. But his goodness turned out not to be *absolutely* constant. His vaunted conscience, seemingly a bulwark against all corruption, was discerning enough to weaken a trifle just when his lifelong quest for status most needed a slight moral lapse. This brief dimming of the lights allowed Darwin to subtly, even unconsciously, pull strings, employing his ample social connections to the detriment of a young and powerless rival.

Some Darwinians have suggested that the conscience can be viewed as the administrator of a savings account in which moral reputation is stored.[72] For decades Darwin painstakingly amassed capital, vast and conspicuous evidence of his scruples; the Wallace episode was a time to risk some of it. Even if he lost a little—even if the affair produced a few suspicious whispers about the propriety of publishing Wallace's paper without his permission—this would still be a risk worth taking, in terms of the ultimate elevation of Darwin's status. Making such judgments about resource allocation is what the human conscience is designed to do, and during the Wallace episode Darwin's did it well.

As it happened, none of Darwin's capital was lost. He came out smelling like a rose. Before the Linnean Society, Hooker and Lyell described what had happened after Darwin received Wallace's paper. "So highly did Mr. Darwin appreciate the value of the views therein set forth, that he proposed, in a letter to Sir Charles Lyell, to obtain Mr. Wallace's consent to allow the Essay to be published as soon as possible. Of this step we highly approved, provided Mr. Darwin did not withhold from the public, as he was strongly inclined to do (in favour of Mr. Wallace), the memoir he had himself written on the same subject, and which, as before stated, one of us had perused in 1844, and the contents of which we had both of us been privy to for many years. . . ."[73]

More than a century later, this sanitized version of events was still the standard version—an utterly scrupulous Darwin virtually coerced into letting his name appear alongside Wallace's. Darwin, one biographer wrote, "seems hardly to have been a free agent in the face of Lyell's and Hooker's pressure for publication."[74]

There is no basis for concluding that Darwin consciously orchestrated his eclipse of Wallace. Consider the judicious appointment of Lyell as "Lord Chancellor." The natural impulse, in times of crisis, to seek the guidance of friends feels perfectly innocent. We don't necessarily think, "I'll call a friend, rather than some stranger, because a friend will share my warped ideas about what I deserve and what my rivals deserve." So too with Darwin's pose of moral anguish: it worked because he didn't know it was a pose—because, in other words, it *wasn't* a pose; he actually felt the anguish.

And not for the first time. Darwin's guilt about asserting priority—pulling rank on Wallace in a quest for still higher rank—

was just the latest in a lifelong series of comparable pangs. (Recall John Bowlby's diagnosis: Darwin suffered "self-contempt for being vain." "Time and again throughout his life his desire for attention and fame is coupled with the deep sense of shame he feels for harbouring such motives.")[75] Indeed, it was the proven authenticity of Darwin's anguish which helped convince Hooker and Lyell that Darwin "strongly" resisted glory and thus helped them convince the world of it. All the moral capital Darwin built up over the years had come at a large psychological cost, but in the end the investment paid dividends.

None of this is meant to imply that Darwin behaved in perfectly adaptive fashion, constantly attuned to the task of genetic proliferation, with every bit of his ample striving and suffering warranted by that end. Given the difference between nineteenth-century England and the environment(s) of our evolution, this sort of functional perfection is the last thing one should expect. Indeed, as we suggested several chapters ago, Darwin's moral sentiments were manifestly more acute than self-interest dictated; he had plenty of capital in his moral savings account without losing sleep over unanswered letters, without crusading on behalf of dead sheep. The claim here is simply that lots of odd and much-discussed things about Darwin's mind and character can make a basic kind of sense when viewed through the lens of evolutionary psychology.

Indeed, his whole career assumes a certain coherence. It looks less like an erratic quest, often stymied by self-doubt and undue deference, and more like a relentless ascent, deftly cloaked in scruples and humility. Beneath Darwin's pangs of conscience lay moral positioning. Beneath his reverence for men of accomplishment lay social climbing. Beneath his painfully recurring self-doubts lay a fevered defense against social assault. Beneath his sympathy toward friends lay savvy political alliance. What an animal!

Part Four: **MORALS OF THE STORY**

Chapter 15: **DARWINIAN (AND FREUDIAN) CYNICISM**

*The possibility of the brain having whole train of thoughts, feel-
ing & perception separate, from the ordinary state of mind, is
probably analogous to the double individuality implied by habit,
when one acts unconsciously with respect to more energetic
self....*

—M Notebook (1938)[1]

The picture of human nature painted thus far isn't altogether
flattering.

We spend our lives desperately seeking status; we are addicted
to social esteem in a fairly literal sense, dependent on the neurotransmitters we get upon impressing people. Many of us claim to be self-
sufficient, to have a moral gyroscope, to hold fast to our values, come
what may. But people truly oblivious to peer approval get labeled
sociopaths. And the epithets reserved for people at the other end
of the spectrum, people who seek esteem most ardently—"self-
promoter," "social climber"—are only signs of our constitutional
blindness. We are all self-promoters and social climbers. The people
known as such are either so effective as to arouse envy or so graceless
as to make their effort obvious, or both.

Our generosity and affection have a narrow underlying purpose.
They're aimed either at kin, who share our genes, at nonkin of the
opposite sex who can help package our genes for shipment to the

next generation, or at nonkin of either sex who seem likely to return the favor. What's more, the favor often entails dishonesty or malice; we do our friends the favor of overlooking their flaws, and seeing (if not magnifying) the flaws of their enemies. Affection is a tool of hostility. We form bonds to deepen fissures.[2]

In our friendships, as in other things, we're deeply inegalitarian. We value especially the affection of high-status people, and are willing to pay more for it—to expect less of them, to judge them leniently. Fondness for a friend may wane if his or her status slips, or if it simply fails to rise as much as our own. We may, to facilitate the cooling of relations, justify it. "He and I don't have as much in common as we used to." Like high status, for example.

It is safe to call this a cynical view of behavior. So what's new? There's nothing revolutionary about cynicism. Indeed, some would call it the story of our time—the by now august successor to Victorian earnestness.[3]

The shift from nineteenth-century earnestness to twentieth-century cynicism has been traced, in part, to Sigmund Freud. Like the new Darwinism, Freudian thought finds sly unconscious aims in our most innocent acts. And like the new Darwinism, it sees an animal essence at the core of the unconscious.

Nor are those the only things Freudian and Darwinian thought have in common. For all the criticism it has drawn in recent decades, Freudianism remains the most influential behavioral paradigm—academically, morally, spiritually—of our time. And to this position the new Darwinian paradigm aspires.

On grounds of this rivalry alone, disentangling Freudian psychology and evolutionary psychology would be worthwhile. But there are other grounds, too, perhaps more important: the forms of cynicism ultimately entailed by the two schools are different, and different in ways that matter.

Both Darwinian and Freudian cynicism carry less bitterness than garden-variety cynicism. Because their suspicion of a person's motives is in large part a suspicion of *unconscious* motives, they view the person—the conscious person, at least—as a kind of unwitting accomplice. Indeed, to the extent that pain is the price paid for the internal subterfuge, the person may be worthy of compassion as well as suspicion. Everyone comes out looking like a victim. It is in describing how and why the victimization takes place that the two schools of thought diverge.

Freud thought of himself as a Darwinian. He tried to look at the human mind as a product of evolution, a fact that—by itself, at least—should forever endear him to evolutionary psychologists. Anyone who sees humans as animals, driven by sexual and other coarse impulses, can't be all bad. But Freud misunderstood evolution in basic and elementary ways.[4] He put much emphasis, for example, on the Lamarckian idea that traits acquired through experience get passed on biologically. That some of these misconceptions were common in his day—and that some were held by Darwin, or at least encouraged by his equivocations—may be a good excuse. But the fact remains that they led Freud to say many things that sound nonsensical to today's Darwinians.

Why would people have a death instinct ("thanatos")? Why would girls want male genitals ("penis envy")? Why would boys want to have sex with their mothers and kill their fathers (the "Oedipus complex")? Imagine genes that specifically encourage any of these impulses, and you're imagining genes that aren't exactly destined to spread through a hunter-gatherer population overnight.

There's no denying Freud's sharp eye for psychic tension. Something resembling the Oedipal conflict between father and son may well exist. But what are its real roots? Martin Daly and Margo Wilson have argued that here Freud conflated several distinct Darwinian dynamics, some of them grounded ultimately in the parent-offspring conflict described by Robert Trivers.[5] For example, when boys reach adolescence, they may, especially in a polygynous society (such as our ancestral environment) find themselves competing with their fathers for the same women. But among those women is *not* the boy's mother; incest often produces deficient offspring, and it's not in the son's genetic interest to have his mother assume the risks and burdens of pregnancy to create a reproductively worthless sibling. (Hence the dearth of boys who try to seduce their mothers.) At a younger age the boy (or for that matter a girl) may have a paternal conflict that *is* over the mother—but not with sex as its goal. Rather, the son and father are fighting over the mother's valuable time and attention. If the struggle has sexual overtones at all, they are only that the father's genetic interest may call for impregnating the mother, while the son's would call for delaying the arrival of a sibling (by, for example, continued breast-feeding, which forestalls ovulation).

These sorts of Darwinian theories are often speculative and, at this early stage in the growth of evolutionary psychology, meagerly tested. But unlike Freud's theories, they are tethered to something firm: an understanding of the process that designed the human brain. Evolutionary psychology has embarked on a course whose broad contours are well-marked and which should, as it proceeds, find continual correction in the dialectic of science.

DARWIN'S KNOBS AND TUNINGS

The path to progress begins by specifying the knobs of human nature—the things that Charles Darwin, for example, shared with all humanity. He cared for his kin, within limits. He sought status. He sought sex. He tried to impress peers and to please them. He tried to be seen as good. He formed alliances and nurtured them. He tried to neutralize rivals. He deceived himself when the preceding goals so dictated. And he felt all the feelings—love, lust, compassion, reverence, ambition, anger, fear, pangs of conscience, of guilt, of obligation, of shame, and so on—that push people toward these goals.

Having located—in Darwin or anyone else—the basic knobs of human nature, the Darwinian next asks: What is distinctive about the tuning of the knobs? Darwin had an unusually active conscience. He nurtured his alliances with unusual care. He worried unusually about the opinions of others. And so on.

Where did these distinctive tunings come from? Good question. Almost no developmental psychologists have taken up the tools of the new paradigm, so there's a shortage of answers. But the route to the answers, at least broadly speaking, is clear. The young, plastic mind is shaped by cues that, in the environment of our evolution, suggested what behavioral strategies were most likely to get genes spread. The cues presumably tend to mirror two things: the sort of social environment you find yourself in; and the sorts of assets and liabilities you bring into that environment.

Some cues are mediated by kin. Freud was right to sense that relatives—parents, in particular—have a lot to say about the shape of the emerging psyche. Freud was also right to sense that parents are not wholly benign, and that deep conflicts between parents and offspring are possible. Trivers's theory of parent-offspring conflict holds that some of the psychic fine-tuning may be for the genetic benefit not of the tunee (the child), but of the tuner (the parent). Disentangling the two types of kin influence—to teach and to exploit—is never easy. And in Darwin's case it's especially hard, for some of his trademark traits—great respect for authority, weighty scruples—are, in addition to being useful in the wider social world, conducive to sacrifice for the family.

If behavioral scientists are to use the new Darwinism to trace mental and emotional development, they will have to abandon one assumption often implicit in the thought of Freud and psychiatrists in general (and, for that matter, just about everyone else): that pain is a symptom of something abnormal, unnatural—a sign that things have gone awry. As the evolutionary psychiatrist Randolph Nesse has stressed, pain is part of natural selection's design (which isn't, of course, to say that it's good).[6] Vast quantities of pain were generated by traits that helped make Darwin an effective animal: his "overactive" conscience, his relentless self-criticism, his "craving for reassurance," his "exaggerated" respect for authority. If indeed Darwin's father, as alleged, encouraged some of this pain, it may be a mistake to ask what demons drove him to do it (unless, perhaps, you then answer: "Genes that were working like a Swiss watch"). What's more, it may be a mistake to assume that the young Darwin didn't himself, at some level, encourage this painful influence; people may well be designed to absorb painful guidance that conduces to genetic proliferation (or would have in the ancestral environment). Many things that look like parental cruelty may *not* be an example of Trivers's parent-offspring conflict.

One condition that may resist comprehension so long as psychologists deem it unnatural is something Darwin suffered from: insecurity. Perhaps over the eons it has made sense for people who couldn't ascend the social hierarchy through classic means (brute force, good looks, charisma) to focus on other routes. One route would be a redoubled commitment to reciprocal altruism—that is, a sensitive, even painfully sensitive, conscience, and a chronic fear of being unliked. The stereotypes of the arrogant, inconsiderate jock and the ingratiating, deferential wimp are no doubt overdrawn, but they may reflect a statistically valid correlation, and they seem to make Darwinian sense. At any rate, they seem to capture Darwin's experience well enough. He was a good-sized boy but awkward and introverted, and at grade school, he wrote, "I could not get up my courage to fight."[7] Though his reserve was misinterpreted by some children as disdain, he was also known as kind—"pleased to do any little acts to gratify his fellows," one schoolmate recalled.[8] Captain FitzRoy would later marvel at how Darwin "makes everyone his friend."[9]

Sharp intellectual self-scrutiny, likewise, might grow out of early social frustration. Children to whom status doesn't come naturally may work harder to become rich sources of information, especially if they seem to have a natural facility with it. Darwin turned his fits of intellectual self-doubt into a series of polished scientific works that both raised his status and made him a valued reciprocal altruist.

If these speculations hold water, then Darwin's two basic kinds of self-doubt—moral and intellectual—are two sides of the same coin, both of them manifestations of social insecurity, and both of them designed as a way to make him a prized social asset when other ways seemed to be failing. Darwin's "acute sensitiveness to praise and blame," as Thomas Huxley put it, can account for his fastidiousness in both realms, and may be rooted in a single principle of mental development.[10] And Darwin's father may have done much—with Darwin's implied consent—to nourish that acute sensitiveness.

When we call people "insecure" we generally mean that they worry a lot: they worry that people don't like them; they worry that they'll lose what friends they have; they worry that they've offended people; they worry that they've given someone bad information. It is common to casually trace insecurity to childhood: rejection on the grade-school playground; romantic failures in adolescence; an un-stable home; the death of a family member; moving around too often to make lasting friends, or whatever. There is a vague and usually unspoken assumption that various kinds of childhood failure or turbulence will lead to adult insecurity.

One can think up reasons (such as those I've just tossed out) why natural selection might have forged some of these links between early experience and later personality. (The early death of Darwin's mother is fertile ground for speculation; in the ancestral environment, complacency was a luxury that a motherless child could not afford.) One can also find, in the data of social psychology, at least loose support for such correlations. Clarity will come when these two sides of the dialectic get in touch with one another: when psychologists start thinking precisely about what kinds of developmental theories make Darwinian sense and then designing research to test those theories.

It is by the same process that we'll start understanding how various other tendencies get forged: sexual reserve or promiscuity, social tolerance and intolerance, high or low self-esteem, cruelty and gentleness, and so on. To the extent that these things are indeed consistently linked to commonly cited causes—the degree and nature of parental love, the number of parents in the household, early romantic encounters, dynamics among siblings, friends, enemies—it is probably because such linkage made evolutionary sense. If psychologists want to understand the processes that shape the human mind, they must understand the process that shaped the human species.[11] Once they do, progress is likely. And unequivocal progress—growing, objective corroboration of ever-more-precise theories—would distinguish the Darwinism of the twenty-first century from the Freudianism of the twentieth.

When the topic turns to the unconscious mind, differences between Freudian and Darwinian thought persist; and again, some of the difference revolves around the function of pain. Recall Darwin's "golden rule": to immediately write down any observation that seemed inconsistent with his theories—"for I had found by experience that such facts and thoughts were far more apt to escape from the memory than favourable ones."[12] Freud cited this remark as evidence of the Freudian tendency "to ward off from memory that which is unpleasant."[13] This tendency was for Freud a broad and general one, found among the mentally healthy and ill alike, and central to the dynamics of the unconscious mind. But there is one problem with this supposed generality: sometimes painful memories are the very *hardest* to forget. Indeed, Freud acknowledged, only a few sentences after citing Darwin's golden rule, that people had mentioned this to him, stressing in particular the painfully persistent "recollection of grievances or humiliations."

Did this mean the tendency to forget unpleasant things wasn't general after all? No. Freud opted for another explanation: it was just that sometimes the tendency to discard painful memories is successful and sometimes it isn't; the mind is "an arena, a sort of tumbling-ground," where opposing tendencies collide, and it isn't easy to say which tendency will win.[14]

Evolutionary psychologists can handle this issue more deftly, because, in contrast to Freud, they don't have such a simple, schematic view of the human mind. They believe the brain was jerry-built over the eons to accomplish a host of different tasks. Having made no attempt to lump the memory of grievances, humiliations, and inconvenient facts under the same rubric, Darwinians don't have to hand out special exemptions to the cases that don't fit. Faced with three questions about remembering and forgetting—(1) why we forget facts inconsistent with our theories; (2) why we remember grievances; (3) why we remember humiliations—they can relax and come up with a different explanation for each one.

We've already touched on the three likely explanations. Forgetting inconvenient facts makes it easier to argue with force and conviction, and arguments often had genetic stakes in the environment of our evolution. Remembering grievances may bolster our haggling in a different way, making us remind people of reparations we're owed; also, a well-preserved grievance may ensure the punishment of our exploiters. As for the memory of humiliations, their uncomfortable persistence dissuades us from repeating behaviors that can lower social status; and, if the humiliations are of sufficient magnitude, their memory may adaptively lower self-esteem (or, at least, lower self-esteem in a way that would have been adaptive in the environment of our evolution).

Thus, Freud's model of the human mind may have been—believe

it or not—insufficiently labyrinthine. The mind has more dark corners than he imagined, and plays more little tricks on us.

THE BEST OF FREUD

What is best in Freud is his sensing the paradox of being a highly social animal: being at our core libidinous, rapacious, and generally selfish, yet having to live civilly with other human beings—having to reach our animal goals via a tortuous path of cooperation, compromise, and restraint. From this insight flows Freud's most basic idea about the mind: it is a place of conflict between animal impulses and social reality.

One biological view of this sort of conflict has come from Paul D. MacLean. He calls the human brain a "triune" brain whose three basic parts recapitulate our evolution: a reptilian core (the seat of our basic drives), surrounded by a "paleomammalian" brain (which endowed our ancestors with, among other things, affection for offspring), surrounded in turn by a "neomammalian" brain. The voluminous neomammalian brain brought abstract reasoning, language, and, perhaps, (selective) affection for people outside the family. It is, MacLean writes, "the handmaiden for rationalizing, justifying and giving verbal expression to the protoreptilian and limbic [paleomammalian] parts of our brains. . . ."[15] Like many neat models, this one may be misleadingly simple; but it nicely captures a (perhaps *the*) critical feature of our evolutionary trajectory: from solitary to social, with the pursuit of food and sex becoming increasingly subtle and elaborate endeavors.

Freud's "id"—the beast in the basement—presumably grows out of the reptilian brain, a product of presocial evolutionary history. The "superego"—loosely speaking, the conscience—is a more recent invention. It is the source of the various kinds of inhibition and guilt designed to restrain the id in a genetically profitable manner; the superego prevents us, say, from harming siblings, or from neglecting our friends. The "ego" is the part in the middle. Its ultimate, if unconscious, goals are those of the id, yet it pursues them with long-term calculation, mindful of the superego's cautions and reprimands.

Congruence between the Freudian and Darwinian views of psychic conflict has been stressed by Randolph Nesse and the psychiatrist Alan T. Lloyd. They see the conflict as a clash among competing advocacy groups, designed by evolution to yield sound guidance, much as the tension among branches of government is designed to yield good governance. The basic conflict—the basic discourse—is "between selfish and altruistic motivation, between pleasure-seeking and normative behavior, and between individual and group interests. The functions of the id match the first half of each of these pairs, while the functions of the ego/superego match the second half." And the basic truth behind the second half of the discourse is the "delayed nature of benefits from social relationships."[16]

In describing this tension between short-term and long-term selfishness, Darwinians have sometimes used the image of "repression." The psychoanalyst Malcolm Slavin suggests that selfish motives may be repressed by children as a way to stay in the good graces of parents—and retrieved moments later, when the need to please passes.[17] Others have stressed the repression of selfish impulses toward friends. We may even repress the memory of a friend's transgressions—an especially wise trick if the friend is of high status or otherwise valuable.[18] The memory could then resurface should the friend see his status plummet or for some other reason merit a more frank appraisal. And, of course, the arena of sex is rife with occasions for tactical repression. Surely a man can better convince a woman of his future devotion if he isn't vividly imagining sexual intercourse with her. That impulse can blossom later, once the ground has been prepared.

As Nesse and Lloyd have noted, repression is just one of the many "ego defenses" that have become part of Freudian theory (largely via Freud's daughter Anna, who wrote the book on ego defenses). And, they add, several other ego defenses are similarly intelligible in Darwinian terms. For example, "identification" and "introjection"—absorbing the values and traits of others, including powerful others—may be a way of cozying up to a high-status person who "distributes status and rewards to those who support his beliefs."[19] And "rationalization," the concoction of pseudoexplanations that conceal our true motives—well, need I elaborate?

All told, Freud's scorecard is not bad: he (and his followers) have identified lots of mental dynamics that may have deep evolutionary roots. He rightly saw the mind as a place of turbulence, much of it subterranean. And, in a general way, he saw the source of the turbulence: an animal of ultimately complete ruthlessness is born into a complex and inescapable social web.

But when he got less general than this, Freud's diagnosis was sometimes misleading. He often depicted the tension at the center of human life as essentially between not self and society but self and civilization. In *Civilization and Its Discontents,* he described the paradox this way: people are pushed together with other people, told to curb their sexual impulses and enter "aim-inhibited relationships of love," and told not just to get along with their neighbors cooperatively but to "love thy neighbor as thyself." Yet, Freud observes, humans are simply not gentle creatures: "[T]heir neighbour is for them not only a potential helper . . . but also someone who tempts them to satisfy their aggressiveness on him, to exploit his capacity for work without compensation, to use him sexually without his consent, to seize his possessions, to humiliate him, to cause him pain, to torture and to kill him. *Homo homini lupus.* [Man is a wolf to man.]" No wonder people are so miserable. "In fact, primitive man was better off in knowing no restrictions of instinct."[20]

This last sentence contains a myth whose correction underlies much of evolutionary psychology. It has been a long, long time since any of our ancestors enjoyed "no restrictions" on these "instincts." Even chimpanzees must weigh their predatory impulses against the fact that another chimp can be "a potential helper," as Freud put it, and thus may be profitably treated with restraint. And male chimpanzees (and bonobos) find their sexual impulses frustrated by females that demand food and other favors in exchange for sex. In our own lineage, as growing male parental investment expanded those demands, males found themselves facing extensive "restrictions" on sexual impulses well before modern cultural norms made life even more frustrating.

The point is that repression and the unconscious mind are the products of millions of years of evolution and were well developed long before civilization further complicated mental life. The new paradigm allows us to think clearly about how these things were designed over those millions of years. The theories of kin selection, parent-offspring conflict, parental investment, reciprocal altruism, and status hierarchy tell us what kinds of self-deception are and aren't likely to be favored by evolution. If present-day Freudians start taking these hints and recast their ideas accordingly, maybe they can save Freud's name from the eclipse it will probably suffer if the task is left to Darwinians.

THE POSTMODERN MIND

All told, the Darwinian notion of the unconscious is more radical than the Freudian one. The sources of self-deception are more numerous, diverse, and deeply rooted, and the line between conscious and unconscious is less clear. Freud described Freudianism as an attempt to "prove to the 'ego' of each one of us that he is not even master in his own house, but that he must remain content with the veriest scraps of information about what is going on unconsciously in his own mind."[21] By Darwinian lights, this wording almost gives too much credit to the "self." It seems to suggest an otherwise clear-seeing mental entity getting deluded in various ways. To an evolutionary psychologist, the delusion seems so pervasive that the usefulness of thinking about any distinct core of honesty falls into doubt.

Indeed, the commonsense way of thinking about the relation between our thoughts and feelings, on the one hand, and our pursuit of goals, on the other, is not just wrong, but backward. We tend to think of ourselves as making judgments and then behaving accordingly: "we" decide who is nice and then befriend them; "we" decide who is upstanding and applaud them; "we" figure out who is wrong and oppose them; "we" figure out what is true and abide by it. To this picture Freud would add that often we have goals we aren't aware of, goals that may get pursued in oblique, even counterproductive, ways—and that our perception of the world may get warped in the process.

But if evolutionary psychology is on track, the whole picture needs to be turned inside out. We believe the things—about morality, personal worth, even objective truth—that lead to behaviors that get our genes into the next generation. (Or at least we believe the kinds of things that, in the environment of our evolution, would have been likely to get our genes into the next generation.) It is the behavioral goals—status, sex, effective coalition, parental investment, and so on—that remain steadfast while our view of reality adjusts to accommodate this constancy. What is in our genes' interests is what seems "right"—morally right, objectively right, whatever sort of rightness is in order.

In short: if Freud stressed people's difficulty in seeing the truth about themselves, the new Darwinians stress the difficulty of seeing truth, period. Indeed, Darwinism comes close to calling into question the very meaning of the word *truth*. For the social discourses that supposedly lead to truth—moral discourse, political discourse, even, sometimes, academic discourse—are, by Darwinian lights, raw power struggles. A winner will emerge, but there's often no reason to expect that winner to be truth. A cynicism deeper than Freudian cynicism may have once seemed hard to imagine, but here it is.

This Darwinian brand of cynicism doesn't exactly fill a gaping cultural void. Already, various avant-garde academics—"deconstructionist" literary theorists and anthropologists, adherents of "critical legal studies"—are viewing human communication as "discourses of power." Already many people believe what the new Darwinism underscores: that in human affairs, all (or at least much) is artifice, a self-serving manipulation of image. And already this belief helps nourish a central strand of the postmodern condition: a powerful inability to take things seriously.[22]

Ironic self-consciousness is the order of the day. Cutting-edge talk shows are massively self-referential, with jokes about cue cards written on cue cards, camera shots of cameras, and a general tendency for the format to undermine itself. Architecture is now *about* architecture, as architects playfully and, sometimes, patronizingly meld motifs of different ages into structures that invite us to laugh along with them. What is to be avoided at all costs in the postmodern age is earnestness, which betrays an embarrassing naïveté.

Whereas modern cynicism brought despair about the ability of the human species to realize laudable ideals, postmodern cynicism doesn't—not because it's optimistic, but because it can't take ideals seriously in the first place. The prevailing attitude is absurdism. A postmodern magazine may be irreverent, but not bitterly irreverent, for it's not purposefully irreverent; its aim is indiscriminate, because everyone is equally ridiculous. And anyway, there's no moral basis for passing judgment. Just sit back and enjoy the show.

It is conceivable that the postmodern attitude has already drawn some strength from the new Darwinian paradigm. Sociobiology, however astringent its reception in academia, began seeping into popular culture two decades ago. In any event, the future progress of Darwinism may strengthen the postmodern mood. Surely, within academia, deconstructionists and critical legal scholars can find much to like in the new paradigm. And surely, outside of academia, one reasonable reaction to evolutionary psychology is a self-consciousness so acute, and a cynicism so deep, that ironic detachment from the whole human enterprise may provide the only relief.

Thus the difficult question of whether the human animal can be a moral animal—the question that modern cynicism tends to greet with despair—may seem increasingly quaint. The question may be whether, after the new Darwinism takes root, the word *moral* can be anything but a joke.

Chapter 16: **EVOLUTIONARY ETHICS**

Our descent, then, is the origin of our evil passions!!—The devil under form of Baboon is our grandfather.

—M Notebook (1838)

It is other question what it is desirable to be taught,—all are agreed general utility.

—"Old and Useless Notes" (undated)[1]

In 1871, twelve years after *The Origin of Species* appeared, Darwin published *The Descent of Man*, in which he set out his theory of the "moral sentiments." He didn't trumpet the theory's unsettling implications; he didn't stress that the very sense of right and wrong, which feels as if heaven-sent, and draws its power from that feeling, is an arbitrary product of our peculiar evolutionary past. But the book did feature, in places, an air of moral relativism. If human society were patterned after bee society, Darwin wrote, "there can hardly be a doubt that our unmarried females would, like the worker-bees, think it a sacred duty to kill their brothers, and mothers would strive to kill their fertile daughters; and no one would think of interfering."[2]

Some people got the picture. The *Edinburgh Review* observed that, if Darwin's theory turned out to be right, "most earnest-minded men will be compelled to give up these motives by which they have attempted to live noble and virtuous lives, as founded on a mistake; our moral sense will turn out to be a mere developed instinct. . . . If these views be true, a revolution in thought is imminent, which will shake society to its very foundations by destroying the sanctity of the conscience and the religious sense."[3]

However breathless this prediction may sound, it wasn't entirely off base. The religious sense has indeed waned, especially among the intelligentsia, the kinds of people who read today's equivalents of the *Edinburgh Review*. And the conscience doesn't seem to carry quite the weight it carried for the Victorians. Among ethical philosophers, there is nothing approaching agreement on where we might turn for basic moral values—except, perhaps, nowhere. It is only a slight exaggeration to say that the prevailing moral philosophy within many philosophy departments is nihilism. A hefty, though unknown, amount of all this can be attributed to the one-two punch Darwin delivered: the *Origin*'s assault on the biblical account of creation, followed by the *Descent*'s doubts about the status of the moral sense.

If plain old-fashioned Darwinism has indeed sapped the moral strength of Western civilization, what will happen when the new version fully sinks in? Darwin's sometimes diffuse speculations about the "social instincts" have given way to theories firmly grounded in logic and fact, the theories of reciprocal altruism and kin selection. And they don't leave our moral sentiments feeling as celestial as they used to. Sympathy, empathy, compassion, conscience, guilt, remorse, even the very sense of justice, the sense that doers of good deserve reward and doers of bad deserve punishment—all these can now be viewed as vestiges of organic history on a particular planet.

What's more, we can't take solace, as Darwin did, in the mistaken belief that these things evolved for the greater good—the "good of the group." Our ethereal intuitions about what's right and what's wrong are weapons designed for daily, hand-to-hand combat among individuals.

It isn't only moral *feelings* that now fall under suspicion, but all of moral discourse. By the lights of the new Darwinian paradigm, a moral code is a political compromise. It is molded by competing interest groups, each bringing all its clout to bear. This is the only discernible sense in which moral values are sent down from on high—

they are shaped disproportionately by the various parts of society where power resides.

So where does this leave us? Alone in a cold universe, without a moral gyroscope, without any chance of finding one, profoundly devoid of hope? Can morality have no meaning for the thinking person in a post-Darwinian world? This is a deep and murky question that (readers may be relieved to hear) will not be rigorously addressed in this book. But we might at least take the trouble to see how Darwin handled the question of moral meaning. Though he didn't have access to the new paradigm, with its several peculiarly dispiriting elements, he definitely caught, as surely as the *Edinburgh Review* did, the morally disorienting drift of Darwinism. Yet he continued to use the words *good* and *bad, right* and *wrong,* with extreme gravity. How did he keep taking morality seriously?

DOOMED RIVALS

As Darwinism was catching on, and the *Edinburgh Review*'s fears were sinking in, a number of thinkers scrambled to avert a collapse of all moral foundation. Many of them skirted evolutionism's threat to religious and moral tradition with a simple maneuver: they redirected their religious awe toward evolution itself, turning it into a touchstone for right and wrong. To see moral absolutes, they said, we need only look to the process that created us; the "right" way to behave is in keeping with evolution's basic direction: we should all go with its flow.

What exactly was its flow? Opinions differed. One school, later called social Darwinism, dwelt on natural selection's pitiless but ultimately creative disposal of the unfit. The moral of the story seemed to be that suffering is the handmaiden of progress, in human as in evolutionary history. The *Bartlett's Familiar Quotations* version of social Darwinism comes from Herbert Spencer, generally regarded as its father: "The poverty of the incapable, the distresses that come upon the imprudent, the starvation of the idle, and those shoulderings aside of the weak by the strong, which leave so many 'in shallows and in miseries,' are the decrees of a large, farseeing benevolence."

Actually, Spencer wrote that in 1851, eight years before the *Origin* appeared. And, for that matter, various people had long had the feeling that gain through pain was nature's way. This was part of the free-market faith that had brought England such rapid material progress. But the theory of natural selection, in the eyes of many capitalists, gave this view an added measure of cosmic affirmation. John D. Rockefeller said that the withering of weak companies in a laissez-faire economy was "the working-out of a law of nature and a law of God."[4]

Darwin found crude moral imputations to his theory laughable. He wrote to Lyell, "I have noted in a Manchester newspaper a rather good squib, showing that I have proved 'might is right' and therefore Napoleon is right and every cheating tradesman is also right."[5] For that matter, Spencer himself would have disavowed that squib. He wasn't as heartless as his more severe utterances imply, nor as heartless as he is now remembered. He put lots of emphasis on the goodness of altruism and sympathy, and he was a pacifist.

How Spencer arrived at these kinder, gentler values illustrates a second approach to figuring out evolution's "flow." The idea was to view evolution's *direction,* not just its dynamics, as a source of guidance; to know how humans should behave, we must first ask toward what end evolution is heading.

There are various ways to answer this question. Today, among biologists, one common answer is that evolution has no discernible end. Spencer, at any rate, believed evolution had tended to move species toward longer and more comfortable lives and the more secure rearing of offspring. Our mission, then, was to nourish these values. And the way to do so was to cooperate with one another, to be nice—to live in "permanently peaceful societies."[6]

All of this now lies in the dustbin of intellectual history. In 1903, the philosopher G. E. Moore decisively assaulted the idea of drawing values from evolution or, for that matter, from *any* aspect of observed nature. He labeled this error the "naturalistic fallacy."[7] Ever since, philosophers have worked hard not to commit it.

Moore wasn't the first to question the inference of "ought" from "is." John Stuart Mill had done it a few decades earlier.[8] Mill's dismissal of the naturalistic fallacy, much less technical and academic than Moore's, was more simply compelling. Its key was to articulate clearly the usually unspoken assumption that typically underlies attempts to use nature as a guide to right conduct: namely, that nature was created by God and thus must embody his values. And, Mill added, not just any God. If, for example, God is not benevolent, then why honor his values? And if he is benevolent, but isn't omnipotent, why suppose that he has managed to precisely embed his values in nature? So the question of whether nature deserves slavish emulation boils down to the question of whether nature appears to be the handiwork of a benevolent and omnipotent God.

Mill's answer was: *Are you kidding?* In an essay called "Nature," he wrote that nature "impales men, breaks them as if on the wheel, casts them to be devoured by wild beasts, burns them to death, crushes them with stones like the first Christian martyr, starves them with hunger, freezes them with cold, poisons them by the quick or slow venom of her exhalations, and has hundreds of other hideous deaths in reserve." And she does all this "with the most supercilious disregard both of mercy and of justice, emptying her shafts upon the best and noblest indifferently with the meanest and worst. . . ." Mill observed, "If there are any marks at all of special design in creation, one of the things most evidently designed is that a large proportion of all animals should pass their existence in tormenting and devouring other animals." Anyone, "whatever kind of religious phrases he may use," must concede "that if Nature and Man are both the works of a Being of perfect goodness, that Being intended Nature as a scheme to be amended, not imitated, by Man."[9] Nor, believed Mill, should we look for guidance to our moral intuition, a device "for consecrating all deep-seated prejudices."[10]

Mill wrote "Nature" before the *Origin* came out (though he published it after), and didn't consider the possibility that suffering is a price paid for organic creation. Still, the question, even then, would remain: If God were benevolent and truly omnipotent, why couldn't he invent a painless creative process? Darwin himself, at any rate, saw the voluminous pain in the world as working against common religious beliefs. In 1860, the year after the *Origin* appeared and long before Mill's "Nature" did, he wrote in a letter to Asa Gray: "I cannot see as plainly as others do, and as I should wish to do, evidence of design and beneficence on all sides of us. There seems to me too much misery in the world. I cannot persuade myself that a beneficent and omnipotent God would have designedly created the Ichneumonidae [parasitic wasps] with the express intention of their feeding within the living bodies of Caterpillars, or that a cat should play with mice."[11]

THE ETHICS OF DARWIN AND MILL

Darwin and Mill not only saw the problem in much the same terms; they also saw the solution in much the same terms. Both believed that, in a universe which for all we know is godless, one reasonable place to find moral guidance is utilitarianism. Mill, of course, did more than subscribe to utilitarianism. He was its premier publicist. In 1861, two years after *On Liberty* and the *Origin* appeared, he published a series of articles in *Fraser's* magazine that are now known by the single title *Utilitarianism* and have become the doctrine's classic defense.

The idea of utilitarianism is simple: the fundamental guidelines for moral discourse are pleasure and pain. Things can be called good to the extent that they raise the amount of happiness in the world and bad to the extent that they raise the amount of suffering. The purpose of a moral code is to maximize the world's total happiness. Darwin quibbled with this formulation. He distinguished between "the general good or welfare of the community," and "the general happiness," and embraced the former, but then conceded that since "happiness is an essential part of the general good, the greatest-happiness principle indirectly serves as a nearly safe standard of right and wrong."[12] He was, for practical purposes, a utilitarian.[13] And he was a great admirer of Mill, both for his moral philosophy and for his political liberalism.

One virtue of Mill's utilitarianism in a post-Darwinian world is its minimalism. If it is harder now to find a grounding for assertions about basic moral values, then, presumably, the fewer and the simpler the foundational assertions, the better. Utilitarianism's foundation consists largely of the simple assertion that happiness, all other things being equal, is better than unhappiness. Who could argue with that?

You'd be surprised. Some people believe that even this seemingly modest moral claim is an unwarranted inference of "ought" from "is"—that is, from the real-world fact that people do like happiness. G. E. Moore himself argued as much (though later philosophers have traced Moore's complaint to a misunderstanding of Mill).[14]

It's true that Mill sometimes worded his argument in a way that invited such criticism.[15] But he never professed to have quite "proven" the goodness of pleasure and the badness of pain; he believed "first principles" beyond proof. His argument followed more modest and pragmatic lines. One of them consisted of saying, basically: Let's face it, we all subscribe at least partly to utilitarianism; some of us just don't use the term.

First of all, we all conduct our *own* lives as if happiness were the object of the game. (Even people who practice severe self-denial typically do so in the name of future happiness, either here or in the hereafter.) And once each of us admits that, yes, we find our own happiness in some basic sense good, something that is not rightly trampled upon without reason, it becomes hard to deny everyone else's identical claim without sounding a bit presumptuous.

Indeed, the point is widely conceded: everyone—except sociopaths, whom the rest of us consider poor moral beacons—agrees that the question of how their acts affect the happiness of others is an important part of moral evaluation. You may believe in any number of absolute rights (freedom, say) or obligations (never cheat). You may consider these things divinely ordained, or unerringly intuited. You may believe that they always override—"trump," as some philosophers say—solely utilitarian arguments. But you don't believe the utilitarian arguments are irrelevant; you implicitly agree that, in the absence of your trump card, they would win.

What's more, when pressed, you probably have a tendency to justify your trump cards in utilitarian terms. You might argue, for example, that even if the occasional isolated act of cheating somehow increased overall welfare in the short run, cheating on a regular basis would erode integrity, so that moral chaos would eventually ensue, to everyone's detriment. Or, similarly, once freedom is denied even to a small group of people, no one will feel secure. This sort of underlying logic—closet utilitarianism—often emerges when the logic behind basic "rights" is teased out. "The greatest-happiness principle," Mill wrote, "has had a large share in forming the moral doctrines even of those who most scornfully reject its authority. Nor is there any school of thought which refuses to admit that the influence of actions on happiness is a most material and even predominant consideration in many of the details of morals, however unwilling to acknowledge it as the fundamental principle of morality, and the source of moral obligation."[16]

The above arguments for "trump cards" illustrate a scantly appreciated fact: utilitarianism can be the basis for absolute rights and obligations. A utilitarian can fiercely defend "inviolable" values, so long as their violation would plausibly lead to big problems in the long run. Such a utilitarian is a "rule" utilitarian, as Mill seems to have been, rather than an "act" utilitarian.[17] Such a person doesn't ask: What is the effect on overall human happiness of my doing such and such today? Instead the question is: What would be the effect if people always did such and such in comparable circumstances, as a rule?

Belief in the goodness of happiness and the badness of suffering isn't just a basic part of moral discourse that we all share. Increasingly it seems to be the only basic part that we all share. Thereafter, fragmentation ensues, as different people pursue different divinely imparted or seemingly self-evident truths. So if a moral code is indeed a code for the entire community, then the utilitarian mandate—happiness is good, suffering bad—seems to be the most practical, if not the *only* practical, basis for moral discourse. It is the common denominator for discussion, the only premise everyone stands on. It's just about all we have left.

Of course, you could dig up a few people who wouldn't go even that far; perhaps citing the naturalistic fallacy, they would insist that there's nothing good about happiness. (My own view is that the goodness of happiness is, in fact, a moral value that remains unscathed by the naturalistic fallacy. Conveniently, space doesn't permit the dissertation-length defense that this claim requires.) Some other people might say that although happiness is a fine thing, they don't think there should be any such thing as a consensually accepted moral code. That's their prerogative. They are free to opt out of moral discourse, and out of any obligations, and benefits, that the resulting code might bring. But if you believe that the idea of a public moral code makes sense, and you want it to be broadly accepted, then the utilitarian premise would seem to be a logical starting point.

Still, the question is a good one: Why *should* we have a moral code? Even accepting the basis of utilitarianism—the goodness of happiness—you might ask: Why should any of us worry about the happiness of others? Why not let everyone worry about their own happiness—which seems, anyway, to be the one thing they can be more or less counted on to do?

Perhaps the best answer to this question is a sheerly practical one: thanks to our old friend non-zero-sumness, everyone's happiness can, in principle, go up if everyone treats everyone else nicely. You refrain from cheating or mistreating me, I refrain from cheating or mistreating you; we're both better off than we would be in a world without morality. For in such a world the mutual mistreatment would roughly cancel itself out anyway (assuming neither of us is a vastly more proficient villain than the other). And, meanwhile, we would each incur the added cost of fear and vigilance.

To put the point another way: life is full of cases where a slight expenditure on one person's part can yield a larger saving on another person's part. For example: holding open a door for the person walking behind you. A society in which everyone holds the door open for people behind them is a society in which everyone is better off (assuming none of us has an odd tendency to walk through doors in front of people). If you can create this sort of system of mutual consideration—a moral system—it's worth the trouble from everyone's point of view.

In this light, the argument for a utilitarian morality can be put concisely: widely practiced utilitarianism promises to make everyone better off; and so far as we can tell, that's what everyone wants.

Mill followed the logic of non-zero-sumness (without using the term, or even being very explicit about the idea) to its logical conclusion. He wanted to *maximize* overall happiness; and the way to maximize it is for everyone to be thoroughly self-sacrificing. You shouldn't hold doors open for people only if you can do so quite easily and thereby save them lots of trouble. You should hold doors open whenever the amount of trouble you save them is even infinitesimally greater than the trouble you take. You should, in short, go through life *considering the welfare of everyone else exactly as important as your own welfare.*

This is a radical doctrine. People who preach it have been known to get crucified. Mill wrote: "In the golden rule of Jesus of Nazareth, we read the complete spirit of the ethics of utility. To do as one would be done by, and to love one's neighbour as oneself, constitute the ideal perfection of utilitarian morality."[18]

DARWIN AND BROTHERLY LOVE

It is surprising to see such a warm, mushy idea—brotherly love—grow out of a word as cold and clinical as "utilitarianism." But it shouldn't be. Brotherly love is implicit in the standard formulations of utilitarianism—maximum *total* happiness, the greatest good for the *greatest number.* In other words: everyone's happiness counts equally; you are not privileged, and you shouldn't act as if you are. This is the second, less conspicuous foundational assumption of Mill's argument. From the beginning he is asserting not only that happiness is good, but that no one person's happiness is special.

It is hard to imagine an assertion that more directly assaults the values implicit in nature. If there's one thing natural selection "wants" us to believe, it's that our individual happiness is special. This is the basic gyroscope it has built into us; by pursuing goals that promise to make us happy, we will maximize the proliferation of our genes (or, at least, would have stood a good chance of doing that in the ancestral environment). Leave aside for the moment that pursuing goals which promise to make us happy, in the long run, often doesn't; leave aside that natural selection doesn't really "care" about our happiness in the end and will readily countenance our suffering if that will get our genes into the next generation. For now the point is that the basic mechanism by which our genes control us is the deep, often unspoken (even unthought), conviction that our happiness is special. We are designed not to worry about anyone else's happiness, except in the sort of cases where such worrying has, during evolution, benefited our genes.

And it isn't just us. Self-absorption is the hallmark of life on this planet. Organisms are things that act as if their welfare were more important than the welfare of all other organisms (except, again, when other organisms can help spread their genes). It may sound innocuous for Mill to say that your happiness is a legitimate goal only so long as it doesn't interfere with the happiness of others, but this is an evolutionary heresy. Your happiness is *designed* to interfere with the happiness of others; the very reason it exists is to inspire selfish preoccupation with it.[19]

Long before Darwin knew about natural selection, long before he could have thought about its "values," his own contrary values were well formed. The ethics embraced by Mill were a Darwin family tradition. Grandfather Erasmus had written about the "greatest happiness principle." And on both sides of the family universal compassion had long been an ideal. In 1788 Darwin's maternal grandfather, Josiah Wedgwood, made hundreds of antislavery medallions showing a black man in chains under the words "AM I NOT A MAN AND A BROTHER?"[20] Darwin sustained the tradition, feeling deeply the anguish of black men who, he observed bitterly, "are ranked by the polished savages in England as hardly their brethren, even in God's eyes."[21]

This sort of simple and deep compassion is what Darwin's utilitarianism ultimately rested on. To be sure, he did, like Mill, pen a rationale for his ethics (a rationale that, oddly, flirts more openly than Mill's with the naturalistic fallacy).[22] But in the end, Darwin was simply a man who empathized boundlessly; and in the end, boundless empathy is what utilitarianism is.

Once Darwin fathomed natural selection, he surely saw how deeply his ethics were at odds with the values it implies. The insidious lethality of a parasitic wasp, the cruelty of a cat playing with a mouse—these are, after all, just the tip of the iceberg. To ponder natural selection is to be staggered by the amount of suffering and death that can be the price for a single, slight advance in organic design. And it is to realize, moreover, that the purpose of this "advance"—longer, sharper canine teeth in male chimpanzees, say—is often to make other animals suffer or die more surely. Organic design thrives on pain, and pain thrives on organic design.

Darwin doesn't seem to have spent much time agonizing over this conflict between natural selection's "morality" and his own. If a parasitic wasp or a cat playing with mice embodies nature's values—well, so much the worse for nature's values. It is remarkable that a creative process devoted to selfishness could produce organisms which, having finally discerned this creator, reflect on this central value and reject it. More remarkable still, this happened in record time; the very first organism ever to see its creator did precisely that. Darwin's moral sentiments, designed ultimately to serve selfishness, renounced this criterion of design as soon as it became explicit.[23]

DARWINISM AND BROTHERLY LOVE

It's conceivable that Darwin's values, ironically, drew a certain strength from his pondering of natural selection. Think of it: zillions and zillions of organisms running around, each under the hypnotic spell of a single truth, all these truths identical, and all logically incompatible with one another: "My hereditary material is the most important material on earth; its survival justifies your frustration, pain, even death." And you are one of these organisms, living your life in the thrall of a logical absurdity. It's enough to make you feel a little alienated—if not, indeed, out and out rebellious.

There is another sense in which Darwinian reflection works against selfishness, a sense Darwin himself could not fully appreciate; there is a sense in which the new Darwinian paradigm can lead one appreciably in the direction of Mill's and Darwin's and Jesus' values.

This is meant informally. I'm not claiming that any moral absolutes *follow* from Darwinism. Indeed, as we've seen, the very idea of moral absolutes has suffered a certain amount of damage at Darwin's hands. But I do believe that most people who clearly understand the new Darwinian paradigm and earnestly ponder it will be led toward greater compassion and concern for their fellow human beings. Or, at least toward the admission, in moments of detachment, that greater compassion and concern would seem to be in order.

The new paradigm strips self-absorption of its noble raiment. Selfishness, remember, seldom presents itself to us in naked form. Belonging as we do to a species (*the* species) whose members justify their actions morally, we are designed to think of ourselves as good and our behavior as defensible, even when these propositions are objectively dubious. The new paradigm, by exposing the biological machinery behind this illusion, makes the illusion harder to buy.

For example, nearly all of us say, and believe, that we don't dislike people without reason. If someone is an object of our wrath, or even of our callous indifference—if we can enjoy his suffering, or easily countenance it—it is because of something he did, we say; he *deserves* to be treated coldly.

Now, for the first time, we understand clearly how humans came to have this feeling that the deserts they dish out are just. And its origins don't inspire great moral confidence.

At the root of this feeling is the retributive impulse, one of the basic governors of reciprocal altruism. It evolved not for the good of the species, or the good of the nation, or even for the good of the tribe, but for the good of the individual. And, really, even this is misleading; the impulse's ultimate function is to get the individual's genetic information copied.

This doesn't necessarily mean the impulse of retribution is *bad*. But it does mean that some of the reasons we've been thinking of it as *good* are now open to question. In particular, the aura of reverence surrounding the impulse—the ethereal sense that retribution embodies some higher ethical truth—is harder to credit once the aura is seen to be a self-serving message from our genes, not a beneficent message from the heavens. Its origin is no more heavenly than that of hunger, hatred, lust, or any of the other things that exist by virtue of their past success in shoving genes through generations.

There is, actually, a defense of retribution that can be cast in moral terms—in utilitarian terms, or in terms of any other morality whose aim is to get people to behave considerately toward one another. Retribution helps solve the "cheater" problem that any moral system faces; people who are seen to take more than they give are thereafter punished, discouraged from always being a door holdee and never a door holder. Even though the retributive impulse wasn't designed for the good of the group, as Mill's moral system is, it can, and often does, raise the sum of social welfare. It keeps people mindful of the interests of others. However lowly its origins, it has come to serve a lofty purpose. This is something to be thankful for.

And it might be enough to exonerate the retributive impulse

except for one fact: the grievances redressed by retribution aren't tallied with the sort of divine objectivity that Mill would prescribe. We don't try to punish only people who truly have cheated or mistreated us. Our moral accounting system is wantonly subjective, informed by a deep bias toward the self.

And this general bias in calculating what we're owed is only one of several departures from clarity of moral judgment. We tend to find our rivals morally deficient, to find our allies worthy of compassion, to gear that compassion to their social status, to ignore the socially marginal altogether. Who could look at all this and then claim with a straight face that our various departures from brotherly love possess the sort of integrity we ascribe to them?

We are right to say that we never dislike people without a reason. But the reason, often, is that it is not in our interests to like them; liking them won't elevate our social status, aid our acquisition of material or sexual resources, help our kin, or do any of the other things that during evolution have made genes prolific. The feeling of "rightness" accompanying our dislike is just window dressing. Once you've seen that, the feeling's power may diminish.*

But wait a minute. Couldn't we similarly discount the sense of rightness accompanying compassion, sympathy, and love? After all, love, like hate, exists only by virtue of its past contribution to genetic proliferation. At the level of the gene, it is as crassly self-serving to love a sibling, an offspring, or a spouse as it is to hate an enemy. If

---

*The argument here is crucially different from other arguments about morality that have been made in this book. Here the contention is not just that the new Darwinian paradigm can help us realize whichever moral values we happen to choose. The claim is that the new paradigm can actually *influence*—legitimately—our choice of basic values in the first place. Some Darwinians insist that such influence can never be legitimate. What they have in mind is the naturalistic fallacy, whose past violation has so tainted their line of work. But what we're doing here doesn't violate the naturalistic fallacy. Quite the opposite. By studying nature—by seeing the origins of the retributive impulse—we see how we have been conned into committing the naturalistic fallacy without knowing it; we discover that the aura of divine truth surrounding retribution is nothing more than a tool with which nature—natural selection—gets us to uncritically accept its "values." Once this revelation hits home, we are less likely to obey this aura, and thus less likely to commit the fallacy.

Herbert Spencer (*below*) is considered (in some ways unfairly) the classic proponent of social Darwinism, which justifies cruelty and oppression on grounds that suffering is evolution's path to "progress." As the 1984 poster at the right suggests, early proponents of the new Darwinian paradigm were accused (almost always unfairly) of making a similarly malicious inference of moral values from nature's workings.

COME AND HEAR

"The quintessential female is an individual specialized for making eggs" — E. O. Wilson

EDWARD O. WILSON
Sociobiologist and

The Prophet of Right Wing Patriarchy

Tuesday, Oct. 9, 1984
4:00 p.m., Auditorium
Medical Sciences Building
BRING NOISEMAKERS

Univ of Toronto

John Stuart Mill: "If there are any marks at all of special design in creation, one of the things most evidently designed is that a large proportion of all animals should pass their existence in tormenting and devouring other animals. . . . If Nature and Man are both the works of a Being of perfect goodness, that Being intended Nature as a scheme to be amended, not imitated, by Man."

Thomas Henry Huxley: "The practice of that which is ethically best—what we call goodness or virtue—involves a course of conduct which, in all respects, is opposed to that which leads to success in the cosmic struggle for existence. In place of ruthless self-assertion it demands self-restraint; in place of thrusting aside, or treading down, all competitors, it requires that the individual shall not merely respect, but shall help his fellows; its influence is directed, not so much to the survival of the fittest, as to the fitting of as many as possible to survive."

George Williams: "Huxley viewed the cosmic process as an enemy that must be combated. I take a similar but more extreme position, based both on the more extreme contemporary view of natural selection as a process for maximizing selfishness, and on the longer list of vices now assignable to the enemy. If this enemy is worse than Huxley thought, there is a more urgent need for biological understanding."

Samuel Smiles, author of the 1859 book *Self-Help*: "The greatest slave is not he who is ruled by a despot, great though that evil be, but he who is in the thrall of his own moral ignorance, selfishness, and vice."

Charles Darwin around 1882: "As man advances in civilisation, and small tribes are united into larger communities, the simplest reason would tell each individual that he ought to extend his social instincts and sympathies to all the members of the same nation, though personally unknown to him. This point being once reached, there is only an artificial barrier to prevent his sympathies extending to the men of all nations and races."

the base origins of retribution are grounds for doubting it, why shouldn't love be doubted too?

The answer is that love should be doubted, but that it survives the doubt in pretty good shape. At least, it survives in good shape by the lights of a utilitarian, or indeed of anyone who considers happiness a moral good. Love, after all, makes us want to further the happiness of others; it makes us give up a little so that others (the loved ones) may have a lot. More than that: love actually makes this sacrifice feel good, thus magnifying total happiness all the more. Of course, sometimes love is hurtful. Witness the woman in Texas who plotted the murder of the mother of her daughter's rival for a cheer-leading slot. Her maternal love, though undeniably intense, doesn't go down on the positive side of the moral ledger. And so too whenever love ends up doing more harm than good. But either way—whether the net result is good or bad—the moral evaluation of love is the same as the evaluation of retribution: we must first clear away the window dressing, the intuitive feeling of "rightness," and then sob-erly assess the effect on overall happiness.

Thus the service performed by the new paradigm isn't, strictly speaking, to reveal the baseness of our moral sentiments; that base-ness, per se, counts neither for nor against them; the ultimate genetic selfishness underlying an impulse is morally neutral—grounds neither for embracing the impulse nor for condemning it. Rather, the par-adigm is useful because it helps us see that the aura of rightness surrounding so many of our actions may be delusional; even when they feel right, they may do harm. And surely hatred, more often than love, does harm while feeling right. That is why I contend that the new paradigm will tend to lead the thinking person toward love and away from hate. It helps us judge each feeling on its merits; and on grounds of merit, love usually wins.

Of course, if you're *not* a utilitarian, sorting these issues out may be more complex. And although utilitarianism was Darwin's and Mill's solution to the moral challenge of modern science, it isn't everyone's. Nor is this chapter intended to *make* it everyone's (al-though, I admit, it's mine). The point, rather, is to show that a Darwinian world needn't be an amoral world. If you accept even the simple assertion that happiness is better than unhappiness (all other things being equal), you can go on to construct a full-fledged morality, with absolute laws and rights and all the rest. You can keep finding laudable some of the things we've always found laudable— love, sacrifice, honesty. Only the most die-hard nihilist, who insists that there's nothing good about the happiness of human beings, could find the word *moral* meaningless in a post-Darwinian world.

ENGAGING THE ENEMY

Darwin was not the only Victorian evolutionist who took a dim view of evolution's "values." Another was his friend and advocate Thomas Huxley. In a lecture titled "Evolution and Ethics," which he delivered at Oxford University in 1893, Huxley took aim at the whole premise of social Darwinism, the idea of deriving values from evolution. Echoing the logic of Mill's essay "Nature," he said that "cosmic evolution may teach us how the good and the evil tendencies of man may have come about; but, in itself, it is incompetent to furnish any better reason why what we call good is preferable to what we call evil than we had before." Indeed, a close look at evolution, with its massive toll in death and suffering, suggested to Huxley that it is rather at odds with what we call good. Let us understand, he said, "once for all, that the ethical progress of society depends, not on imitating the cosmic process, still less in running away from it, but in combating it."[24]

Peter Singer, one of the first philosophers to take the new Darwinism seriously, has noted, in this context, that "the more you know about your opponent, the better your chances of winning."[25] And George Williams, who did so much to define the new paradigm, has embraced both Huxley's and Singer's points, and stressed how strongly the new paradigm underscores them. His revulsion at natural selection's values, he writes, is even greater than Huxley's, "based both on the more extreme contemporary view of natural selection as a process for maximizing selfishness, and on the longer list of vices now assignable to the enemy." And if the enemy is indeed "worse than Huxley thought, there is a more urgent need for biological understanding."[26]

Biological understanding to date suggests some basic rules for engaging the enemy. (My enumeration of them isn't meant to imply that I'm notably successful in following them.) A good starting point would be to generally discount moral indignation by 50 percent or so, mindful of its built-in bias, and to be similarly suspicious of moral indifference to suffering. We should be especially vigilant in certain situations. We seem, for example, prone to grow indignant about the behavior of distinct groups of people (nations, say) whose interests conflict with a distinct group to which we belong. We also tend to be inconsiderate of low-status people and exceedingly tolerant of high-status people; making life somewhat easier on the former at the expense of the latter is probably warranted, at least by utilitarian lights (and the lights of other egalitarian moralities).

This isn't to say that utilitarianism is mindlessly egalitarian. A powerful person who uses his or her station humanely is a valuable social asset, and thus may merit special treatment, so long as the treatment facilitates such conduct. A famous example in the annals of utilitarian writing is the question of whether you would first save an archbishop or a chambermaid if the two were trapped in a burning building. The standard answer is that you should save the archbishop—even if the chambermaid is your mother—since he will do more good in the future.[27]

Well, maybe so, if the high-status person is an archbishop (and even then, perhaps, it depends on the archbishop). But most high-status people aren't. And there is little evidence that high-status people have any particular proclivity toward conscience or sacrifice. Indeed, the new paradigm stresses that they have attained their status not for "the good of the group" but for themselves; they can be expected to use it accordingly, just as they can be expected to pretend otherwise.[28] Status merits much less indulgence than it generally gets. It is only human nature to extend deference to Mother Teresa and Donald Trump; in the second case, this part of human nature is perhaps unfortunate.

Of course, these prescriptions assume a utilitarian premise—that the happiness of other people is the object of a moral system. What about the nihilists? What about people who insist that not even happiness is a good thing, or that only *their* happiness is a good thing, or that for some other reason the welfare of others shouldn't concern them? Well, for one thing, they probably go around acting as if it did. For the pretense of selflessness is about as much a part of human nature as is its frequent absence. We dress ourselves up in tony moral language, denying base motives and stressing our at least minimal consideration for the greater good; and we fiercely and self-righteously decry selfishness in others. It seems fair to ask that even people who don't buy the stuff about utilitarianism and brotherly love at least make one minor adjustment in light of the new Darwinism: be consistent; either start subjecting all that moral posturing to skeptical scrutiny or quit the posturing.

For people who choose the former, the simplest single source of guidance is to bear in mind that the feeling of moral "rightness" is something natural selection created so that people would employ it selfishly. Morality, you could almost say, was designed to be misused by its own definition. We've seen what may be the rudiments of self-serving moralizing in our close relatives the chimpanzees as they pursue their agendas with righteous indignation. Unlike them, we can distance ourselves from the tendency long enough to see it—long enough, indeed, to construct a whole moral philosophy that consists essentially of attacking it.

Darwin, on grounds such as this, believed that the human species is a moral one—that, in fact, we are the only moral animal. "A moral being is one who is capable of comparing his past and future actions or motives, and of approving or disapproving of them," he wrote. "We have no reason to suppose that any of the lower animals have this capacity."[29]

In this sense, yes, we are moral; we have, at least, the technical capacity for leading a truly examined life; we have self-awareness, memory, foresight, and judgment. But the last several decades of evolutionary thought lead one to emphasize the word *technical*. Chronically subjecting ourselves to a true and bracing moral scrutiny, and adjusting our behavior accordingly, is not something we are designed for. We are potentially moral animals—which is more than any other animal can say—but we aren't naturally moral animals. To be moral animals, we must realize how thoroughly we aren't.

Chapter 17: **BLAMING THE VICTIM**

As all men desire their own happiness, praise or blame is bestowed on actions and motives, according as they lead to this end.
—*The Descent of Man* (1871)

We acquire many notions unconsciously, without abstracting them & reasoning on them (as justice?? . . .)
—N Notebook (1838)[1]

In the mid-1970s, the book *Sociobiology* gave the new Darwinian paradigm its first burst of publicity. It also gave its author, E. O. Wilson, his first burst of public abuse. He was called a racist, a sexist, a capitalist imperialist. His book was characterized as a right-wing plot, a blueprint for the continued oppression of the oppressed.

It may seem odd that such fears would persist many decades after the unmasking of the "naturalistic fallacy" and the crumbling of social Darwinism's intellectual foundation. But the word *natural* has more than one application to moral questions. If a man cheating on his wife, or exploiting the weak, excuses himself by saying it's "only natural," he doesn't necessarily mean it's divinely ordained. He may just mean that the impulse runs so deep as to be practically irresistible; what he's doing may not be good, but he can't much help it.

For years, the "sociobiology debate" subsisted largely on this one issue. Darwinians were accused of "genetic determinism" or "biological determinism"—which, it was said, left no room for "free will." They then accused their accusers of confusion; Darwinism, rightly understood, posed no threat to lofty political and moral ideals.

It is true that the accusations were often confused (and that the charges directed specifically against Wilson were gratuitous). But it's also true that some fears on the left have a firm grounding even after the confusion is dispelled. The question of moral responsibility in the view of evolutionary psychology is a large one, and dicey. In fact, it is large enough, properly understood, to alarm the right as well as the left. There are deep and momentous issues lying out there, going largely unaddressed.[2]

As it happens, Charles Darwin addressed the deepest of them more than a century ago in thoroughly acute and humane fashion. But he didn't tell the world. As aware as any modern Darwinian of how explosive a truly honest analysis of moral responsibility might be, he never published his thoughts. They have remained in obscurity, in the darkest recesses of his private writings—a grab bag of papers that he labeled, with typically emphatic modesty, "Old & USELESS notes about the moral sense & some metaphysical points." Now, with the biological basis of behavior coming rapidly to light, is a good time to excavate Darwin's treasure.

REALITY REARS ITS UGLY HEAD

The occasion for Darwin's analysis is a conflict between ideal and real. Brotherly love is great in theory. In practice, however, problems arise. Even if you could somehow convince lots of people to pursue brotherly love—reality problem number one—you would run into reality problem number two: brotherly love tends to make society fall apart.

After all, true brotherly love is unconditional compassion; it harbors utter doubt about the validity of harming anyone, however repugnant their behavior. And in a society where no one gets punished for anything, repugnant behavior will grow.

This paradox lurks in the background of utilitarianism, especially John Stuart Mill's rendering of it. Mill may say that a good utilitarian is someone who loves unconditionally, but until the day when everyone *does* love unconditionally, the realization of utilitarianism's goal—maximum overall happiness—will entail highly conditional love. Those who haven't seen the light must be encouraged to act nice. Murder must be punished, altruism praised, and so on. People must be held accountable.[3]

Remarkably, Mill didn't confront this tension anywhere in his basic text on the subject, *Utilitarianism.* A few dozen pages after embracing the universal love taught by Jesus, he endorsed the principle "of giving to each what they deserve, that is, good for good as well as evil for evil."[4] This is an irreconcilable difference—between saying "Do unto others as you would have them do unto you" and saying, "Do unto others as they have done to you"; between saying "Love your enemies" or "Turn the other cheek" and saying "An eye for an eye, a tooth for a tooth."[5]

Maybe Mill can be excused for taking a charitable view of the sense of justice, the governor of reciprocal altruism.[6] As we've noted, the machinery of reciprocal altruism is, for a utilitarian, a real evolutionary godsend; by dishing out a steady stream of tits for tats, it provides the sticks and carrots that keep people in touch with the needs of others. Given that human nature didn't evolve to elevate the community's welfare, it does a none too shabby job of it. Lots of non-zero-sum fruits get reaped.

Still, thanking the retributive impulse for services rendered isn't the same as thanking it for light shed. Whatever its practical value, there is no reason to believe that the inherent sense of justice—the sense that people *deserve* punishment, that their suffering is a good thing *in and of itself*—reflects a higher truth. The new Darwinian paradigm, indeed, reveals the sense of rightness surrounding retribution to be mere genetic expediency, and to be warped accordingly. This unmasking was part of the basis for my suggestion in the previous chapter that the new paradigm will tend to steer people toward compassion.

There is a second powerful reason that the idea of retributive punishment looks dubious from the standpoint of modern Darwinism. Evolutionary psychology professes to be the surest path to a complete explanation of human behavior, good and bad, and of the underlying psychological states: love, hate, greed, and so on. And to know all is to forgive all. Once you see the forces that govern behavior, it's harder to blame the behaver.

This has nothing to do with a supposedly right-wing doctrine of "genetic determinism." To begin with, the question of moral responsibility has no exclusive ideological character. Though some on the far right might be thrilled to hear that businessmen can't help but exploit laborers, they would be less happy to hear that criminals can't help but commit crimes. And neither Bible-thumpers in the "moral majority" nor feminists especially want to hear male philanderers say they're slaves to their hormones.

More to the point: the phrase "genetic determinism" exudes ignorance as to what the new Darwinism is about. As we've seen, everyone (including Darwin) is a victim not of genes, but of genes and environment together: knobs and tunings.

Then again, a victim is a victim. A stereo has no more control over its tunings than over the knobs it was born with; whatever importance you attach to the two factors, there's no sense in which the stereo is to blame for its music. In other words: though the fears of "genetic determinism" that were current in the 1970s were unfounded, the fears of "determinism" weren't. Yet that's also the good news—more reason to doubt impulses of blame and censure and extend our compassion beyond its natural confines of family and friends. Then again, that's also the bad news: this philosophically valid endeavor has some pernicious real-world effects. The situation, in short, is a mess.

Of course, you can argue with the proposition that all we are is knobs and tunings, genes and environment. You can insist that there's something . . . something *more*. But if you try to visualize the form this something would take, or articulate it clearly, you'll find the task impossible, for any force that is not in the genes or the environment is outside of physical reality as we perceive it. It's beyond scientific discourse.

This doesn't mean it doesn't exist, of course. Science may not tell the whole story. But just about everyone on both sides of the sociobiology debate in the 1970s professed to be scientifically minded. That's what was so ironic about all the anthropologists and psychologists who complained of sociobiology's "genetic determinism." The then-reigning philosophy of the social sciences was "cultural determinism" (as anthropologists put it) or "environmental determinism" (as psychologists put it). And when it comes to free will, and thus to blame and credit, determinism is determinism is determinism. As Richard Dawkins has noted, "Whatever view one takes on the question of determinism, the insertion of the word 'genetic' is not going to make any difference."[7]

DARWIN'S DIAGNOSIS

Darwin saw all of this. He didn't know about genes, but he certainly knew about the concept of heredity, and he was a scientific materialist; he didn't think any nonphysical forces were needed to explain human behavior, or anything else in the natural world.[8] He saw that all behavior must therefore boil down to heredity and environment. "[O]ne doubts existence of free will," he wrote in his notebooks, because "every action determined by heredetary [sic] constitution, example of others or teaching of others."[9]

What's more, Darwin saw how these forces have their combined effect: by determining a person's physical "organization," which in turn determines thought and feeling and behavior. "My wish to improve my temper, what does it arise from but organization," he asked in his notebook. "That organization may have been affected by circumstances & education, & by choice which at that time organization gave me to will."[10]

Here Darwin is making a point that even today often goes ungrasped: *all* influences on human behavior, environmental as well as hereditary, are mediated biologically. Whatever combination of things has given your brain the exact physical organization it has at this moment (including your genes, your early environment, and your assimilation of the first half of this sentence), that physical organization is what determines how you will respond to the second half of this sentence. So, even though the term *genetic determinism* is confused, the term *biological determinism* isn't—or, at least, it wouldn't be if people would realize that it's not a mere synonym for *genetic determinism*. Then again, if they realized that, they'd realize they could drop the word "biological" without losing anything. The sense in which E. O. Wilson is a "biological determinist" is the sense in which B. F. Skinner was a "biological determinist"—which is to say, he was a determinist.[11] The sense in which evolutionary psychology is "biologically determinist" is the sense in which all psychology is "biologically determinist."

As for why, if all behavior is determined, we "feel" as if we're making free choices, Darwin had a strikingly twentieth-century explanation: our conscious mind isn't privy to all the motivating forces. "The general delusion about free will obvious.—because man has power of action, & he can seldom analyse his motives (originally mostly INSTINCTIVE, & therefore now great effort of reason to discover them: this is important explanation) he thinks they have none."[12]

Darwin doesn't seem to have suspected what the new Darwinism suggests: that some of our motives are hidden from us not incidentally but by design, so that we can credibly act as if they aren't what they are; that, more generally, the "delusion about free will" may be an adaptation. Still, he got the basic idea: free will is an illusion, brought to us by evolution. All the things we are commonly blamed or praised for—ranging from murder to theft to Darwin's eminently Victorian politeness—are the result not of choices made by some immaterial "I" but of physical necessity. "This view should teach one profound humility, one deserves no credit for anything," Darwin wrote in his notes. "[N]or ought one to blame others."[13] Here Darwin has unearthed the most humane scientific insight of all—and, at the same time, one of the most dangerous.

Darwin saw the danger in the forgiveness brought by understanding; he saw that determinism, by eroding blame, threatens society's moral fiber. But he wasn't too worried about this doctrine spreading. However compelling the logic seemed to a thoughtful scientific materialist, most people aren't thoughtful scientific materialists. "This view will not do harm, because no one can be really *fully* convinced of its truth, except man who has thought very much, & he will know his happiness lays in doing good & being perfect, & therefore will not be tempted, from knowing every thing he does is independent of himself to do harm."[14] In other words: So long as this knowledge is confined to a few English gentlemen, and doesn't infect the masses, everything will be all right.

The masses are now getting infected. What Darwin didn't realize is that the technology of science would eventually make the case for determinism vivid. He saw that "thought, however unintelligible it may be, seems as much function of organ, as bile of liver," but he probably didn't dream that we would start pinpointing specific connections between the organ and the thoughts.[15]

Today these connections regularly make headlines. Scientists link crime to low serotonin. Molecular biologists try—with slight but growing success—to isolate genes that incline the brain toward mental illness. A natural chemical called oxytocin is found to underlie love. And an unnatural chemical, the drug Ecstasy, induces a deeply benign state of mind; now anyone can be Gandhi for a day. People are getting the sense—from news in genetics, molecular biology, pharmacology, neurology, endocrinology—that we are all machines, pushed and pulled by forces that we can't discern but that science can.

This picture, though utterly biological, has no special connection with *evolutionary* biology. Genes, neurotransmitters, and the various other elements of mind control are being studied, for the most part, without special inspiration from Darwinism.

But Darwinism will increasingly frame this picture and give it narrative force. We will see not only *that,* for example, low serotonin encourages crime, but *why:* it seems to reflect a person's perception of foreclosed routes to material success; natural selection may "want" that person to take alternate routes. Serotonin and Darwinism together could thus bring sharp testament to otherwise vague complaints about how criminals are "victims of society." A young inner-city thug is pursuing status by the path of least resistance, no less than you; and he is compelled by forces just as strong and subtle as the ones that have made you what you are. You may not reflect on this when he kicks your dog or snatches your purse, but afterwards, on reflection, you may. And you may then see that you would have been him had you been born in his circumstances.

The landslide of news about the biology of behavior is just beginning. People, by and large, haven't succumbed to it and concluded that we're all mere machines. So the notion of free will lives on. But it shows signs of shrinking. Every time a behavior is found to rest on chemistry, someone tries to remove it from the realm of volition.

That "someone" is typically a defense lawyer. The most famous example is the "Twinkie defense." A lawyer convinced a California jury that a junk-food diet had left his client with a "diminished capacity" to think clearly, and that full "premeditation" of his crime—murder—was thus impossible. Other examples abound. In both British and American courts, women have used premenstrual syndrome to partly insulate themselves from criminal responsibility. As Martin Daly and Margo Wilson rhetorically asked in their book *Homicide,* can a "high-testosterone" defense of male murderers be far behind?[16]

Of course, psychology was eroding culpability even before biology came along to help it. "Posttraumatic stress disorder" is a defense lawyer's favorite malady—said to encompass everything from "battered-woman syndrome" to "depression-suicide syndrome" (which purportedly leads people not only to commit crimes, but to bungle them, with the unconscious goal of being caught). The disorder was originally couched in purely psychological terms, with little reference to biology. But work is constantly under way to link such maladies to biochemistry, because physical evidence is what really gets a jury's attention. Already, an expert witness touting a conjectured posttraumatic stress disorder subcategory called "action-addict syndrome" (a dependency on the thrill of danger) has traced the problem to endorphins, which the criminal desperately craves, and obtains via crime.[17] And compulsive gamblers, it turns out, have abnormally high levels of endorphins in their blood when they gamble. Thus (the argument goes) gambling is a disease.

Well, we all like our endorphins, and we all do things to get them, ranging from jogging to sex. And when we do those things, our endorphin levels are abnormally high. No doubt rapists feel good at some point during or after their crimes; no doubt that pleasure has a biochemical basis; and no doubt this basis will come to light. If defense lawyers get their way and we persist in removing biochemically mediated actions from the realm of free will, then within decades that realm will be infinitesimal. As, indeed, it should be—on strictly intellectual grounds, at least.

There are at least two ways to respond to the growing body of evidence that biochemistry governs all. One is to use the data, perversely, as proof of volition. The argument runs as follows: Of course all these criminals have free will, regardless of the state of their endorphins, blood-sugar levels, and everything else. Because if biochemistry negated free will, then *none* of us would have free will! And we know *that's* not the case. Right? (Pause.) *Right?*

This sort of whistling in the dark is often heard in the books and articles that bemoan crumbling culpability. It was also implicit in the referendum that finally removed the "diminished capacity" defense from California law. Presumably the voters sensed that if something as natural as sugar could indeed turn you into a robot, that would mean everyone's a robot, and no one deserves punishment. Precisely.

The second response to dehumanizing biochemical data is Darwin's—complete surrender. Give up on free will; no one really deserves blame or credit for anything; we are all slaves of biology. We must view a wicked man, Darwin wrote in his notes, "like a sickly one." It would "be more proper to pity than to hate & be disgusted."[18]

In short: brotherly love is a valid doctrine. The hatred and revulsion that send people to jail and to the gallows—and, in other contexts, lead to arguments, fights, and wars—are without intellectual foundation. Of course, they may have a *practical* foundation. Indeed, that's the problem: blame and punishment are as practically necessary as they are intellectually vacuous. That's why Darwin took comfort in the hope that his insights would never become common.

DARWIN'S PRESCRIPTION

What to do? If Darwin knew that the cat, alas, was out of the bag, that the material underpinnings of behavior were on public display, what would he suggest? How should society respond to creeping knowledge of our robotic nature? There are hints in his notes. To begin with, we should try to disentangle punishment from the visceral impulses that drive it. This will sometimes mean narrowing its use, restricting it to the cases where it actually does some good. "[I]t is right to punish criminals; but solely to *deter* others," Darwin wrote.

This is very much in the spirit of the time-honored utilitarian prescription. We should punish people only so long as that will raise overall happiness. There is nothing good, in itself, about retribution; the suffering inflicted on wrongdoers is just as sad as the suffering of everyone else, and counts equally in the grand utilitarian calculus. It is warranted only when outweighed by the growth it brings in the welfare of others, through the prevention of future crime.[19]

This idea strikes many people as reasonable and not terribly radical, but taking it seriously would mean overhauling legal doctrine. In American law, punishment has several explicit functions. Most are strictly practical: keeping the criminal off the streets, discouraging him from crime after his release, discouraging others who witness his fate, rehabilitating him—all of which a utilitarian would applaud. But one of the stated functions of punishment is strictly "moral": retribution, pure and simple. Even if punishment serves no discernible purpose, it is supposedly good. If on some desert island you happen upon a ninety-five-year-old prison escapee whose very existence was long ago forgotten, you will serve the cause of justice by somehow making him suffer. Even if you don't enjoy dishing out the punishment, and if no one back on the mainland ever hears about it, you can rest assured that, somewhere in the heavens, the God of Justice is smiling.

The doctrine of retributive justice doesn't play the prominent role it once played in the courts. But there is discussion these days, especially among conservatives, of reemphasizing it. And even now it is one reason courts spend so much time deciding whether people "volitionally" committed a crime—as opposed to being "insane" or "temporarily insane" or having "diminished capacity," or whatever. If utilitarians ran the world, messy words like "volition" would never enter the picture. The courts would ask two questions: (a) Did the defendant commit the crime? and (b) What is the practical effect of punishment—on the criminal's own future behavior, and on the behavior of other would-be criminals?

Thus, when a woman who has been beaten or raped by her husband kills or mutilates him, the question wouldn't be whether she has a "disease" called battered-woman syndrome. And when a man kills his wife's lover, the question wouldn't be whether jealousy is "temporary insanity." The question, in both cases, would be whether punishment would prevent these people, and similarly situated people, from committing crimes in the future. This question is impossible to answer precisely, but it's less messy than the question of volition, and it has the added virtue of not being rooted in an outmoded worldview.

Of course, the two questions have a certain amount in common. The courts tend to recognize "free will," and hence justifiable "blame," in the kinds of acts that can be deterred by the anticipation of punishment. Thus, neither a utilitarian nor an old-fashioned judge would send an out-and-out psychotic to jail (though both might institutionalize him if he seemed likely to repeat the crime). As Daly and Wilson write, "The enormous volume of mystico-religious bafflegab about atonement and penance and divine justice and the like is the attribution to higher, detached authority of what is actually a mundane, pragmatic matter: discouraging self-interested competitive acts by reducing their profitability to nil."[20]

All told, then, "free will" has been a fairly useful fiction, a rough proxy for utilitarian justice. But all the time-wasting debates now in progress (Is alcoholism a disease? Are sex crimes an addiction? Does premenstrual syndrome nullify volition?) suggest that it is beginning to outlive its usefulness. After another decade or two of biological research, it may be more trouble than it's worth; and in the meantime, the scope of "free will" may have shrunk considerably. We will then face (at least) two choices: either (a) artificially restore free will to robustness by redefining it (proclaim, for example, that the existence of a biochemical correlate has no bearing on whether a behavior is volitional); or (b) dispense with volition altogether and adopt explicitly utilitarian criteria of punishment. Both of these options amount to roughly the same thing: as the biological (that is, environmental-genetic) underpinnings of behavior come into view, we must get used to the idea of holding robots responsible for their malfunctions—so long, at least, as this accountability will do some good.

Dispensing with the idea of volition might strip the legal system of some emotional support. Jurors so readily mete out punishment in part because of their vague sense that it's an inherently good thing. Still, this vague sense is a stubborn sense, unlikely to be extinguished by a change of legal doctrine. And even where it weakens, the practical value of punishment will likely remain clear enough to keep jurors doing their jobs.

THOROUGHLY POSTMODERN MORALITY

The truly formidable threat posed by scientific enlightenment is in the moral, not the legal, realm. The problem here isn't that the sense of justice, the governor of reciprocal altruism, will break down entirely. Even people of extreme detachment and humanity, if they feel cheated, lied to, or otherwise mistreated, manage to summon enough indignation for utilitarian purposes. Darwin believed in everyone's ultimate blamelessness, but he could conjure up anger when pressed. He found himself "burning with indignation" at the behavior of his bitter critic, Richard Owen. Writing to Huxley, Darwin said, "I believe I hate him more than you do."[21]

As a rule, if we all worked toward the ideal of universal compassion and forgiveness, drawing on all the enlightenment modern science has to offer, the meager progress we made would hardly bring civilization tumbling down around us. Few of us are anywhere near overkill in the brotherly-love department. And it is unlikely that all the demystifying logic of modern biology will get us there. The hard animal core of TIT FOR TAT is secure against the ravages of truth.

The real moral danger is less direct. Moral systems draw their strength not just from the principles behind TIT FOR TAT—aggrieved parties punishing offenders—but from society at large punishing offenders. Charles Dickens was afraid to take up publicly with his mistress not because his wife would have punished him. (He had already left her; and how much power did she have anyway?) He was afraid, rather, of infamy.

And so it is whenever a strong animal impulse is consistently thwarted by a moral code: violation would bring low repute, the avoidance of which is also a strong animal impulse. Effective moral codes fight fire with fire.

Indeed, they fight fire with an elaborate fire-making machine. Robert Axelrod, whose computer tournament so nicely supported the theory of reciprocal altruism, has also studied the ebb and flow of norms. He finds that robust moral codes rest not just on norms but on "metanorms": society disapproves not only of the code's violators but also of those who tolerate violators by failing to disapprove.[22] Had Dickens gone public with his adultery, his friends might well have had to cut ties with him or else suffer punishment themselves for failing to punish.

It is in the world of norms and metanorms, with its oblique and diffuse retaliation, that modern science takes its toll on moral fiber. We needn't worry about creeping determinism muting a victim's rage. But the rage of spectators may wane as they come to believe that, for example, male philandering is "natural," a biochemical compulsion—and that, anyway, the wife's retributive furor is an arbitrary product of evolution. Life—the life, at least, of those other than ourselves, our kin, and our close friends—becomes a movie that we watch with the bemused detachment of an absurdist. This is the specter of a thoroughly postmodern morality. Darwinism isn't its only source, nor is biology more broadly, but together the two could do much to feed it.

The basic paradox here—the intellectual groundlessness of blame, and the practical need for it—is something few people seem eager to acknowledge. One anthropologist has made the following two statements about divorce: (a) "I do not want to encourage someone saying, 'Well, it's programmed in and I can't help it.' We can help it. While these behaviors may be powerful, many people in fact resist them quite successfully"; and (b) "[T]here are men and women walking the streets today saying to themselves, 'I'm a failure! I've had two marriages, and neither of them has worked.' Well, that's probably a natural human behavior pattern, and they feel a little better when they hear what I have to say. I don't think people need to feel failure following a divorce."[23]

Each of these statements is defensible, but you can't have it both ways. It's accurate, on the one hand, to say that any given divorce was inevitable, driven by a long chain of genetic and environmental forces, all mediated biochemically. Still, to stress this inevitability is to affect public discourse, and thus to affect future environmental forces and future neurochemistry, rendering inevitable future divorces that otherwise wouldn't have been. To call things in the past inexorable makes more things in the future inexorable. To tell people they're not to blame for past mistakes is to make future mistakes more likely. The truth is hardly guaranteed to set us free.

Or, to put the point another, perhaps more upbeat, way: the truth depends on what we say the truth is. If men are told that the impulse to philander is deeply "natural," essentially irrepressible, then the impulse—for those men, at least—may indeed be so. In Darwin's day, though, men were told something else: that animal impulses are formidable foes but can, with constant and arduous effort, be defeated. This then became, for many men, the truth. Free will was, in an important sense, created by their belief in it.

In the same sense, one might argue, their "successful" belief in free will justifies our own belief in it. But not belief in the *metaphysical* doctrine of free will. There is nothing in the behavior of self-disciplined Victorians that upsets the doctrine of determinism; they were just products of their environment, of a time and place where belief in the possibility of self-control was in the air—as were (therefore) stiff moral sanctions against those who failed at the task. Still, these men represent, in a sense, an argument for putting the same influences in our air. At least, these men are evidence that the influences can work; they are cause to consider the doctrine of free will "true" in a sheerly pragmatic sense of the word.[24] But whether such pragmatism can outweigh *real* truth—whether a self-fulfilling "belief" in free will can survive the ever-more-manifest dubiousness of free will as a metaphysical doctrine—is another question altogether.

And, anyway, even if this artifice succeeds, and the idea of "blame" remains conveniently robust, we are back to the challenge of confining it to useful proportions: blaming people only when blame serves the greater good, not letting self-righteousness get carried away (as it naturally tends to do). And, meanwhile, we will still face the deeper challenge of reconciling necessary moral sanction with the limitless compassion that is always, in fact, appropriate.

MILL AS A PURITAN

Launching a war against divorce, complete with harsher sanctions against philanderers, and zero tolerance for their claims that philandering is "natural," may or may not be worth the miscellaneous costs. This is a question about which reasonable people may disagree. But creeping determinism is, in any case, a problem, because moral codes of *some* sort are surely desirable. Morality, after all, is the only way to harvest various fruits of non-zero-sumness—notably those fruits that aren't harvested by kin-selected altruism or reciprocal altruism. Morality makes us mindful of the welfare of people other than family and friends, raising society's overall welfare. You don't have to be a utilitarian to think that's a good thing.

Actually, morality isn't the *only* way to harvest these particular fruits. But it's the cheapest way, and the least creepy. If no one drinks before driving, society is better off. And most of us would rather see compliance enforced by an internalized moral code than by a ubiquitous police force. This is the rigorous answer to people who ask why terms like *morality* and *values* should be taken seriously. Not because tradition is a good thing in itself. But because of what a strong moral code is uniquely able to offer: the more elusive benefits of non-zero-sumness, without lots of police.

John Stuart Mill felt that moral codes could be as stifling and eerie as ubiquitous police. He complained, in *On Liberty,* of living "under the eye of a hostile and dreaded censorship."[25] So it may seem ironic, at the very least, to pen this ode to moral toughness right after penning an ode to Mill's ethical philosophy, utilitarianism.

But Mill's real complaint wasn't about strong moral codes; it was about strong and mindless moral codes. Specifically: codes banning behaviors that wouldn't have harmed anyone—codes, in other words, that weren't sound from a utilitarian standpoint. In those days, various statistically aberrant lifestyles, such as homosexuality, were considered grave crimes against humanity, even though it was hard to find a human they hurt. And divorce was fairly scandalous even if both husband and wife wanted it and were childless.

But not all rules looked so absurd to Mill. In fact, he pointedly did not embrace a general right to leave a marriage.[26] Couching his views on marital responsibility in almost unrecognizably abstract terms, he wrote: "When a person, either by express promise or by conduct, has encouraged another to rely upon his continuing to act in a certain way—to build expectations and calculations, and stake any part of his plan of life upon that supposition—a new series of moral obligations arises on his part towards that person, which may possibly be overruled, but cannot be ignored." And as for leaving a marriage after having children: "[I]f the relation between two contracting parties . . . as in the case of marriage, has even called third parties into existence, obligations arise on the part of both the contracting parties toward those third persons, the fulfillment of which, or at all events the mode of fulfillment, must be greatly affected by the continuance or disruption of the relation between the original parties to the contract."[27] In other words: it's bad to walk out on your family.

Mill's gripe in *On Liberty* is with Victorian moral gravity, not with moral gravity itself. There had been a time in the distant past, he wrote, when "the element of spontaneity and individuality was in excess, and the social principle had a hard struggle with it. . . ." Back then, the difficulty was "to induce men of strong bodies or minds to pay obedience to any rules which required them to control their impulses." But, "society has now fairly got the better of individuality; and the danger which threatens human nature is not the excess, but the deficiency, of personal impulses and preferences."[28] It isn't clear that if Mill were around today he would make the same judgment.

Certainly Mill would attack residues of mindless Victorianism, such as homophobia. But he might well not favor the sort of hedonism that, in the late 1960s, was identified with the left (hallucinogenic drugs and sex) nor the sort that, in the 1980s, was identified with the right (nonhallucinogenic drugs and BMWs).

In fact, Mill considered hedonism fair game for moral judgment even when it hurt no one except the hedonist. We shouldn't *punish* people for ceding their long-term welfare to the animal within, Mill wrote; still, they can only expect that, since they are hazardous models for emulation, we may choose not to associate with them, and indeed may warn our friends against doing so. "A person who shows rashness, obstinacy, self-conceit—who cannot live within moderate means—who cannot restrain himself from hurtful indulgences—who pursues animal pleasures at the expense of those of feeling and intellect—must expect to be lowered in the opinion of others, and to have a less share of their favourable sentiments. . . ."[29]

Here John Stuart Mill, libertarian, meets Samuel Smiles, puritan. Though Mill ridiculed the idea of a "radically corrupt" human nature that must be suffocated in the name of spiritual progress, he also doubted that the higher sentiments, which yield morality, would flower without cultivation. "The truth is," he wrote, "that there is hardly a single point of excellence belonging to human character, which is not decidedly repugnant to the untutored feelings of human nature."[30] Smiles himself couldn't have said it better; a not altogether rosy view of human nature underlay his emphasis in *Self-Help* on strenuous self-restraint.

Indeed, notwithstanding the seemingly opposite drifts of Smiles's and Mill's 1859 books, the two men saw eye-to-eye quite broadly. Both (along with Darwin) embraced the left-of-center political reforms of the day, as well as their philosophical framework; Smiles was a big fan of utilitarianism, which was known in those days as "philosophical radicalism."

Mill's position on human nature accords well enough with modern Darwinism. Surely it would be an exaggeration to say that we are innately evil—that, as Mill's caricature of Calvinism would have it, we cannot be good without ceasing to be human. Indeed, the ingredients of morality, from empathy to guilt, have a deep basis in human nature. At the same time, these ingredients don't spontaneously coalesce into a mind that is truly benevolent; they were not designed for the greater good. Nor do these ingredients reliably promote our *own* happiness. Our happiness was never high among natural selection's priorities, and even if it had been, happiness wouldn't naturally arise in an environment so different from the context of our evolution.

DARWINISM AND IDEOLOGY

There is thus a sense in which the new paradigm lends itself to morally conservative use. By showing that the "moral sentiments" aren't naturally deployed morally, it suggests that a strong moral code may be needed if people are to respect the greater good. Marvelous though it is how often the mutual pursuit of self-interest leads two or more human beings to find common benefit, much common benefit will go unfound unless we take morality seriously.

Does this sort of moral conservatism have a deep connection with *political* conservatism? Not really. True, political conservatives spend more time than their opposites championing moral austerity. But they also tend to think that the strong moral code we should all obey is the one they espouse ex cathedra—or, at least, the one that has the blessing of "tradition." A Darwinian, by contrast, looks at time-honored moral codes with deep ambivalence.

On the one hand, codes that have long endured must have a kind of compatibility with human nature, and probably do serve the interests of at least someone. But of whom? The molding of a moral code is a power struggle, and power in human societies is usually distributed complexly and unequally. Figuring out which agendas are served can be tricky.

The dissection of moral codes—determining who pays for them and who benefits, and the costs and benefits of alternative codes— is best done with the tools of the new paradigm. And it is best done with care. We should, in the end, dispense with those norms that don't make practical sense, but in the meanwhile we should recognize that norms often do make practical sense; they have grown out of an informal give and take that, though never purely democratic, is some-times roughly pluralistic. What's more, this implicit negotiation prob-ably took into account some (perhaps harsh) truths about human nature that may not at first be apparent. We should look at moral axioms the way a prospector looks at shiny rocks—with great respect and great suspicion, a healthy ambivalence pending further, and ur-gent, inspection.

The result of such appraisal will be too diverse to characterize with a simple label. It may be called conservative, so long as that refers to a tentative respect for tradition and not an undying love for it. Then again, the result of the analysis may be called liberal, so long as liberalism isn't equated with hedonism or with moral laissez-faire. If liberalism's moral philosophy is what the (in his day) "radical"

John Stuart Mill laid out in *On Liberty,* then it includes a healthy appreciation of the dark side of human nature and the need for self-restraint, even for moral censure.

As for the effects of creeping biological determinism—which is to say, creeping determinism—they also defy ideological pigeonholing. On the one hand, by stressing that incarceration is always a moral tragedy, if a practical necessity, determinism accents the urgency of erasing the social conditions, such as poverty, that lead to punishable behavior. Darwin saw this. In his notes, after professing his determinism and recognizing the philosophical vacuousness of retribution, he wrote: "Believer in these views will pay great attention to Education." Animals, he noted, "do attack the weak & sickly as we do the wicked.—we ought to pity & assist & educate by putting contingencies in the way to aid motive power."[31]

Yet, Darwin wrote, if a wicked man is "incorrigably bad nothing will cure him."[32] Indeed. Though the new paradigm stresses the mental plasticity that liberals have long stressed, it also suggests—as does casual observation—that this plasticity is not infinite, and certainly not eternal; many mechanisms of mental development seem to have their essential effects during the first two or three decades of life. It's not yet clear how concrete various aspects of the character then become. (Can a man become a nearly incorrigible rapist, or at least incorrigible until his testosterone level drops, near middle age?) But the answers may at times be favored on the political right, by those who argue for locking 'em up and throwing away the key.

Progress in evolutionary psychology will plainly affect—*legitimately* affect—moral and political discourse for decades to come. But no simple ideological label will summarize the effects. Once everyone understands this, there will be no horde of critics on the left, or on the right, for Darwinians to fend off. Enlightenment can then proceed apace.

Chapter 18: **DARWIN GETS RELIGION**

In my journal I wrote that whilst standing in the midst of the grandeur of a Brazilian forest, "it is not possible to give an adequate idea of the higher feelings of wonder, admiration, and devotion which fill and elevate the mind." I well remember my conviction that there is more in man than the mere breath of his body. But now the grandest scenes would not cause any such convictions and feelings to rise in my mind. It may be truly said that I am like a man who has become colour-blind. . . .

—*Autobiography* (1876)[1]

When the HMS *Beagle* left England, Darwin was an orthodox and earnest Christian. He would later recall "being heartily laughed at by several of the officers (though themselves orthodox) for quoting the Bible as an unanswerable authority on some point of morality." But he was beginning to harbor quiet doubts. He was troubled by the Old Testament's "manifestly false history of the world" and its depiction of God as "a revengeful tyrant." He wondered about the New Testament too; though he found the moral teachings of Jesus beautiful, he saw that their "perfection depends in part on the interpretation which we now put on metaphors and allegories."

Darwin longed to regain certainty. He daydreamed about the unearthing of ancient manuscripts that would corroborate the Gospels. It didn't help. "Disbelief crept over me at a very slow rate."[2]

Having lost his Christian faith, Darwin held for many years to a vague theism. He believed in a "First Cause," a divine intelligence that had set natural selection in motion with some end in mind. But then he began to wonder: "[C]an the mind of man, which has, as I fully believe, been developed from a mind as low as that possessed by the lowest animal, be trusted when it draws such grand conclusions?"[3] Darwin finally settled into a more or less stable agnosticism. He might in upbeat moments entertain theistic scenarios; but for long periods of his life, upbeat moments weren't common.

In one sense, however, Darwin always remained a Christian. Like others of his time and place, he was steeped in the moral austerity of Evangelicism. He lived by the tenets that echoed in English churches and found secular expression in Samuel Smiles's *Self-Help:* that a man, by exercising his "powers of action and self-denial" could stay "armed against the temptation of low indulgences." This, as we've seen, was for Darwin the "highest stage in moral culture"— recognizing "that we ought to control our thoughts, and 'not even in inmost thought to think again the sins that made the past so pleasant to us.' "[4]

But if Darwin was in this sense an Evangelical Christian, he could, with almost equal accuracy, be called a Hindu or Buddhist or Muslim. The theme of strict self-governance, the control of animal appetites, appears again and again in the world's great religions. Also widespread, if a bit less so, is the doctrine of brotherly love that Darwin found so beautiful. Six centuries before Jesus, Lao-tzu had said, "It is the way of the Tao . . . to recompense injury with kindness."[5] Buddhist scriptures call for "an all embracing love for all the universe . . . unmarred by hate within, not rousing enmity."[6] Hinduism has the doctrine of "ahimsa," the absence of all harmful intent.

What does a Darwinian make of this striking recurrence of themes? That various men at various times have been privy to the divine revelation of several universal truths? Not exactly.

The Darwinian line on spiritual discourse is much like the Darwinian line on moral discourse. People tend to say and believe things that are in their evolutionarily ingrained interests. This doesn't mean that harboring these ideas always gets their genes spread. Some religious doctrines—celibacy, for example—may dramatically fail to do that. The expectation, rather, is simply that the doctrines people latch on to will have a kind of harmony with the mental organs natural selection has designed. "Harmony," admittedly, is a pretty broad term. These doctrines may, on the one hand, slake some deep psychological thirst (belief in an afterlife gratifies the will to survive); or they may, on the other hand, suppress some thirst so unslakable as to be a burden (lust, for example). But in one sense or another, the beliefs people subscribe to should be explicable in terms of the evolved human mind. Thus when diverse sages manage to sell the same themes, the themes may say something about the contours of that mind, about human nature.

Does this mean common religious teachings have some sort of timeless value as rules to live by? Donald T. Campbell, one of the first psychologists to get enthusiastic about modern Darwinism, has suggested as much. In an address to the American Psychological Association, he spoke of "the possible sources of validity in recipes for living that have been evolved, tested, and winnowed through hundreds of generations of human social history. On purely scientific grounds, these recipes for living might be regarded as better tested than the best of psychology's and psychiatry's speculations on how lives should be lived."[7]

Campbell said this in 1975, just after the publication of Wilson's *Sociobiology* and before Darwinian cynicism had fully crystallized. Today many Darwinians would be less sanguine. Some have noted that, while ideas must by definition have a kind of harmony with the brains they settle into, that doesn't mean they're good for those brains in the long run. Some ideas, indeed, seem to parasitize brains—they are "viruses," as Richard Dawkins puts it.[8] The idea that injecting heroin is fun keeps infecting people by appealing to myopic cravings, rarely to the ultimate advantage of those people.

Besides, even if an idea does spread by serving people's long-term interests, the interests may be those of its sellers, not its buyers. Religious leaders tend to have high status, and it is not beyond the pale to see their preachings as a form of exploitation, a subtle bending of the listener's will to the speaker's goals. Certainly Jesus' teachings, and the Buddha's teachings, and Lao-tzu's teachings had the effect of amplifying the power of Jesus and Buddha and Lao-tzu, raising their stature within a growing group of people.

Still, it's not as though religious doctrines were always *forced* on people. Granted, the Ten Commandments had a certain totalitarian authority, conveyed by the political leadership and carrying God's own signature. Jesus too, though lacking political office, regularly invoked God's endorsement. But the Buddha, for one, didn't stress supernatural authority. And, though born to a noble station, he is said to have abandoned the trappings of status to roam the world and teach; his movement started, apparently, from scratch.

The fact is that many people at various times have bought various religious doctrines under no great external coercion. Presumably, there was some psychological payoff. The great religions are at some level ideologies of self-help. It would indeed be wasteful, as Campbell suggests, to throw out eons of religious tradition without inspecting it first. The sages may have been self-serving, like the rest of us, but that doesn't mean they weren't sages.

DEMONS

One great theme of the great religions is demonic temptation. Time and again we see an evil being that tries, in the guise of innocence, to entice people into seemingly minor but ultimately momentous wrongdoing. In the Bible and the Koran there is Satan. In Buddhist scripture there is the arch-tempter Mara, who insidiously deploys his daughters, Rati (Desire) and Raga (Pleasure).

Demonic temptation may not sound like an especially scientific doctrine, but it captures nicely the dynamics by which habits are acquired: slowly but surely. For example, natural selection "wants" men to have sex with an endless series of women. And it realizes this goal with a subtle series of lures that can begin, say, with the mere contemplation of extramarital sex and then grow steadily more powerful and ultimately inexorable. Donald Symons has observed, "Jesus said, 'Whosoever looketh on a woman to lust after her hath committed adultery with her already in his heart' because he understood that the function of the mind is to cause behavior."[9]

It is no coincidence that demons and drug dealers often use the same opening line ("Just try a little; it will feel good"), or that religious people often see demons in drugs. For habituation to any goal—sex or power, say—is literally an addictive process, a growing dependence on the biological chemicals that make these things gratifying. The more power you have, the more you need. And any slippage will make you feel bad, even if it leaves you at a level that once brought ecstasy. (One habit that natural selection never "meant" to encourage was drug addiction itself. This miracle of technology is an unanticipated biochemical intervention, a subversion of the reward system. We were meant to get our thrills the old-fashioned way, from a hard day's work: eating, copulating, undermining rivals, and so on.)

Demonic temptation connects almost seamlessly with the more basic notion of evil. Both ideas—a malign being, and a malign force—lend emotional power to spiritual counsel. When the Buddha tells us to "dig up the root of thirst" so that "Mara, the tempter, may not crush you again and again," we are supposed to steel ourselves for the battle to come; those are fightin' words.[10] Warnings that drugs or sex or a belligerent dictator are "evil" bring much the same effect.

The concept of "evil," though less metaphysically primitive than, say, "demons," doesn't fit easily into a modern scientific worldview. Still, people seem to find it useful, and the reason is that it is metaphorically apt. There is indeed a force devoted to enticing us into various pleasures that are (or once were) in our genetic interests but do not bring long-term happiness to us and may bring great suffering to others. You could call that force the ghost of natural selection. More concretely, you could call it our genes (*some* of our genes, at least). If it will help to actually use the word *evil*, there's no reason not to.

When the Buddha urges digging up the "root of thirst," he isn't necessarily counseling abstinence. Certainly there is talk in many religions of abstinence from various things, and certainly abstinence is one way to short-circuit the addictiveness of vice. But the Buddha put his emphasis not so much on a laundry list of proscriptions as on a generally austere attitude, a cultivated indifference to material rewards and sensory pleasure: "Cut down the whole forest of desires, not a tree only!"[11]

This fundamental defiance of human nature is encouraged in some measure by other religions. In the Sermon on the Mount, Jesus said, "Lay not up for yourselves treasures upon earth"; and "Take no thought for your life, what ye shall eat, or what ye shall drink; nor yet for your body, what ye shall put on."[12] The Hindu scriptures, like the Buddhist, dwell at more length, and more explicitly, on withdrawal from the realm of pleasure. The spiritually mature man is one who "abandons desires," who "has lost desire for joys," who "withdraws, as a tortoise his limbs from all sides, his senses from the objects of sense."[13] Hence the ideal man as depicted in the Bhagavad Gita: a man of discipline, who acts without worrying about the fruits of his action, a man who is unmoved by acclaim and by criticism. This was the image that inspired Gandhi to persevere without "hope of success or fear of failure."

That Hinduism and Buddhism sound so much alike is not shocking. The Buddha was born a Hindu. But he carried the theme of sensory indifference further, boiling it down to a severe maxim—life is suffering—and placing it at the very center of his philosophy. If you accept the inherent misery of life, and follow the teachings of the Buddha, then you can, oddly enough, find happiness.

In all these assaults on the senses there is a great wisdom—not only about the addictiveness of pleasures but about their ephemerality. The essence of addiction, after all, is that pleasure tends to dissipate and leave the mind agitated, hungry for more. The idea that just one more dollar, one more dalliance, one more rung on the ladder will leave us feeling sated reflects a misunderstanding about human nature—a misunderstanding, moreover, that is built into human nature; we are designed to feel that the next great goal will bring bliss, and the bliss is designed to evaporate shortly after we get there. Natural selection has a malicious sense of humor; it leads us along with a series of promises and then keeps saying "Just kidding." As the Bible puts it, "All the labour of man is for his mouth, and yet the appetite is not filled."[14] Remarkably, we go our whole lives without ever really catching on.

The advice of the sages—that we refuse to play this game—is nothing less than an incitement to mutiny, to rebel against our creator. Sensual pleasures are the whip natural selection uses to control us, to keep us in the thrall of its warped values system. To cultivate some indifference to them is one plausible route to liberation. While few of us can claim to have traveled far on this route, the proliferation of this scriptural advice suggests it has been followed *some* distance with *some* success.

There is also a more cynical explanation for that proliferation. One way to reconcile poor people to their plight is to convince them that material pleasures aren't fun anyway. Exhortations to forswear indulgence could be simply an instrument of social control, of oppression. So too with Jesus' assurance that in the afterlife the "first shall be last and the last shall be first"[15]—it sounds a bit like a way of recruiting low-status people to his growing army, a recruitment that may come at their own expense, as they cease to struggle for worldly success. Religion, in this view, has always been the opiate of the masses.

Maybe so. But it remains true that pleasure *is* ephemeral; that its constant pursuit is *not* a reliable source of happiness (as not only Samuel Smiles but also John Stuart Mill noted); that we are built *not* to easily grasp this fact; and that the reasons for all this are clearer in light of the new Darwinian paradigm.

There are scattered hints in the ancient scriptures of an understanding that human striving—after pleasure, after wealth, after status—is yoked to self-deception. The Bhagavad Gita teaches that men "devoted to enjoyment and power" are "robbed of insight." To pursue the fruits of action is to live in a "jungle of delusion."[16] The Buddha said that "the best of virtues [is] passionlessness; the best of men he who has eyes to see."[17] In Ecclesiastes it is written: "Better is the sight of the eyes than the wandering of the desire."[18]

Some of these utterances are, in context, ambiguous, but there is no doubt about the clarity with which sages have seen one particular human delusion: the basic moral bias toward self. The idea recurs in Jesus' teaching—"He that is without sin among you, let him first cast a stone"; "Thou hypocrite, first cast out the beam out of thine own eye; and then shalt thou see clearly to cast out the mote out of thy brother's eye."[19] The Buddha put it in plainer language: "The fault of others is easily perceived, but that of one's self is difficult to perceive."[20]

The Buddha saw, in particular, that much delusion grows out of the human penchant for one-upmanship. In warning his followers against dogmatic squabbling, he said:

> The senses' evidence,
> and works, inspire such scorn
> for others, and such smug
> conviction *he* is right,
> that all his rivals rank
> as "sorry, brainless fools."[21]

This grasp of our naturally skewed perspective is bound up with exhortations toward brotherly love. For a premise of these exhortations is that we are deeply inclined *not* to view everyone with the charity we extend to our kin and ourselves. Indeed, if we weren't so deeply inclined, if we didn't buttress this inclination with all the moral and intellectual conviction at our disposal, you wouldn't have to start a whole religion to correct the imbalance.

The renunciation of sensory pleasure is also tied to brotherly love. Acting with generosity and consideration is tricky unless you somehow escape the human preoccupation with feeding the ego. Taken as a whole, some bodies of religious thought are a fairly coherent program for maximizing non-zero-sumness.

THEORIES OF BROTHERLY LOVE

The question remains: How did these bodies get started? Why has the doctrine of brotherly love so thrived? Leave aside for the moment that it is honored mainly in the breach, that even those who most diligently pursue it may manage to dilute their self-love only slightly, that organized religions have often been vehicles for violating the doctrine on a spectacular scale. The mere fact that the idea lives on in this species is curious. In light of Darwinian theory, everything about the idea of brotherly love seems paradoxical except for the rhetorical power of the term *brotherly*. And this alone, surely, hasn't been enough to sell the idea.

Proposed solutions to this mystery range from the highly cynical to the mildly inspiring. At the more inspiring end of the spectrum is a theory by the philosopher Peter Singer. His book *The Expanding Circle* asks how the range of human compassion grew beyond its primitive bounds—the family, or perhaps the band. Singer notes that human nature, and the structure of human social life, long ago got people in the habit of publicly justifying their actions in objective terms. When we urge respect for our interests, we talk as if we are asking for no more than we would give anyone else in our shoes. Singer believes that once this habit is established (by the evolution of reciprocal altruism, among other things), the "autonomy of reasoning" takes over. "The idea of a disinterested defense of one's conduct" grew out of self-interest, "but in the thought of reasoning beings, it takes on a logic of its own which leads to its extension beyond the bounds of group."

This extension has grown impressively. Singer recounts how Plato urged his fellow Athenians to adopt what at the time was a major moral advance: "He argued that Greeks should not, in war, enslave other Greeks, lay waste their lands or raze their houses; they should do these things only to non-Greeks."[22] The growth of moral concern to the bounds of the nation-state has long since become the norm. Eventually, Singer believes, it may reach global proportions: starvation in Africa will seem as scandalous to Americans as starvation in America. Pure logic will have brought us truly in touch with the great religious teachings of the ages—the fundamental moral equality of everyone. Our compassion will, as it should, spread evenly across humanity. Darwin shared this hope. He wrote in *The Descent of Man:* "As man advances in civilisation, and small tribes are united into larger communities, the simplest reason would tell each individual that he ought to extend his social instincts and sympathies to all the members of the same nation, though personally unknown to him. This point being once reached, there is only an artificial barrier to prevent his sympathies extending to the men of all nations and races."[23]

In a sense, Singer is saying that our genes have been too clever by half. They long ago began cloaking raw selfishness in the lofty language of morality, using it to exploit the various moral impulses natural selection created. Now this language, as harnessed to pure logic, impels the brains they built to behave with selflessness. Natural selection designed two things for narrow self-interest—cold reason and warm moral impulses—and somehow, when combined, they take on a life of their own.

Enough inspiration. The most cynical explanation of why so many sages have urged an expanded moral compass is the one set out near the beginning of this chapter: a large compass expands the power of the sages doing the urging. The Ten Commandments, with their bans on lying, stealing, and murder, made Moses' flock more manageable. And the Buddha's warnings about dogmatic squabbling kept his power base from splintering.

Supporting this cynicism is the fact that the universal love espoused in many scriptures doesn't emerge from scrutiny looking truly universal. The odes to selflessness in the Bhagavad Gita come in a somewhat ironic context: Lord Krishna is spurring the warrior Arjuna toward self-discipline so that he will more effectively slaughter an enemy army—an army, no less, that contains some of his own kin.[24] And in Paul's Epistle to the Galatians, after singing the praises of love, peace, gentleness, and goodness, he says, "[L]et us do good unto all men, *especially* unto them who are of the household of faith."[25] These are wise words indeed, coming from the head of the household. The case has been made that even Jesus didn't really preach universal love, that his injunctions to love your "enemies," when appraised carefully, are seen to apply only to Jewish enemies.[26]

In this light, Singer's "expanding circle" seems an extension less of moral logic than of political reach. As social organization goes beyond the level of the hunter-gatherer band—to the tribe, the city-state, the nation-state—religious organization on an ever larger scale is feasible. So sages take the opportunity to expand their power—which means preaching a commensurately broad tolerance. Thus, appeals for brotherly love are comparable to a politician's self-serving appeals to patriotism. In fact, appeals to patriotism *are*, in a way, appeals for brotherly love on a national scale.[27]

There is a third theory that stands near the middle of the cynicism spectrum. Yes, it holds, the Ten Commandments may have made Moses' flock more manageable. But presumably many of the sheep benefited too, since mutual restraint and consideration bring non-zero-sum benefits. In other words, religious leaders, however self-interested, haven't been simply foisting their interests on the masses. They've been finding overlap between their interests and the masses' interests, and the overlap has gotten larger; as the scope of social and economic organization has grown, and with it the zone of non-zero-sumness, the self-interest of people has lain in behaving with at least minimal decency toward larger and larger numbers of people. Religious leaders are more than happy to have their stature rise commensurately.

There has been a change not just in the *scope* of social organization, but in its nature. The moral sentiments were designed for a particular environment—or, more precisely, for a particular series of environments, including hunter-gatherer villages and other, earlier, societies that are lost in the mists of prehistory. It is safe to say that these societies didn't have an elaborate judicial system and a large police force. Indeed, the strength of the retributive impulse is testament to a time when, if you didn't stand up for your interests, no one else would.

At some point, things began to change, and the value of these impulses began to wane. Today, most of us waste great quantities of time and energy indulging our indignation. We rail ineffectually at careless drivers; we spend a day working with police to find a purse-snatcher, even though the purse contained what we earn in three hours' work and catching the thief won't change the odds of being victimized in the future; we smolder at the fortune of professional rivals, even though we are powerless to bring them misfortune and would profit from treating them with greater civility.

When exactly in human history some of the moral sentiments began to obsolesce is hard to say. But it is worth pondering Donald Campbell's insight that it is the religions of the ancient *urban* civilizations—"independently developed in China, India, Mesopotamia, Egypt, Mexico, and Peru"—that reliably produced the familiar elements of modern religions: the curbing of "many aspects of human nature," including "selfishness, pride, greed, . . . covetousness, . . . lust, wrath."

Campbell believes this curbing was needed for "optimal social coordination."[28] Whether he means optimal for the ruler or optimal for the ruled he doesn't say. But we can take heart in the fact that, though the two are sometimes at odds, they aren't mutually exclusive.

What's more, the "social coordination" in question may extend beyond the scope of any single nation. It is by now trite to say that the peoples of the world are more interdependent than ever. Trite but true. Material progress has greatly deepened economic integration, and various technologies have brought threats that humanity can forestall only in concert, such as environmental degradation and nuclear proliferation. There may have been a time when it was commonly in the interests of political leaders to stoke their people's intolerance and bigotry to the point of international strife. This time is passing.

The Hindu scriptures teach that a single universal soul resides in everyone; the wise man "sees himself in all and all in him."[29] As a metaphor for a great philosophical truth—the equal sacredness (read: utilitarian worth) of every human sphere of consciousness—this teaching is profound. And as the basis for a practical rule of living— that the wise man refrains from harming others so that "he harms not himself"[30]—this teaching is prescient. The ancient sages pointed—however ambiguously, however selfishly—to a truth that was not just valid, and not just valuable, but destined to grow in value as history advanced.

TODAY'S SERMONETTE

In illustrating the "puritan conscience" of Victorian England, Walter Houghton described a man who wrote down all his "sins and errors" and habitually detected "selfishness . . . in every effort and resolve."[31] The idea goes back at least as far as Martin Luther, who said a saint is someone who understands that everything he does is egotistical.

This definition of sainthood reflects favorably on Darwin. Here is a characteristic utterance: "But what a horridly egotistical letter, I am writing; I am so tired, that nothing short of the pleasant stimulus of vanity & writing about one's own dear self would have sufficed."[32] (Needless to say, this sentence followed a passage that would strike few people today as egotistical. He had been voicing anxiety, not confidence, about how his work aboard the *Beagle* would be received.)

Whether or not Darwin, by Luther's measure, fully qualifies as a saint, it is certainly true that Darwinism, by this measure, can help make a person saintly. No doctrine heightens one's consciousness of hidden selfishness more acutely than the new Darwinian paradigm. If you understand the doctrine, buy the doctrine, and apply the doctrine, you will spend your life in deep suspicion of your motives.

Congratulations! That is the first step toward correcting the moral biases built into us by natural selection. The second step is to keep this newly learned cynicism from poisoning your view of everyone else: to pair harshness toward self with leniency toward others; to somewhat relax the ruthless judgment that often renders us conveniently indifferent to, if not hostile to, their welfare; to apply liberally the sympathy that evolution has meted out so stingily. If this operation is inordinately successful, it might result in a person who takes the welfare of others markedly, but at least not massively, less seriously than his own.

Darwin did a reasonable job of this. Though fairly attuned to, and disdainful of, other people's vanity, his general attitude toward others was one of great moral seriousness; he reserved most of his mockery for himself. Even when he couldn't help but hate people, he tried to keep his hate in perspective. Regarding archenemy Richard Owen, he wrote to his friend Hooker, "I am become quite demoniacal about Owen" and "I mean to try to get more angelic in my feelings."[33] The point isn't whether he succeeded. (He didn't.) The point is that to half-jokingly apply the word "demoniacal" to one's hatreds is to show more moral self-doubt, and less self-importance, than most of us usually manage. (This is all the more impressive as Darwin's feelings were hardly eccentric; Owen, though a particular threat to Darwin's status by virtue of his disbelief in natural selection, was also a spiteful and widely disliked man.)[34] Darwin came fairly close to the nearly impossible and highly commendable: a detached, thoroughly modern (if not postmodern) cynicism toward self, paired with Victorian earnestness toward others.

Another thing Martin Luther said is that chronic moral torment is a sign of God's grace. If so, Darwin was a walking grace repository. Here was a man who could lie guiltily awake at night because he hadn't yet answered some bothersome piece of fan mail.[35]

We might ask what is so gracious about filling someone with anguish. One answer is that other people can benefit from it. Perhaps what Luther should have said is that a morally tormented person is a *medium* for God's grace. And this (metaphorically speaking, at least) Darwin sometimes was: he was a utilitarian magnifier. Through the magic of non-zero-sumness, he turned his minor sacrifices into other people's major gains. By spending a few minutes writing a letter, he could markedly brighten the day, and perhaps the week, of some unknown soul. This is not what the conscience was designed for, since these people were usually in no position to reciprocate, and were often too remote to help Darwin's moral reputation. As we've seen, a good conscience, in the most demanding, most moral sense of the term, is one that doesn't work only as natural selection "intended."

Some people worry that the new Darwinian paradigm will strip their lives of all nobility. If love of children is just defense of our DNA, if helping a friend is just payment for services rendered, if compassion for the downtrodden is just bargain-hunting—then what is there to be proud of? One answer is: Darwin-like behavior. Go above and beyond the call of a smoothly functioning conscience; help those who aren't likely to help you in return, and do so when nobody's watching. This is one way to be a truly moral animal. Now, in the light of the new paradigm, we can see how hard this is, how right Samuel Smiles was to say that the good life is a battle against "moral ignorance, selfishness, and vice"; these are indeed the enemies, and they are tenacious by design.

Another antidote to despair over the ultimate baseness of human motivation is, oddly enough, gratitude. If you don't feel thankful for the somewhat twisted moral infrastructure of our species, then consider the alternative. Given the way natural selection works, there were only two possibilities at the dawn of evolution: (a) that eventually there would be a species with conscience and sympathy and even love, all grounded ultimately in genetic self-interest; (b) that no species possessing these things would ever exist. Well, *a* happened. We do have a foundation of decency to build on. An animal like Darwin can spend lots of time worrying about other animals—not just his wife, children, and high-status friends, but distant slaves, unknown fans, even horses and sheep. Given that self-interest was the overriding criterion of our design, we are a reasonably considerate group of organisms. Indeed, if you ponder the utter ruthlessness of evolutionary logic long enough, you may start to find our morality, such as it is, nearly miraculous.

DARWIN'S END

Darwin himself would have been among the last to see God's grace in his anguish, or in anything else. He reported, near the end of his life, that his typical frame of mind was agnostic. When he declared, the day before he died, "I am not the least afraid of death," it was almost surely in anticipation of relief from his earthly suffering, not in hope of anything better to come.[36]

Darwin had pondered the meaning of life for "a man who has no assured and ever present belief in the existence of a personal God or of a future existence with retribution and reward." He believed such a man would find "in accordance with the verdict of all the wisest men that the highest satisfaction is derived from following certain impulses, namely the social instincts. If he acts for the good of others, he will receive the approbation of his fellow men and gain the love of those with whom he lives; and this latter gain undoubtedly is the highest pleasure on this earth." Still, "his reason may occasionally tell him to act in opposition to the opinion of others, whose approbation he will then not receive; but he will still have the solid satisfaction of knowing that he has followed his innermost guide or conscience."[37]

Maybe this last sentence was a loophole, designed for a man who had spent his life building a theory that lacked the universal "approbation of his fellow men," a theory that, though true, might not tend toward "the good of others." Certainly it is a theory with which our species has yet to make its peace.

Having crafted a moral measuring stick, Darwin gave his life a passing grade. "I believe that I have acted rightly in steadily following and devoting my life to science." Still, while feeling "no remorse from having committed any great sin," he had "often regretted that I have not done more direct good to my fellow creatures. My sole and poor excuse is much ill-health and my mental constitution, which makes it extremely difficult for me to turn from one subject or occupation to another. I can imagine with high satisfaction giving up my whole time to philanthropy, but not a portion of it; though this would have been a far better line of conduct."[38]

It's true that Darwin didn't live the optimally utilitarian life. No one ever has. Still, as he prepared to die, he could rightly have reflected on a life decently and compassionately lived, a string of duties faithfully discharged, a painful, if only partial, struggle against the currents of selfishness whose source he was the first man to see. It wasn't a perfect life; but human beings are capable of worse.

Acknowledgments

A number of people were nice enough to read and comment on drafts of parts of this book: Leda Cosmides, Martin Daly, Marianne Eismann, William Hamilton, John Hartung, Philip Hefner, Ann Hulbert, Karen Lehrman, Peter Singer, Donald Symons, John Tooby, Frans de Waal, and Glenn Weisfeld. I know all of them had better things to do, and I'm grateful.

A few people actually summoned enough grim self-discipline to read a draft of the whole book: Laura Betzig, Jane Epstein, John Pearce, Mickey Kaus (who has also improved many of my other writings over the years), Mike Kinsley (who, while editor of *The New Republic* and since, has improved even more of them), and Frank Sulloway (who was also kind enough to lend various other aid, including the use of his photo archives). Gary Krist gave me trustworthy feedback on an even earlier, messier version of the whole book, and also provided sound advice and vital moral support later in the game. Each of these people deserves a medal.

Marty Peretz gave me an extended leave of absence from *The New Republic,* in keeping with his general, and rare, policy of letting people explore things that interest them. I am lucky to work for someone who genuinely respects ideas. During that leave, Henry and Eleanor O'Neill provided a winter's free lodging in Nantucket, allowing me to write a part of this book under some of the most beautiful conditions imaginable.

Edward O. Wilson, by writing *Sociobiology* and *On Human Nature,* got me interested in this stuff, and has been helpful since then. John Tyler Bonner, James Beniger, and Henry Horn, who cotaught a seminar on sociobiology while I was in college, sustained my interest. While an editor at *The Sciences* magazine in the mid-1980s, I had the privilege of editing Mel Konner's column, "On Human Nature." I learned a lot from the column, and from my conversations with Mel, about this view of life.

Thanks to Bill Strobridge (for encouraging me to become a writer), Ric Aylor (for steering me toward B. F. Skinner's writings while I was still in high school), Bill Newlin (for early advice), Jon Weiner, Steve Lagerfeld, and Jay Tolson (for later advice), Sarah O'Neill (for timely babysitting and other acts of altruism), and my brother, Mike Wright (for fueling my fascination with this book's subject in ways he doesn't know, including being such a moral animal himself). Several colleagues at *The New Republic* whom I've already mentioned—Ann Hulbert, Mickey Kaus, and Mike Kinsley—deserve a curtain call for providing advice and commiseration on a day-to-day basis. I feel privileged to have known and worked with them over the past few years. John McPhee, who as my college teacher did much to shape the direction of my life, also gave me valued advice during this project. This isn't a very McPheeesque book, but it is guided by some of his values (e.g., it's all true, so far as I know, and I didn't choose the subject with income maximization in mind).

Various scholars (including many of those mentioned above, especially in the first paragraph) have let me interrogate them, formally or informally: Michael Bailey, Jack Beckstrom, David Buss, Mildred Dickemann, Bruce Ellis, William Irons, Elizabeth Lloyd, Kevin MacDonald, Michael McGuire, Randolph Nesse, Craig Palmer, Matt Ridley, Peter Strahlendorf, Lionel Tiger, Robert Trivers, Paul Turke, George Williams, David Sloan Wilson, and Margo Wilson. A number of people provided me with reprints of their papers, the answers to nagging questions, etcetera: Kim Buehlman, Elizabeth Cashdan, Steve Gangestad, Mart Gross, Elizabeth Hill, Kim Hill, Gary Johnson, Debra Judge, Bobbi Low, Richard Marius, and Michael Raleigh. I'm sure I'm forgetting people, including a lot of members of the

Human Behavior and Evolution Society whom I've buttonholed at their meetings.

My editor, Dan Frank, is rare among contemporary editors in the amount and quality of attention he gives to manuscripts. A number of other people at Pantheon, including Marge Anderson, Altie Karper, Jeanne Morton, and Claudine O'Hearn, have also been helpful. My agent, Rafe Sagalyn, has been generous with his time and sound in his advice.

Finally, to my wife, Lisa, I owe the largest debt. I still remember when she first read the first draft of the first part of this book and explained to me—yet without using this word—that it was bad. She has read the manuscript in various forms since then and often presented similarly penetrating judgments in similarly diplomatic fashion. Often, when I was faced with conflicting advice, or otherwise befuddled, her reaction served as my guiding light. In addition, she has done all kinds of other things that allowed me to write this book without going totally crazy. I could not have asked for more (although, as I recall, I did on a few occasions).

Lisa disagrees with parts of this book. I'm sure that everyone else I've mentioned does too. That's the way things are in a young science that is morally and politically charged.

Appendix: Frequently Asked Questions

In 1859, after Darwin sent his brother Erasmus a copy of *The Origin of Species,* Erasmus replied with a letter of praise. The theory of natural selection was so logically compelling, he said, that the fossil record's failure to document incremental evolutionary change didn't much bother him. "In fact the a priori reasoning is so entirely satisfactory to me that if the facts won't fit in, why so much the worse for the facts is my feeling."

This sentiment is more widely shared by evolutionists than some of them would admit. The theory of natural selection is so elegant and powerful as to inspire a kind of faith in it—not *blind* faith, really, since the faith rests on the theory's demonstrated ability to explain so much about life. But faith nonetheless; there is a point after which one no longer entertains the possibility of encountering some fact that would call the whole theory into question.

I must admit to having reached this point. Natural selection has now been shown to plausibly account for so much about life in general and the human mind in particular that I have little doubt that it can account for the rest. Still, "the rest" is no trivial chunk of terrain. There is much about human thought, feeling, and behavior that still puzzles and challenges a Darwinian—and much else that may not strike a confirmed Darwinian as so puzzling but does strike the layperson that way. It would be quite un-Darwinesque of me not to mention a few prominent examples. Darwin was nearly preoccupied with his theory's real and apparent shortcomings, and his insistence on confronting them is one thing that makes the *Origin* so persuasive. The shortcoming that Erasmus alluded to came from the chapter Darwin called "Difficulties on Theory." In later editions, Darwin added another chapter called "Miscellaneous Objections to the Theory of Natural Selection."

What follows is hardly an exhaustive list of the puzzles and apparent puzzles that surround the new Darwinian paradigm as applied to the human mind. But it conveys their nature and suggests some prospects for solving them. It also addresses some of the most commonly asked questions about evolutionary psychology and, I hope, helps dispel some common misconceptions.

1. What about homosexuals? One wouldn't expect natural selection to create people who are disinclined to do the things (for example, heterosexual intercourse) that get their genes transported into the next generation. At the dawn of sociobiology, some evolutionists thought that the theory of kin selection might solve this paradox. Homosexuals, perhaps, were like sterile ants: rather than spend their energy trying to get their genes sent directly to the next generation, they use oblique conduits; rather than invest in children of their own, they invest in siblings, nieces, nephews.

In principle this explanation could work, but reality doesn't seem to favor it. First of all, how many homosexuals spend an inordinate amount of time helping siblings, nephews, and nieces? Second, look at what many of them do spend their time doing: pursuing homosexual union about as ardently as heterosexuals seek heterosexual union. What's the evolutionary logic in that? Sterile ants don't spend lots of time caressing other sterile ants, and if they did it would constitute a puzzle.

It is notable that bonobos, our near kin, exhibit bisexuality (though apparently not exclusive homosexuality). They engage in genital rubbing, for example, as a sign of friendliness, a way to defuse tension. This points to a general principle: once natural selection has created a form of gratification—genital stimulation, in this case—that form can come to serve other functions; it can either be adapted to these other functions via genetic evolution or can come to serve them via sheerly cultural change. Thus, ancient Greece developed a cultural tradition whereby boys sometimes pleased men with sexual stimulation. (And, in sheerly Darwinian terms, it's quite debatable who was exploiting whom; boys who used this technique to cultivate mentors were at least getting their status raised in the process; the men—again, in sheerly Darwinian terms—seem to have been wasting their time.)

In this view, the fact that some people's sexual impulses get diverted from typical channels is just another tribute to the malleability of the human mind. Given a particular set of environmental influences, it may do any number of things. (Prison is an extreme example of such an environmental influence; when heterosexual gratification is impossible, the sexual urge—especially the relatively strong and indiscriminate *male* sexual urge—may seek the closest substitute.)

Is there a "gene" for homosexuality? There is evidence suggesting that some genes are more likely to lead to homosexuality than others. But that doesn't mean there's a "gay gene"—a gene that drives one inexorably to homosexuality, regardless of environment; and it certainly doesn't mean that the genes in question were selected by natural selection *by virtue* of their contribution to homosexuality. (Some genes no doubt make a person more likely to enter, say, banking, or professional football, than other genes; but there's no "banker gene" or "pro football gene"—no gene that was selected *by virtue* of its contribution to one's banking or football playing. Just genes conducive to, say, facility with numbers, or to physical strength.) Indeed, once you rule out the kin-selection theory of homosexual inclination, it is very hard to imagine a gene being selected *by virtue* of its leading to exclusive homosexuality. If there is a "gay gene" that has spread to a sizable part of the population, it probably was having some effect other than homosexual inclination in the environment in which it spread.

One reason some people are so concerned about the "gay gene" question is that they want to know if homosexuality is "natural," a question that—to them, at least—seems to have moral consequence. They think it matters greatly whether (a) there is a gene (or combination of genes) conducive to homosexuality that indeed was selected by virtue of that effect; or (b) there is a gene (or combination of genes) conducive to homosexuality that was selected for some other reason but, in some environments, has the effect of encouraging homosexuality; or (c) there is a gene (or combination of genes) conducive to homosexuality that is a fairly recent arrival on the human scene and hasn't yet gotten a strong endorsement from natural selection for any particular property; or (d) there is no "gay gene."

But who cares? Why should the "naturalness" of homosexuality in any way affect our moral judgment of it? It is "natural," in the sense of being "approved" by natural selection, for a man to kill someone he finds sleeping with his wife. Rape may, in the same sense, be "natural." And seeing that your children are fed and clothed is surely "natural." But most people rightly judge these things by their consequences, not their origins. What is plainly true about homosexuality is the following: (1) some people are born with a combination of genetic and environmental circumstance that impels them strongly toward a homosexual lifestyle; (2) there is no inherent contradiction between homosexuality among consenting adults and the welfare of other people. For moral purposes (I believe) that should be the end of the discussion.

2. *Why are siblings so different from one another?* If genes are so important, why do people who have so many genes in common so often turn out so unlike one another? In a certain sense, this question isn't a logical one to ask an evolutionary psychologist. After all, mainstream evolutionary psychology doesn't study how different genes lead to different behaviors, but how the genes common to the human species can lead to various behaviors—sometimes different, sometimes similar. In other words, evolutionary psychologists typically analyze behavior without regard to an individual's peculiar genetic constitution. Still, the answer to this question about siblings sheds much light on a puzzle that *is* central to evolutionary psychology: If the main genetic influences on human behavior come from genes that all people share, why do people *in general* behave so differently from one another? We've addressed this question from various angles in this book, but the matter of siblings sheds a new kind of light on it.

Consider Darwin. He was the second youngest of six children. As such, he conforms to a striking pattern that has only recently come to light: people who initiate or support scientific revolutions are exceedingly unlikely to be firstborn children. Frank Sulloway (see Sulloway [in press]), who has documented this pattern with voluminous data, has also found that people who lead or support *political* revolutions are very unlikely to be firstborn children.

How to explain this pattern? Presumably, Sulloway notes, it has something to do with the fact that younger children often find themselves in competition with older siblings—authority figures—for resources. Indeed, they may find themselves in conflict not just with these particular authorities, but with a whole establishment. After all, firstborn children, having higher reproductive value than their younger siblings (see chapter seven), should, in theory, tend to be favored by parents, all other things being equal. So there may often be a natural commonality of interest, an alliance, between parents and older siblings that younger siblings find themselves combatting. The establishment lays down the law, the younger sibling challenges it. It could be adaptive for children who find themselves thus situated to become good at questioning received rules. That is: a species-typical developmental program may tend to steer children with older siblings toward radical thought.

The larger point here is about "nonshared environment," whose importance geneticists have grasped only over the last decade (see Plomin and Daniels [1987]). People who doubt environmental determinism like to point to two brothers, raised side-by-side, and ask why one of them became, say, a criminal and the other a district attorney. If environment is so important, they ask, why did these people turn out so differently? Such questions misconstrue the meaning of "environment." Though two brothers do share some aspects of their environment (the same parents, same school) a large part of their environment is "nonshared" (who their first-grade teacher was, who their friends are, and so on).

Paradoxically, as Sulloway (see Sulloway [in preparation]) points out, siblings may, by virtue of being siblings, have *particularly* disparate "nonshared" environments. For example, while you and your next-door neighbor may both be firstborn children—and thus "share" this environmental influence—the same can't possibly be true of both you and your sibling. What's more: Sulloway believes that one sibling, by virtue of occupying a certain strategic "niche" within the family ecology, may push other siblings toward other niches in their struggle to compete for resources. Thus, a younger sibling may find that another sibling has already won great favor through, say, conscientious sacrifice for the parents; in response, he or she may seek another "niche"—excellence in school, say—rather than try to compete in the already crowded sacrifice market.

3. *Why do people choose to have few or no kids?* This is sometimes cited as a great evolutionary "mystery." Academics have puzzled over the "demographic transition" that lowered birthrates in industrialized societies, trying to explain it in Darwinian terms. Some theorize, for example, that in a modern environment, having what was once considered an average-sized family can be bad for your genetic legacy. Maybe you will wind up with more grandchildren if you have two children, both of whom you can afford to educate at expensive private schools, than if you have five children who get educated at cheaper schools and find themselves unable to support children themselves. Thus, in having fewer children, people are behaving adaptively.

There is a simpler solution: natural selection's primary means of getting us to reproduce hasn't been to instill in us an overwhelming, conscious desire to have children. We are designed to love sex and then to love the consequences that materialize nine months later— not necessarily to anticipate loving the consequences. (Witness the Trobriand Islanders, who according to Malinowski hadn't grasped the connection between sex and childbirth but, nonetheless, had managed to keep reproducing.) Only in the wake of contraceptive technology has this design faltered.

The choice of family size is one of many cases where we have outsmarted natural selection; through conscious reflection—seeing, for example, that children, however lovable, can be quite burdensome in certain quantities—we can choose to short-circuit the ultimate goals that natural selection "intended" us to pursue.

4. Why do people commit suicide? Again, one can try to construct scenarios in which this sort of behavior might be adaptive. Maybe a person in the ancestral environment who had become a burden on his family would actually maximize inclusive fitness by taking himself out of the picture. Maybe, for example, food is so scarce that by continuing to eat he would deprive more reproductively valuable relatives of nutrients to the point of endangering their lives.

This explanation is not entirely implausible, but there are some problems with it. One is that, in the modern environment at least, people who commit suicide seldom belong to families near starvation. And, really, being near starvation is just about the only circumstance in which suicide could make much Darwinian sense. Given fairly abundant available food, almost anyone—except the seriously handicapped, or the extremely old and infirm—could, by staying alive, contribute substantially to their reproductively valuable relatives: gather berries, tend children, teach children, etcetera. (And, anyway, even if you have become an unjustifiable burden on your family, would out-and-out suicide be the genetically optimal path? Wouldn't it be better for, say, a depressed man's genes if he just wandered away from the village, hoping to find better luck elsewhere—hoping, perhaps, to encounter some strange woman he can try to seduce, if not rape?)

A likely resolution of the suicide paradox lies in remembering that the behavioral "adaptations" designed by natural selection aren't the behaviors per se but the underlying mental organs. And mental organs that were adaptive enough, in one environment, to become part of human nature may, in another environment, lead to behaviors that are maladaptive. We've seen, for example, why feeling bad about yourself may sometimes be adaptive (chapter thirteen). But, alas, the mental organ designed to make you feel bad about yourself can misfire; feeling bad about yourself for too long, without relief, may lead to suicide.

Modern environments seem more likely than some previous environments to lead to this sort of malfunctioning. They permit, for example, a degree of social isolation that was unknown to our ancestors.

5. *Why do people kill their own children?* Infanticide is no mere product of a modern environment. It has happened abundantly in hunter-gatherer cultures and agrarian cultures. Is it, then, the result of an adaptation—a mental organ that implicitly calculates when killing a newborn will maximize genetic fitness? Quite possibly. Not only are unhealthy and handicapped babies more likely to be killed; so are babies born under various other kinds of inauspicious circumstances—when, say, the mother already has young children and has no husband.

Of course, in the modern environment, it is harder to explain the killing of offspring as a sound genetic stratagem. But as we've seen (chapter four), many cases of supposed offspring murder are in fact the murder of a stepoffspring. Many of the rest, I suspect, are committed by husbands who may in fact be the natural father but have begun to doubt—consciously or unconsciously—that they are. And in those relatively few cases where a mother kills her own newborn baby, it is often amid the sort of environmental cues that, in the ancestral environment, might have meant that infanticide would be genetically profitable: relative poverty, no reliable source of male parental investment, etcetera.

6. *Why do soldiers die for their country?* Jumping on a hand grenade—or, in the ancestral environment, suicidally leading a defense against club-wielding invaders—may make Darwinian sense if you're in the presence of close relatives. But why die for a bunch of people who are just friends? That's one favor you'll never have the pleasure of seeing repaid.

First, it is worth remembering that in the ancestral environment, in a small hunter-gatherer village, the average degree of kinship to a comrade in arms was not negligibly low—and, indeed, depending on patterns of marriage, could be fairly high (see Chagnon [1988]). In discussing the theory of kin selection in chapter seven, we focused on mental organs that identify close kin and treat them with special generosity; and, we suggested, genes conducive to such discrimination will tend to flourish at the expense of genes that bestow altruism more diffusely. But there may be a few circumstances that don't permit such fine discernment. One such circumstance is a collective threat. If, say, a whole hunter-gatherer band, including your immediate family and many near relatives, is under dire attack, inordinate bravery could make straightforward genetic sense by virtue of kin selection. Men in modern war may sometimes act under the influence of a tendency to bestow just such indiscriminate altruism in warlike situations.

Another difference between modern war and ancestral war is that the genetic payoff of victory is now lower. It is reasonable to suspect—based on observation of preliterate societies—that the rape or abduction of women was once a common feature of war. Thus the rewards were large enough, in Darwinian terms, to justify substantial risk (though not plainly suicidal behavior). And it is likely that the men who demonstrated the most valor during war were rewarded most richly.

In sum, the best guess about valor in wartime is that it is the product of mental organs that once served to maximize inclusive fitness and may no longer do so. But the organs persist, ready to be exploited by, among others, political leaders who profit from war (see Johnson [1987]).

Human behavior poses many other Darwinian mysteries. What are the functions of humor and laughter? Why do people make deathbed confessions? Why do people take vows of poverty and chastity—and even, occasionally, keep them? What is the exact function of grief? (Surely it signifies, as we assumed in chapter seven, the degree of emotional investment in the deceased, and surely the emotional investment itself made genetic sense while the person was alive. But now that the person is gone, how does grieving serve the genes?)

The solution to such mysteries is one of the great challenges in contemporary science. Often the route to solution will involve these themes: (1) distinguishing between the behavior and the mental organ governing it; (2) remembering that the mental organ, not the behavior, is what was actually designed by natural selection; (3) remembering further that, though these organs must have led to adaptive behavior in the environment of their design (since that's the only reason natural selection ever designs a mental organ), they may no longer do so; (4) remembering that the human mind is incredibly complex, that it was designed to yield a large array of behavior, depending on all kinds of subtleties of circumstance, and that the array of behaviors it yields is tremendously expanded by the unprecedented diversity of circumstance in the modern social environment.

Notes

ABBREVIATIONS

Autobiography: Nora Barlow, ed. (1959) *The Autobiography of Charles Darwin*, New York: Harcourt Brace and Co.

CCD: Frederick Burkhardt and Sydney Smith, eds. (1985–91) *The Correspondence of Charles Darwin*, 8 vols., Cambridge: Cambridge University Press.

Descent: Charles Darwin (1871) *The Descent of Man, and Selection in Relation to Sex*, Princeton, N.J.: Princeton University Press, 1981.

ED: Henrietta Litchfield, ed. (1915) *Emma Darwin: A Century of Family Letters, 1792–1896*, 2 vols., New York: D. Appleton and Co.

Expression: Charles Darwin (1872) *The Expression of the Emotions in Man and Animals*, Chicago: University of Chicago Press, 1965.

LLCD: Francis Darwin, ed. (1888). *Life and Letters of Charles Darwin*, 3 vols., New York: Johnson Reprint Corp., 1969.

Notebooks: Paul H. Barrett et al., eds. (1987) *Charles Darwin's Notebooks, 1836–1844*, Ithaca, N.Y.: Cornell University Press.

Origin: Charles Darwin (1859) *The Origin of Species*, New York: Penguin Books, 1968.

Papers: Paul H. Barrett, ed. (1977) *The Collected Papers of Charles Darwin*, Chicago: University of Chicago Press.

Voyage: Janet Browne and Michael Neve, eds. (1989) Charles Darwin's *Voyage of the Beagle*, New York: Penguin Books.

INTRODUCTION: DARWIN AND US

[1] *Origin*, p. 458.
[2] Greene (1963), pp. 114–15.
[3] Tooby and Cosmides (1992), pp. 22–25, 43.

[4] See Tooby and Cosmides (1992).

[5] Some of those who avoid the label insist that they do so because of deep doctrinal differences between Wilson's formulation of the field and theirs. There are indeed differences, and the conceptual sophistication of the field has certainly grown since 1975. But these differences almost surely would not have kept many people from using Wilson's word had the word not acquired unfortunate political connotations within academia.

[6] See Brown (1991) and the last chapter of Pinker (1994).

[7] Smiles (1859), pp. 16, 332–33.

[8] Mill (1859), pp. 50, 62.

[9] *Autobiography*, p. 21.

[10] LLCD, vol. 3, p. 200; *Autobiography*, pp. 73–74.

[11] Clark (1984), p. 168.

[12] Bowlby (1991), pp. 74–75; Smiles (1859), p. 17.

CHAPTER 1: DARWIN COMES OF AGE

[1] CCD, vol. 1, p. 460.

[2] Marcus (1974), pp. 16–17.

[3] See Stone (1977), p. 422; Himmelfarb (1968), p. 278; Young (1936), pp. 1–5; Houghton (1957).

[4] Young (1936), pp. 1–2.

[5] Houghton (1957), pp. 233–34.

[6] Houghton (1957), pp. 62, 238; Young (1936), pp. 1–4.

[7] *Descent*, vol. 1, p. 101.

[8] *Autobiography*, pp. 46–56.

[9] See Gruber (1981), pp. 52–59; for a modern reply to Paley, see Dawkins (1986).

[10] *Autobiography*, pp. 56–57, 59.

[11] *Autobiography*, p. 85. On Darwin's conversion to evolutionism and his formulation of natural selection, see Sulloway (1982) and Sulloway (1984).

[12] Clark (1984), p. 6.

[13] *Autobiography*, pp. 27–28, 58, 67.

[14] Clark (1984), p. 3.

[15] Himmelfarb (1959), p. 8.

[16] Clark (1984), p. 137.

[17] *Origin*, p. 263.

[18] For discussions of the conceptual foundations of evolutionary psychology, see Cosmides and Tooby (1987), Tooby and Cosmides (1992), Symons (1989), and Symons (1990).

[19] See Humphrey (1976), Alexander (1974), p. 335, and Ridley (1994).

[20] Some Darwinians now contest this use of the term *random*. They argue that the process of generation yields traits more likely to be useful than a literally random process would yield. Some believe the process of generating traits has itself evolved through natural selection—that the genes governing the process have been selected to favor the generation of useful genes. See, for example, Wills (1989). This is an important ongoing debate, but it has little relevance to this book; though the outcome will shed light on the speed with which evolution occurs, it won't change our expectations about what sorts of traits evolution will tend to produce.

[21] ED, vol. 1, pp. 226–27.

[22] Desmond and Moore (1991), pp. 51, 54, 89.

[23] But see Brent (1983), pp. 319–20, for highly circumstantial evidence of premarital sexual activity.

[24] Marcus (1974), p. 31.

CHAPTER 2: MALE AND FEMALE

[1] *Descent*, vol. 2, pp. 396–97.

[2] *Descent*, vol. 1, p. 273.

[3] *Descent*, vol. 1, p. 342, vol. 2, pp. 240–41; Wilson (1975), pp. 318–24.

[4] *Descent*, vol. 2, pp. 32–37, 97, 252–55.

[5] One theory is that evolution initially endowed females with a fondness for simple signs of robustness in the male—slightly brighter-than-average color, say—as these signs bespoke robust offspring; but this female preference, once ingrained, meant that slightly more colorful males had a reproductive advantage regardless of whether the color actually signified health; so genes for male color began faring well. This very reproductive success of male color, in turn, further amplified, by natural selection, the female preference for color, since females that preferred bright males would tend to have bright, sexually successful male offspring. Hence a vicious circle: the more females liked color, the more colorful males got, and vice versa. This theory has in recent years been challenged from various corners (though the "alternative" theories are not always incompatible with it). For good overviews of this unresolved issue, see Ridley (1994) and Cronin (1991).

[6] See *Descent*, vol. 1, p. 274. Darwin was, in one sense, on the right track. He did trace male eagerness back to the relative largeness of the female sex cell. Because of this size, he reasoned, it is relatively easy to transport the male sex cell to the female sex cell. Thus, for example, in marine animals with little power of locomotion, the sperm would be more likely to float to the eggs than vice versa. But floating is a pretty random process, and Darwin reasoned that males would have profited in evolutionary terms by seeking out the egg and depositing the sperm. This tendency of males to seek females would then persist even in higher terrestrial animals, in the form of "strong passions." Among this theory's several problems is its failure to account for the muting of male eagerness and the amplification of female eagerness in species where the imbalance in parental investment is the opposite of the norm, as discussed later in this chapter.

[7] Quoted in Hrdy (1981), p. 132.

[8] The term was coined by John Bowlby, the eminent psychiatrist who was also a biographer of Darwin (Bowlby [1991]).

[9] There has been an animated debate about the relative importance of studying the EEA and studying the contribution of traits to fitness in a modern, or at least very recent, environment. (There has also been a debate about how to define the EEA.) *Ethology and Sociobiology* devoted a whole issue—vol. 11, no. 4/5, 1990—to arguments about the significance of the EEA.

[10] See Tooby and Cosmides (1990b).

[11] Bateman (1948), p. 365.

[12] Richard Dawkins's brasher, more accessible, and more popularly known book,

The Selfish Gene, was wholly steeped in Williams's worldview, and Dawkins acknowledged in the first chapter that he had been "heavily influenced by G. C. Williams's great book . . ."

[13] Williams (1966), pp. 183–84.

[14] Williams (1966), p. 184.

[15] Trivers (1972), p. 139.

[16] Whether Trivers gave this theoretical extension its *final* form is debatable. In 1991, Timothy Clutton-Brock and A.C.J. Vincent suggested that, rather than focus on parental "investment," which is hard to quantify, we should focus strictly on the potential rate of reproduction of each sex. Looking across species, they showed that which sex has the higher potential rate of reproduction is an extremely strong predictor of which sex will compete more intensely for access to the other sex. I've noticed that many people find relative potential rate of reproduction an intuitively clearer explanation of female coyness than relative parental investment. In fact, in introducing the subject earlier in this chapter, I focused on relative potential rate of reproduction, telling the story much as Clutton-Brock and Vincent would tell it. See Clutton-Brock and Vincent (1991).

[17] Cited in Buss and Schmitt (1993), p. 227. It's possible, of course, that many of the women feared for their physical safety, since the men were unknown to them.

[18] Cavalli-Sforza et al. (1988).

[19] Malinowski (1929), pp. 193–94.

[20] See Symons (1979), p. 24, for a discussion of Trobriand jealousy in this context.

[21] Malinowski (1929), pp. 313–14, 319.

[22] Malinowski (1929), p. 488.

[23] For a discussion of the precise sense in which species-typical mental adaptations are "universally" evident, see Tooby and Cosmides (1989).

[24] Trivers (1985), p. 214.

[25] *Descent,* vol. 2, p. 30.

[26] Trivers (1985), p. 214.

[27] *Notebooks,* p. 370. (Darwin later inserted "often" between "argument" and "pursued.")

[28] Williams (1966), pp. 185–86. Gronell (1984) studies a pipefish species that "only partly . . . substantiates" the prediction of greater female assertiveness, but both pipefish species reviewed by Clutton-Brock and Vincent (1991) more clearly confirm the prediction.

[29] V. C. Wynne-Edwards, quoted in West-Eberhard (1991), p. 162.

[30] Trivers (1985), pp. 216–18; Daly and Wilson (1983), p. 156; Wilson (1975), p. 326. Clutton-Brock and Vincent (1991) review 16 species in which females can reproduce at a higher rate than males owing to high male parental investment. In 14 species, females compete markedly more intensely for mates than males, as predicted. The authors argue that in the two "possible exceptions"—seahorses (close relatives of the pipefish) and rheas—the theoretically higher reproductive rate of females may in practice be nullified by special ecological circumstances.

[31] De Waal (1989), p. 173.

[32] It's not a settled matter that *Homo erectus* is our direct ancestor.

[33] See Wrangham (1987) for a reconstruction of a proto-ape based only on the African apes—not including the orangutan.

[34] Rodman and Mitani (1987).

[35] Stewart and Harcourt (1987).

[36] De Waal (1982); Nishida and Hiraiwa-Hasegawa (1987).
[37] Badrian and Badrian (1984), Susman (1987), de Waal (1989), Nishida and Hiraiwa-Hasegawa (1987), and Kano (1990).
[38] Wolfe (1991), pp. 136–37; Stewart and Harcourt (1987).
[39] De Waal (1982), p. 168.
[40] Goodall (1986), pp. 453–66.
[41] Wolfe (1991), p. 130.
[42] Leighton (1987).

CHAPTER 3: MEN AND WOMEN

[1] *Descent*, vol. 2, p. 362.
[2] Morris (1967), p. 64.
[3] Murdock (1949), pp. 1–3. In some societies, the maternal uncle plays a larger role in child-rearing than the father. Richard Alexander argues that this tends to occur in societies in which sexual habits make it quite uncertain that the average husband is in fact the father of his wife's children. In such a context, the theory of kin selection (which we'll come to in chapter 7) suggests that a man might be better off, in Darwinian terms, investing in his sister's children than in his wife's. See Alexander (1979), pp. 169–75.
[4] Trivers (1972), p. 153.
[5] Benshoof and Thornhill (1979), Tooby and DeVore (1987).
[6] The Mangaians of Polynesia, for example, were long noted for their lack of romantic love. Symons (1979), p. 110, repeated this conventional view but he now sees it as mistaken (personal communication). A graduate student under his supervision, Yonie Harris (University of California, Santa Barbara), has studied the Mangaians extensively, and she believes romantic love to be a human universal (personal communication). More generally, see Jankowiak and Fisher (1992).
[7] This isn't to say that the calculations made by the female are simple. See Cronin (1991) and Ridley (1994) for good overviews of the complex arguments about the relative importance of the various things females may "seek" in a male: "good genes" that will raise fitness in either male or female offspring; genes that will raise fitness only in male offspring (e.g., genes for long antlers) by raising their chances of obtaining a mate; an absence of pathogens (which cuts the risk of catching a disease during sex and may also signify pathogen-resistant genes); some form of investment (as with the hanging fly, as described in this paragraph of the text); etc.
[8] Thornhill (1976).
[9] Buss (1989) and Buss (1994), chapter 2.
[10] Women placed greater importance on ambition and industriousness in twenty-eight of the twenty-nine cultures with statistically significant differences between the sexes. See Tooby and Cosmides (1989) for a discussion of how specific behaviors or preferences can be less than universal yet still signify the deeper existence of species-typical mental organs.
[11] Trivers (1972), p. 145.
[12] Tooke and Camire (1990). Tooke and Camire speculate that the noted sensitivity of females to nonverbal cues may grow out of this arms race.
[13] For Trivers's 1976 observation, and elaboration on its logic, see chapter 13, "De-

ception and Self-Deception." The first person to apply this logic in the context of mating seems to have been Joan Lockard (1980). Tooke and Camire (1990) provide tentative evidence of male self-deception.

[14] I've heard men who admitted to deceiving a woman say the deception lay not so much in what they said (e.g., "I'll stay with you forever") as in what they didn't say (e.g., "I'm pretty sure I won't"). It is as if they were walking a fine line, committing every deception *but* the kind of overt falsehood that might make subsequent reprisal easy to justify. In such cases the question of how consciously the deception was executed may be extremely murky, even if the deceiver is quite conscious of the deception in retrospect.

[15] Kenrick et al. (1990); see Buss and Schmitt (1993) for evidence that men are, in other respects as well, less selective about short-term mates than women are.

[16] Trivers (1972), pp. 145–46.

[17] Malinowski (1929), p. 524.

[18] See Buss (1989) and *New York Times,* June 13, 1989, p. C1.

[19] See Buss (1994), p. 59. Of course, men, in choosing a woman they'll invest in, do have some things to worry about aside from fertility, such as her likely fidelity. But this consideration seems more or less matched by the woman's concern with the man's likelihood of sticking around to invest in the children.

[20] Trivers (1972), p. 149.

[21] Symons (1979) had noted that men are more concerned about a mate's adultery than women but hadn't noted the specific female concern with emotional infidelity as a sign of resource diversion.

[22] See Daly, Wilson, and Weghorst (1982) for a summary of some of these studies. That male jealousy is focused more on sex and female jealousy on "loss of time and attention" had been noted in Teismann and Mosher (1978). That wives of all social levels are more likely to accept the sexual infidelity of their husbands than vice versa had been noted in the Kinsey Report, cited in Symons (1979), p. 241.

[23] Buss et al. (1992).

[24] Symons (1979), pp. 138–41; see also Badcock (1990), pp. 142–60.

[25] Shostak (1981), p. 271.

[26] On gorillas, see Stewart and Harcourt (1987), pp. 158–59. On langurs, see Hrdy (1981).

[27] Daly and Wilson (1988), p. 47.

[28] Hill and Kaplan (1988), p. 298; Hill (personal communication).

[29] Hrdy (1981), pp. 153–54, 189.

[30] Symons (1979), pp. 138–41. Much the same theory was separately advanced in Benshoof and Thornhill (1979).

[31] See, e.g., Hill (1988) and Daly and Wilson (1983), p. 318. This issue remains unsettled.

[32] Hill and Wenzl (1981); Grammer, Dittami, and Fischmann (1993).

[33] Baker and Bellis (1993). There are other ways a woman might unconsciously discriminate against her regular mate. According to Baker and Bellis, the female orgasm, as long as it doesn't come well before ejaculation, raises the amount of sperm retained and thus the chances of conception. So the kind of men who thrill a woman to the point of ecstasy may be the kind of men natural selection "wants" to father her childen. And Baker and Bellis did find a *somewhat* greater tendency

for unfaithful women to have such orgasms with their outside lovers than with their regular mates. (But the data are, at this point, tentative, and the methodology that produced them is less than straightforward.) These two tactics for controlling conception—the timing of the copulation, and the orgasm—could, in theory, sponsor various sorts of female discernment. In the early stages of a relationship, the orgasm could gauge a man's likely commitment; if his growing familiarity to a woman makes her more likely to have a well-timed orgasm, then she is less likely to conceive a child by an unfamiliar, and perhaps uncommitted, man. But if the man is "sexy" enough—if he has brute strength and other signs of "good genes"—orgasm may come earlier in the testing period, by design. Incidentally, Baker and Bellis also found that "double matings"—two different sex partners within a five-day period— are more common around ovulation. Baker and Bellis see this second finding as evidence for a theory of "sperm competition": maybe one Darwinian purpose of infidelity is to let the sperm of various men fight mano-a-mano within the uterus; then the egg will likely be penetrated by hearty, combative sperm, and, if the resulting offspring is a son, it will often be a son with heartier-than-average, more successfully competitive sperm.

[34] See Betzig (1993a).

[35] Daly and Wilson (1983), p. 320.

[36] Harcourt et al. (1981). See also Wilson and Daly (1992).

[37] Baker and Bellis (1989). Many nonhuman species long thought quite monogamous, notably birds, are turning out to feature some quite randy females, now that molecular biology can match offspring with their real fathers. See Montgomerie (1991).

[38] Wilson and Daly (1992), pp. 289–90.

[39] Symons (1979), p. 241.

[40] Though Buss's study of thirty-seven cultures found males expressing a stronger preference for virgins than did females in all twenty-three cultures that had a statistically significant difference between the sexes, fourteen cultures showed no such difference. Most are modern European cultures in which virginity is a rare thing for either sex to find in a mate. Still, we know that in at least some of these cultures, such as Sweden, women with reputations for extreme promiscuity are not considered desirable wives. See Buss (1994), chapter 4.

[41] Mead (1928), p. 105.

[42] Mead (1928), pp. 98, 100.

[43] Freeman (1983), pp. 232–36, 245. The term I've depicted as "whore" Freeman translates as "prostitute," but he notes that it has stronger connotations of shame than his translation conveys.

[44] Mead (1928), p. 98; Freeman (1983), p. 237.

[45] Mead (1928), p. 107.

[46] In Mangaia, Yonie Harris (personal communication) says, promiscuous women are called "sluts," though the fact that the English word is used makes her unsure whether the concept predates Western influence. On the Ache, see Hill and Kaplan (1988), p. 299.

[47] William Jankowiak (personal communication).

[48] Buss and Schmitt (1993), table on p. 213.

[49] See Tooby and Cosmides (1990a).

[50] Maynard Smith (1982), pp. 89–90.

[51] The basic idea of frequency-dependent selection had been less fully articulated in 1930 by the British biologist Ronald Fisher, who used it to explain why the ratio of newborn males to newborn females hovers near fifty-fifty. The reason is not, as people tend to imagine, that such a ratio is "good for the species." Rather, if genes favoring the birth of either sex begin to dominate the population, then genes favoring the birth of the other sex will grow in reproductive value, and thus grow in number until the balance is restored. See Maynard Smith (1982), pp. 11–19 and Fisher (1930), pp. 141–43.

[52] Dawkins (1976), pp. 162–65.

[53] Unpublished work by Mart Gross, University of Toronto (personal communication).

[54] Dugatkin (1992).

[55] The "sexy son" hypothesis was proposed by Gangestad and Simpson (1990). The authors have intriguing data suggesting, indirectly, that sexually uninhibited women have a disproportionate number of male offspring—which makes perfect sense if their strategy is geared to having "sexy sons." (Even though sperm cells, not egg cells, determine the child's sex, the mother could alter the birth ratio by, for example, selectively destroying fertilized eggs.) Still, there is no reason why alteration of the sex ratio couldn't come in response to environmental cues. There is evidence, after all, that some mammals who find themselves in adverse environmental conditions tend to give birth to females more than males. (See chapter 7, "Families," for the likely explanation.)

[56] Tooby and Cosmides (1990a) stress the "noise" explanation. In particular, they say, the genetic variation may be selected as a way to frustrate co-evolving pathogens, and only incidentally affect personality. For a Darwinian view of genetics and personality that is more open to the possibility of genetically distinct personality "types," see Buss (1991).

[57] Trivers (1972), p. 146.

[58] Walsh (1993) found the inverse correlation between a woman's estimation of her own attractiveness and number of sex partners. The data finding no correlation were gathered by Steve Gangestad (personal communication); in this case (perhaps significantly), the rating of female attractiveness wasn't done by the female but by observers.

[59] See, e.g., Chagnon (1968).

[60] Cashdan (1993). Of course, the causality could work in either direction: women who wear sexy clothes and are willing to engage in sex frequently may, by virtue of these habits, tend to find themselves in the company of males not willing to invest paternally—at least, not in them. The causality probably works in both directions.

[61] Gangestad (personal communication). See Simpson et al. (1993).

[62] Buss and Schmitt (1993), esp. pp. 214, 229.

[63] Quoted in Thornhill and Thornhill (1983), p. 154. This paper was the first extensive analysis of human rape from within the new Darwinian paradigm. See also Palmer (1989).

[64] See Barret-Ducrocq (1989).

[65] The anthropologists Patricia Draper and Henry Harpending have suggested that an adolescent girl's (or boy's) approach to sex may depend heavily on whether there is a father around the house. They argue that during evolution, the presence or

absence of a father correlated with the kinds of strategies being practiced by men generally, and thus offered a cue to the sort of courtship environment in which a woman would find herself. The upshot is that children reared in father-absent homes are likely to be, among other things, followers of short-term sexual strategies. (One criticism of this theory is that the presence or absence of a father in a girl's household would seem to be one of the less reliable cues available—compared to, say, direct observation of male mating habits in the girl's generation.) See Draper and Harpending (1982) and Draper and Harpending (1988).

[66] Buehlman, Gottman, and Katz (1992). The study actually found two equally strong predictors of divorce: statements by the husband indicating disappointment with the marriage and his "withdrawal" during discussion of the marriage—his failure, for example, to respond expansively when asked to describe how the couple met. But this withdrawal is, to some extent, merely a second manifestation of dissatisfaction with the marriage (and, indeed, the two indices were strongly correlated). In any event, along both dimensions, the husband's sentiments were markedly stronger predictors of divorce than the wife's.

[67] Charmie and Nsuly (1981), esp. pp. 336–40.

[68] Symons (1979) was among the first to question the pair-bonding thesis and to note problems with the human-gibbon comparison. See also Daly and Wilson (1983). On gibbon behavior, see Leighton (1987).

[69] Alexander et al. (1979).

[70] It's also possible to argue, more speculatively, that some human sexual dimorphism reflects not intra-male combat but the importance of hunting in human evolution.

[71] These numbers come from a computerized database derived from G. P. Murdock's *Ethnographic Atlas* and were gathered courtesy of Steven J. C. Gaulin. Note that six of the 1,154 societies—about one-half of one percent—are polyandrous; that is, women have multiple spouses. But all these societies are polygynous as well, so either sex can have more than one spouse. And the polyandrous marriages often turn out to be not polyandry strictly speaking (more than one husband in a single household) but a kind of serial monogamy that presumably allows husbands some confidence of paternity. See Daly and Wilson (1983), pp. 286–88, for a discussion of polyandry.

[72] Morris (1967), pp. 10, 51, 83.

CHAPTER 4: THE MARRIAGE MARKET

[1] *Descent,* vol. 1, p. 182.

[2] Gaulin and Boster (1990).

[3] Alexander (1975), Alexander et al. (1979).

[4] The "polygyny threshold" model of polygyny in birds was fully developed in Orians (1969). See also Daly and Wilson (1983), pp. 118–23, and Wilson (1975), p. 328. Gaulin and Boster (1990) make the connection between the polygyny threshold model and Alexander's terminology.

[5] Gaulin and Boster (1990). I use the word *nonpolygynous* because the authors break the data down not as polygamous vs. monogamous, but as polygynous vs. nonpolygynous, where nonpolygynous includes the minuscule number of known polyandrous societies.

[6] If the lawyer's previous fiancée doesn't want to share him, he can presumably find another woman who will. And, as more highly ranked males do find two or three willing wives, it will be harder and harder for highly ranked females to insist on monogamy without sacrificing their selectivity. By the way, if you're wondering why we don't assume that *men* move up the scale and agree to share a high-ranking wife, one reason is that men, for reasons that should by now be clear, generally find spouse-sharing a much more abhorrent prospect than women do. Genuine polyandry has been found only in a handful of cultures, always accompanied by polygyny. See Daly and Wilson (1983), pp. 286–88.

[7] I'm assuming a woman could agree to marry a man on the legally binding condition that he never acquire a second wife. But men would be free to refuse to marry under that condition, which is why some women would not insist on that condition.

[8] This thesis—institutionalized monogamy as an implicit compromise among men in relatively egalitarian societies—grows out of the work of various scholars. Richard Alexander (1975), p. 95, points in this direction, especially as paraphrased by Betzig (1982), p. 217. The first time I encountered the thesis as stated here was in a 1990 conversation with Kevin MacDonald, who has not, so far as I know, put it quite like this in writing. But for related work, see MacDonald (1990). See also Tucker (1993), who stresses the connection between monogamy and democratic values.

[9] Zulu: Betzig (1982); Inca: Betzig (1986).

[10] Stone (1985), p. 32.

[11] See MacDonald (1990).

[12] See Daly and Wilson (1988), Daly and Wilson (1990a).

[13] Daly and Wilson (1990a). The difference persists for men older than thirty-five. But, curiously, there is no sharp difference for men under twenty-four. Daly and Wilson suggest one explanation: that men who mature physically at an early age are more likely to be both delinquent and sexually active (and hence, presumably, more likely to marry). These data come from Detroit (1972) and Canada (1974–83).

[14] On risk, crime, etc., see Daly and Wilson (1988), esp. pp. 178–79; Thornhill and Thornhill (1983); Buss (1994); and Pedersen (1991). Buss and Pedersen discuss these things as products of a high sex ratio—the ratio of marriage-age males to marriage-age females. But polygyny, including de facto polygyny (i.e., serial monogamy) produces what is in many ways the rough practical equivalent of a high sex ratio.

[15] This point is made by, e.g., Tucker (1993), who also notes that serial monogamy encourages male violence.

[16] Saluter (1990), p. 2. For both men and women, the fraction of the population never married actually dropped between 1960 and 1990. This may seem inconsistent with the idea that serial monogamy tends to leave socially disadvantaged men mateless, but it isn't necessarily. As the divorce rate grows, the total number of people ever married should tend to rise; but the amount of time that the average person spends married may decline, and perhaps for disadvantaged men this decline is especially steep. Census data such as these don't shed light on that question. The main point here is that these data do suggest that females are divided among males less equitably now than before: in 1960, 7.5 percent of women forty or older, and 7.6 percent of men forty or older, had never been married. In 1990, the figures had diverged: 5.3 percent for women, and 6.4 percent for men. It is interesting that, for both women and men between thirty-nine and forty-five, the fraction of never married has actually grown between 1960 and 1990. It now stands at 8.0 and 10.5

percent, respectively. So too for the thirty to thirty-four bracket: for women the number has grown from 6.9 to 16.4, and for men from 11.9 to 27.0. All these numbers are somewhat ambiguous; surely many of the never married, especially in the younger age brackets, could have married but instead have spent much time living with a mate out of wedlock. In the absence of statistics on all monogamous relationships—including cohabitation without marriage as well as mutual fidelity without cohabitation—a very clear quantitative analysis is impossible.

[17] See Symons (1982).

[18] Daly and Wilson (1988), p. 83.

[19] Daly and Wilson (1988), pp. 89–91. Such patterns can be misleading, since stepparent families may have troubles that predate the stepparenthood. But, as Daly and Wilson note (p. 87), stepparent families, unlike single-parent families, are not especially likely to live in poverty.

[20] Laura Betzig (personal communication).

[21] See chapter 12, "Social Status."

[22] Wiederman and Allgeier (1992).

[23] Tooby and Cosmides (1992), p. 54.

[24] Tooby and Cosmides (1992), p. 54.

CHAPTER 5: DARWIN'S MARRIAGE

[1] CCD, vol. 2, pp. 117–18; *Notebooks*, p. 574.

[2] See Stone (1990), pp. 18, 20, 325, 385, 424, and, in general, his chapters 7, 10, and 11. In addition to privately arranged separations, there was the option of "judicial separation," but this was almost never pursued. See Stone (1990), p. 184.

[3] See, e.g., Whitehead (1993).

[4] ED, vol. 2, p. 45.

[5] LLCD, vol. 1, p. 132.

[6] CCD, vol. 1, pp. 40, 209.

[7] CCD, vol. 1, pp. 425, 429, 439. Even much smaller age gaps were cause for comment. Emma Darwin, before her marriage, had written to her mother about an engagement that had transpired "to our great surprise, as she is 24," whereas the groom was twenty-one. See ED, vol. 1, p. 194.

[8] CCD, vol. 1, p. 72.

[9] Marriage among cousins was not uncommon in nineteenth-century England. Sex between closer relatives—siblings, or parent and child—brings great danger of genetic pathology in the offspring. In view of these genetic consequences, it is not surprising that people around the world abhor incest. Specifically, the abhorrence appears to depend on an innate mechanism for learning who close relatives are. The mechanism is most conspicuous when it misfires; children reared sibling-like with unrelated children—as on an Israeli kibbutz—seem disinclined toward romance with them even though there's no cultural sanction against it. See Brown (1991), chapter 5.

[10] CCD, vol. 1, p. 190.

[11] CCD, vol. 1, pp. 196–97.

[12] CCD, vol. 1, p. 211.

[13] CCD, vol. 1, p. 220.

[14] CCD, vol. 1, p. 254.

[15] CCD, vol. 1, p. 229.

[16] ED, vol. 1, p. 255; Desmond and Moore (1991), p. 235; Wedgwood (1980), p. 219–21. On Erasmus's interest, see CCD, vol. 1, p. 318.

[17] ED, vol. 1, p. 272.

[18] CCD, vol. 2, pp. 67, 79, 86.

[19] *Papers*, vol. 1, pp. 49–53.

[20] CCD, vol. 2, pp. 443–45.

[21] CCD, vol. 2, pp. 443–44.

[22] See *Notebooks*, pp. 157, 237.

[23] See Sulloway (1979b), p. 27, and Sulloway (1984), p. 46.

[24] *Autobiography*, p. 120.

[25] *Notebooks*, p. 375.

[26] See Buss (1994).

[27] ED, vol. 2, p. 44.

[28] CCD, vol. 2, p. 439; ED, vol. 2, p. 1.

[29] ED, vol. 2, pp. 1, 7.

[30] ED, vol. 2, p. 6.

[31] CCD, vol. 2, p. 126.

[32] I have presupposed the common view that Darwin's marriage memorandum was written without any particular woman in mind. But this view may well be wrong. It rests on such phrases as "without one's wife was better than an angel & had money," which suggest ignorance as to who the wife will be. But it would be like Darwin to be a little coy; he habitually played his cards close to the vest. In any event, even aside from the striking coincidence that Emma was indeed an angel who had money, there is one other complication with the conventional view: after proposing to Emma in November, Darwin wrote to Lyell that he had decided to do so during his previous visit to her, which was in late July. Some find it hard to imagine that the memorandum was written only a few weeks before the choice of mate, and without an inkling of it.

[33] CCD, vol. 2, p. 119; Himmelfarb (1959), p. 134.

[34] CCD, vol. 2, p. 133.

[35] CCD, vol. 2, pp. 132, 150, 147.

[36] Daly and Wilson (1988), p. 163. They're talking specifically about a growing propensity toward risk in males who have failed utterly to mate, but the logic applies equally to other manifestations of intensified pursuit of sex, including plain old passion.

[37] But see Brent (1983), pp. 319–20, for highly circumstantial evidence of premarital sexual activity.

[38] Bowlby (1991), p. 166.

[39] *Notebooks*, p. 579.

[40] Brent (1983), p. 251.

[41] CCD, vol. 2, pp. 120, 169.

[42] See ED, vol. 2, p. 47.

[43] CCD, vol. 2, p. 172. She is actually addressing Darwin here, playfully referring to him in the third person. The date of the letter is uncertain, but according to Darwin she wrote it "shortly after our marriage." The editors of Darwin's correspondence date the letter as circa February; the Darwins had been married on January 29, 1839.

[44] *Notebooks*, p. 619.

[45] Houghton (1957), p. 341, and, in general, his chapter 13.

[46] Houghton (1957), pp. 354–55.

[47] Quoted in Houghton (1957), pp. 380–81.

[48] See Rasmussen (1981).

[49] See, e.g., Betzig (1989).

[50] Desmond and Moore (1991), p. 628.

[51] See Thomson and Colella (1992) and Kahn and London (1991). Kahn and London argue that the greater risk of divorce faced by women who aren't virgins going into marriage is due to preexisting differences between the two types of women, and conclude that divorce therefore isn't causally linked to premarital sex. Thomson and Colella focus on the positive correlation between premarital cohabitation and the likelihood of divorce. Like Kahn and London, they provide evidence that the link may not be causal, but (see p. 266) they concede that this evidence is itself ambiguous.

[52] Laura Betzig (personal communication). See Short (1976). Many Victorian women used wet nurses, but this was much less common among lower-class women, and less common generally than it had been in pre-Victorian times.

[53] Symons (1979), pp. 275–76, describes these and other factors that often make males "the partner with the lesser emotional investment."

[54] CCD, vol. 2, pp. 140–41.

[55] Irvine (1955), p. 60.

[56] Rose (1983), pp. 149, 181, 169.

CHAPTER 6: THE DARWIN PLAN FOR MARITAL BLISS

[1] *Autobiography*, pp. 96–97.

[2] CCD, vol. 4, p. 147.

[3] Himmelfarb (1959), p. 133.

[4] *Autobiography*, p. 97.

[5] See Ellis and Symons (1990).

[6] Kenrick, Gutierres, and Goldberg (1989).

[7] Some studies have found an *inverse* correlation between number of children and satisfaction with marriage. But that may reflect a tendency of marriages that produced few or no children—and for that reason might have been unhappy—to dissolve early and thus not be included in the studies. Or, to put it another way: couples that stay together in spite of having few or no children may tend to be extraordinarily compatible; thus their happiness compared to couples with many children isn't necessarily a *result* of having few or no children.

[8] About half of all American couples who marry will divorce, and the chances of divorce are significantly higher than this if no children are born. See, e.g., Essock-Vitale and McGuire (1988), p. 230, and Rasmussen (1981).

[9] Brent (1983), p. 249.

[10] A 1985 survey found that 26 percent of all men who had ever married either were separated or had gotten divorced. Of the ones who married a second time, 25 percent were either separated or had gotten divorced. (That doesn't imply a 75 percent success rate for marriage; the survey included men of all ages, and some of the younger men will eventually get divorced, thus lowering the success rate.) Of course, since the twice-married are older on average than the once-married, second marriages will have, upon death of husband, a higher success rate than first marriages. But that

doesn't mean their survival prospects are better on a year-by-year basis. These numbers are the result of my calculations, performed on raw survey data from the U.S. Census Bureau.

[11] Rose (1983), p. 108. Randolph Nesse (1991a) would agree with Mill. He notes that marital harmony is often falsely assumed to be the norm, so that many people are "dissatisfied with marriages that are, in fact, better than average" (p. 28).

[12] Mill (1863), pp. 278, 280–81.

[13] *Los Angeles Times,* Jan. 5, 1991.

[14] Saluter (1990), p. 2.

[15] Calculations made from June 1985 Census Bureau report: "Age at survey date by current marital status, number of times married, how the first and second marriages ended, race, Spanish origin, and sex, for persons 15 years and over at survey date: United States."

[16] And, of course, some of these never-married men are themselves serial monogamists; that they've never tied the knot legally doesn't make them any less of a threat to less fortunate, mateless men.

[17] *Washington Post,* Jan. 1, 1991, p. Z15; *Washington Post,* Oct. 20, 1991, p. W12.

[18] Rose (1983), pp. 107–9.

[19] Stone (1991), p. 384.

[20] *New York Times Book Review,* Nov. 4, 1990, p. 12.

[21] All these numbers come from Roper surveys, summarized in Crispell (1992).

[22] On the impact of no-fault divorce on women, see, e.g., Levinsohn (1990).

[23] On biological determinism, see, e.g., Fausto-Sterling (1985); on sex differences, see, e.g., Gilligan (1982).

[24] Shostak (1981), p. 238.

[25] The phrase "male bonding" was coined by Lionel Tiger (1969).

[26] *New York Times,* Feb. 12, 1992, p. C10.

[27] See, for example, Lehrman (1994).

[28] Cashdan (1993).

[29] Kendrick, Gutierres, and Goldberg (1989) has at least some relevance here.

[30] For example, a declining birthrate leads, two decades later, to a greater number of twenty-one-year-olds than eighteen-year-olds, a greater number of twenty-two-year-olds than nineteen-year-olds, and so on. Since men tend to marry younger women, this means a surplus of males in the marriage market. Men might respond to the mate shortage by showing a greater commitment to monogamy and less wanderlust, and women, sensing their heightened market value, might show a lower tolerance for low-commitment sex. There is very tentative evidence that this dynamic helped stop the growth of divorce rates in the mid-1980s. See Pedersen (1991) and Buss (1994), chapter 9.

[31] Stone (1977), p. 427.

[32] Colp (1981).

[33] Colp (1981), p. 207.

[34] CCD, vol. 1, p. 524.

[35] Marcus (1974), p. 18.

[36] See, e.g., Alexander (1987). Alexander has done as much as anyone to shape modern Darwinian thinking about morality.

[37] Kitcher (1985), pp. 5, 9.

1 *Origin*, p. 258; CCD, vol. 4, p. 422.

2 See Trivers (1985), pp. 172–73 and Wilson (1975), chapters 5, 20.

3 *Origin*, p. 257; on the possible role of the insect-sterility puzzle in Darwin's delay, see Richards (1987), pp. 140–56.

4 *Origin*, p. 258.

5 Hamilton (1963), pp. 354–55. The fuller, more commonly cited version of the theory and its application to insect societies are in Hamilton (1964).

6 Haldane (1955), p. 44. See also Trivers (1985), chapter 3. Kin selection was also, as Hamilton notes in his 1964 paper, foreshadowed in Fisher (1930).

7 Trivers (1985), p. 110.

8 Other possible mechanisms include innately recognized chemical signals, comparable to those in social insects, and "phenotypic matching," in which an individual might identify as kin organisms that resemble it (visually, olfactorily) or resemble another organism previously identified as kin. See Wilson (1987), Wells (1987), Dawkins (1982), chapter 8, and Alexander (1979).

9 Hamilton (1963), pp. 354–55.

10 CCD, vol. 4, p. 424. This note and the ones on pp. 425 and 426 are written in Emma's hand. Presumably she was transcribing Darwin's words; when he was feeling especially unwell, she sometimes did.

11 Darwin did, though, see that what we call kin selection could apply to human beings in some contexts. See *Descent*, vol. 1, p. 161, on successful inventors: "Even if they left no children, the tribe would still include their blood-relations. . . ."

12 Trivers (1974), p. 250.

13 Trivers (1985), pp. 145–46.

14 CCD, vol. 4, pp. 422, 425.

15 CCD, vol. 4, pp. 426, 428.

16 Robert M. Yerkes, quoted in Trivers (1985), p. 158.

17 LLCD, vol. 1, p. 137; CCD, vol. 4, p. 430.

18 Trivers (1974), p. 260.

19 CCD, vol. 2, p. 439.

20 Trivers (1985), p. 163. Trivers here also suggests that the psychological mechanisms by which the child keeps track of its own interests and its parents' expressed interests, and reconciles the two, may bear some correspondence to Freud's distinction among id, superego, and ego, respectively.

21 Trivers (1974), p. 260.

22 From *The Communist Manifesto*.

23 In some situations parents might get more bang per buck by investing heavily in a relatively *dis*advantaged child—if, for example, one child is assured of reproductive success, and there is a fairly clear ceiling on that success, while another child is less equipped for success but could become better equipped with modest investment. This logic may lead to an overall parental bias toward investment in sons. After all, the things that make males sought-after mates (ambition, skill in the accumulation of resources) may be more readily imbued than some of the assets that are most valued in females (youth, beauty). This is one possible explanation for the much-noted tendency of teachers to pay special attention to their male students—which

doesn't, of course, mean that the tendency is beyond correcting through conscious effort.

[24] See Trivers and Willard (1973).

[25] Trivers and Willard (1973).

[26] Perhaps the clearest example of varied sex ratio is found in the red deer. The key variable isn't the mother's physical fitness per se, but her place in the status hierarchy, which correlates strongly with the reproductive success of male offspring. High-status mothers have mostly sons and low-status mothers mostly daughters. See Trivers (1985), p. 293. On pack rats, see Daly and Wilson (1983), p. 228.

[27] Desmond and Moore (1991), p. 449. Huxley was actually commenting not on natural selection but on a certain baseness he had noticed in particular kinds of animals.

[28] Dickemann (1979).

[29] Hrdy and Judge (1993). On the tendency of wealthy families to favor sons in the inheritance of wealth, see, e.g., Smith, Kish, and Crawford (1987) and Hartung (1982). Hartung found that the more polygynous a society, the more its patterns of inheritance tend to comply with Trivers-Willard logic.

[30] Betzig and Turke (1986).

[31] This point—that rich male offspring outproduce rich female offspring—has itself been documented by Boone (1988) in a study of fifteenth- and sixteenth-century Portuguese nobility. (Still, there's no strong reason to expect the pattern to hold up generally in environments quite different from the ancestral environment, especially now that contraception is commonly used.)

[32] Gaulin and Robbins (1991). Data are taken from graphs and thus may be slightly off, but not significantly.

[33] On the other hand, for some of the findings that support the Trivers-Willard hypothesis, conscious calculation cannot be ruled out. Affluent parents may note, for example, that a son, more than a daughter, can use money to broaden his range of prospective mates. For a review of the Trivers-Willard evidence that notes this and other weak links in the existing data, see Hrdy (1987). There are some studies that find no Trivers-Willard effect in human populations, but I'm aware of no studies that find an effect in the opposite direction, and would thus tend to negate the many studies that do find a Trivers-Willard effect. See Ridley (1994) for additional studies that find a Trivers-Willard effect.

[34] Hamilton (1964), p. 21.

[35] Freeman (1978), p. 118.

[36] See Daly and Wilson (1988), pp. 73–77, for their analysis of shifting reproductive value and their use of it to explain patterns of parental homicide.

[37] Crawford, Salter, and Lang (1989). The first correlation was .64, the second an extremely high .92 (out of 1.0).

[38] Bowlby (1991), p. 247; ED, vol. 2, p. 78. Emma did add: "though it will be long indeed before we either of us forget that poor little face."

[39] *New York Times*, Oct. 7, 1993, p. A21. See Mann (1992) for evidence that mothers favor healthy infants over their low-birth-weight twins.

[40] CCD, vol. 7, p. 521.

[41] Bowlby (1991), p. 330.

[42] CCD, vol. 4, pp. 209, 227. Bowlby (1991), pp. 272, 283, 287, contends that Darwin's father's illness and death moved Charles deeply, sending him into a spell

of bad health. There's no doubt that Darwin was distraught immediately after the death; he failed to attend the funeral, almost surely because of his mental, as well as his physical, state. Bowlby also notes that one of the perfunctory references to his father's death comes after an apology for not writing sooner, even to acknowledge sympathy notes—"but all the autumn and winter I have been much dispirited and inclined to do nothing but what I was forced to." Yet Bowlby admits that in the months after his father's death there is in Darwin's letters no "evidence of active mourning . . . no references to being sad, to missing his father or to reminiscing about him." Thus Bowlby is forced to argue that Darwin's mourning was "inhibited" and manifested in physiological illness.

[43] LLCD, vol. 1, pp. 133–34.

[44] CCD, vol. 4, p. 143. See Freeman (1983), p. 70, and Desmond and Moore (1991), p. 375.

[45] CCD, vol. 4, p. 225.

[46] CCD, vol. 5, p. 32 (footnote); *Autobiography*, pp. 97–98.

[47] LLCD, vol. 3, p. 228. Darwin did later write that his tears over Annie had "lost that unutterable bitterness of former days"—Desmond and Moore (1991), p. 518—but this came in the course of consoling a close friend after the death of a daughter.

CHAPTER 8: DARWIN AND THE SAVAGES

[1] *Descent*, vol. 1, p. 71.

[2] CCD, vol. 1, pp. 306–7; *Voyage*, pp. 173, 178.

[3] CCD, vol. 1, pp. 303–4; see p. 306 (footnote 5) on evidence that the cannibalism stories were apocryphal.

[4] *Descent*, vol. 2, pp. 404–5.

[5] *Voyage*, p. 172.

[6] *Voyage*, pp. 172–73.

[7] CCD, vol. 1, p. 380.

[8] See Alland (1985), p. 17.

[9] *Descent*, vol. 1, p. 93.

[10] Malinowski (1929), p. 501.

[11] *Descent*, vol. 1, p. 99.

[12] In *Descent*, vol. 1, p. 109, he writes of "the variability or diversity of the mental faculties in men of the same race, not to mention the greater differences between the men of distinct races." In vol. 2, p. 327, he refers to "the lower races."

[13] *Descent*, vol. 1, pp. 99, 164.

[14] *Descent*, vol. 1, pp. 96–99.

[15] *Descent*, vol. 1, pp. 88, 94–95.

[16] *Descent*, vol. 1, p. 94.

[17] Darwin, *Voyage of the Beagle*, p. 277 of unabridged edition published by Anchor/Doubleday (1962).

[18] *Descent*, vol. 1, pp. 75–78.

[19] *Expression*, p. 213.

[20] *Descent*, vol. 1, pp. 72, 88.

[21] *Descent*, vol. 1, p. 80.

[22] *Descent*, vol. 1, p. 166. On the subject of Darwin's group selectionism, and indeed his thinking about the moral sentiments more generally, see Cronin (1991).

[23] *Descent*, vol. 1, p. 163.

[24] See D. S. Wilson (1989), Wilson and Sober (1989), and Wilson and Sober (in press).

[25] Williams (1966), p. 262. See Wilson (1975), p. 30, for the view that Williams's position is too dogmatic.

CHAPTER 9: FRIENDS

[1] *Expression*, p. 216.

[2] *Descent*, vol. 1, pp. 163–64.

[3] Williams (1966), p. 94.

[4] See, e.g., *Descent*, vol. 1, p. 80.

[5] Game theory was developed formally by John Von Neumann and Oskar Morgenstern in *Theory of Games and Economic Behavior* (Princeton University Press, 1953), though Von Neumann had first used game theory in the 1920s.

[6] This point, with slightly different emphasis, is made by Maynard Smith (1982), p. vii. He notes: "In seeking the solution of a game, the concept of human rationality is replaced by that of evolutionary stability. The advantage here is that there are good theoretical reasons to expect populations to evolve to stable states, whereas there are grounds for doubting whether human beings always behave rationally."

[7] For an extremely clear dissection of the logic of the prisoner's dilemma, see Rapoport (1960), p. 173.

[8] See Rothstein and Pierotti (1988), though in my view their criticisms of the model are far from devastating.

[9] One could make a fine technical distinction between "cooperation" and "reciprocal altruism," but it wouldn't matter for our purposes. I'll use the terms interchangeably.

[10] *Voyage*, p. 183.

[11] Cosmides and Tooby (1989), p. 70. See also Barkow (1992).

[12] See Cosmides and Tooby (1989).

[13] Bowlby (1991), p. 321.

[14] TIT FOR TAT never got a chance to spread throughout the population. The end of time, in this universe, was 1,000 generations—the blink of an eye, in evolutionary terms. But TIT FOR TAT had become the most populous creature by the second generation, and after 1,000 generations was still growing more rapidly than any other.

[15] Williams (1966), p. 94.

[16] Years later, Axelrod was shown to be wrong about this; a population of TIT FOR TATs can in fact be successfully invaded. But the moral of the story—that cooperation breeds cooperation—remains intact, for the strategy that can make inroads against TIT FOR TAT turns out to be even nicer than TIT FOR TAT. It is like TIT FOR TAT, only it is sporadically "forgiving"; every so often it repays apparent badness with good, responding to a perceived cheater by cooperating on the next encounter. This strategy is most successful in environments that, like real life, are "noisy": a player occasionally misperceives or misremembers the deeds of another. See Lomborg (1993) and *New York Times*, April 15, 1992, p. C1.

[17] Axelrod (1984), p. 99.

[18] Actually, this reciprocal-altruism gene might get a second boost too: in many cases it would be helping copies of *itself* directly, and thus would spread for reasons that

are technically distinguishable from the other reason it spreads via kin selection. See Rothstein and Pierotti (1988).

[19] See Singer (1984), p. 146. For examples of social-exchange theory, see Gergen, Greenberg, and Willis (1980).

[20] Trivers's 1971 paper was bold in the range of animals to which it applied the theory, and some of the more extreme applications—in birds, in some species of fish—haven't gained much supporting evidence. But the applications that matter for our purposes—to mammals, and, more specifically, primates and humans—continue to hold up well. Williams had stressed in 1966 that it is mainly among mammals—clearly capable of recognizing individuals and keeping a mental record of their past behavior—that reciprocal altruism should be expected to evolve most readily.

[21] Reciprocal altruism for dolphins seems to be more thoroughly documented than for porpoises. For references on both, see Taylor and McGuire (1988).

[22] Wilkinson (1990). See also Trivers (1985), pp. 363–66.

[23] See de Waal (1982), de Waal and Luttrell (1988), and Goodall (1986).

[24] On the emotional infrastructure of reciprocal altruism, see Nesse (1990a).

[25] See Cosmides and Tooby (1992) and Cosmides and Tooby (1989). For good brief summaries of the cheater-detection experiments, see Cronin (1991), pp. 335–40, and Ridley (1994), chapter 10.

[26] Trivers (1971), p. 49.

[27] Wilson and Daly (1990). Many of these murders are no doubt motivated, directly or indirectly, by sexual rivalry. But the same principle holds: the fierceness of such fights is at least partly explained by the fact that in the ancestral environment, reputations for fierce fighting spread quickly.

[28] See Trivers (1971), p. 50.

[29] There are other Darwinian explanations—as in Frank (1990)—for altruism toward strangers (tipping waiters we've never seen before and will never see again, etc.), though in my view they tend to be needlessly complex. Some of them depict such generosity as designed to convince the altruist of his goodness, after which he can better convince others of it.

[30] For more frankly group-selectionist scenarios, see the work of David Sloan Wilson, a biologist who feels that the aversion to group selectionism has grown indiscriminate, fostering an overly cynical view of human motivation. If Wilson is right, then the pendulum will eventually swing at least some distance back in the direction of group selectionism. Still, the place of individual selectionism in intellectual history is secure. It is what brought the breakthrough; rejection of convoluted group selectionist theories is what started and sustained the last three decades of radical progress in understanding human nature.

[31] Daly and Wilson (1988), p. 254. See also Wilson and Daly (1992).

CHAPTER 10: DARWIN'S CONSCIENCE

[1] *Descent*, vol. 1, pp. 165–66.

[2] Bowlby (1991), pp. 74–76.

[3] Bowlby (1991), p. 60.

[4] CCD, vol. 1, pp. 39, 507; CCD, vol. 3, p. 289.

[5] LLCD, vol. 1, pp. 119, 124.

[6] LLCD, vol. 3, p. 220; Desmond and Moore (1991), p. 329. For an analysis of the

implications of Darwinism for the animal-rights movement, see James Rachels (1990).

[7] See Nesse (1991b) on the general subject of adaptive psychic pain.

[8] See MacDonald (1988b), esp. p. 158. See also Schweder, Mahapatra, Miller (1987), pp. 10–14.

[9] Loehlin (1992). Personality psychologists don't define "conscientiousness" as broadly as I use the term in this chapter, but there's much overlap; for example, Darwin's compulsive attention to social obligations and to the most detailed aspects of his work would be covered. Presumably evolutionary psychologists will someday show the amorphous thing we call the "conscience" to actually consist of various adaptations (or subadaptations) designed for various functions. In this chapter I use the term somewhat loosely and informally.

[10] *Autobiography,* pp. 26, 45.

[11] *Autobiography,* p. 22.

[12] Brent (1983), p. 11. Brent dissents from the view he is here characterizing.

[13] *Autobiography,* p. 22.

[14] *Autobiography,* p. 28; ED, vol. 2, p. 169.

[15] LLCD, vol. 1, p. 11.

[16] Trivers (1971), p. 53.

[17] See Cosmides and Tooby (1992).

[18] Piaget (1932), p. 139.

[19] *New York Times,* May 17, 1988, p. C1. See also Vasek (1986). And see Krout (1931) for an extremely insightful analysis of childhood lying which stresses that children often lie to gain status and attention. Krout also notes Darwin's own childhood lies and debunks the turn-of-the-century view that some children are "natural-born prevaricators" while others aren't.

[20] *Autobiography,* p. 23; CCD, vol. 2, p. 439.

[21] *New York Times,* May 17, 1988, C1.

[22] Vasek (1986), p. 288.

[23] CCD, vol. 2, p. 439.

[24] Smiles (1859), p. 372.

[25] ED, vol. 2, p. 145.

[26] Smiles (1859), pp. 399, 401.

[27] See, e.g., *Washington Post,* Jan. 5, 1986. The contrast between "personality" and "character" is related to the distinction between "inner-directed" and "other-directed" orientation famously made by Riesman (1950).

[28] Smiles (1859), pp. 397–400.

[29] Smiles (1859), pp. 401–2.

[30] Brent (1983), p. 253.

[31] See Cosmides and Tooby (1989).

[32] Smiles (1859), p. 407.

[33] Less anecdotally: there is evidence that people who live in cities, or at least spend their adolescence in cities, take an especially "Machiavellian" approach to social interactions. See Singer (1993), p. 141.

[34] Daly and Wilson (1988), p. 168.

[35] Smiles (1859), pp. 415, 409.

[36] For another example of a Darwinian explanation of how early environment can shape character, see Draper and Belsky (1990).

[37] See Wilson (1975), p. 565.

[38] This is one of the many defensible claims for which E. O. Wilson was excoriated during the "sociobiology controversy" of the 1970s. See Wilson (1975), p. 553.

[39] Houghton (1957), p. 404. On hypocrisy as testament to a moral code's robustness, see Himmelfarb (1968), pp. 277–78.

[40] James Lincoln Collier, *The Rise of Selfishness in America,* quoted in *New York Times,* Oct. 15, 1991, p. C17.

[41] Desmond and Moore (1991), pp. 333, 398.

[42] ED, vol. 2, p. 168.

CHAPTER 11: DARWIN'S DELAY

[1] CCD, vol. 2, p. 298.

[2] For a summary of Darwin's symptoms and treatment, see prologue of Bowlby (1991).

[3] CCD, vol. 3, p. 397.

[4] CCD, vol. 3, pp. 43–46, 345. On Darwin's workweek, see Bowlby (1991), pp. 409–11.

[5] Bowler (1990).

[6] Gruber (1981), p. 68.

[7] Himmelfarb (1959), p. 210.

[8] Himmelfarb (1959), p. 212.

[9] Debate continues over whether the dynamics of natural selection make the generation of more complex life, and highly intelligent life, quite likely or even virtually inevitable. (See, e.g., Williams [1966], chapter 2, Bonner [1988], and Wright [1990].) The point here is that natural selection entails no mystical force that makes this trend literally inevitable.

[10] *Notebooks,* p. 276. Gruber (1981) puts particular emphasis on passages like this one in explaining Darwin's delay.

[11] CCD, vol. 3, p. 2.

[12] CCD, vol. 2, pp. 47, 430–35.

[13] CCD, vol. 2, p. 150.

[14] Clark (1984), pp. 65–66. Other emotionally taxing influences implicated in Darwin's ill health include his mother's early death, which Bowlby (1991) sees as a contributing factor to "hyperventilation syndrome," his own favored diagnosis of Darwin's malady.

[15] CCD, vol. 3, p. 346.

[16] See LLCD, vol. 1, p. 347, or CCD, vol. 4, pp. 388–409.

[17] See Richards (1987), p. 149.

[18] See Gruber (1981), pp. 105–6.

[19] As for sexual recombination: before genetics, sex seemed in some ways an obstacle for Darwin's theory. The natural way to think of sexual reproduction is as a "mixture" of the mother's and father's traits. And a simple "mixture" of traits won't broaden the available range. Mix two vats of water, one hot and one cold, and you get something in between. Of course, we have only to look at two tall parents who have an even taller child to suspect that this analogy doesn't hold. But that's because we know that the parents' life experiences don't affect their hereditary material and

that the environmental effects on a child's height are typically modest. Darwin didn't know the former.
[20] LLCD, vol. 2, p. 54.

CHAPTER 12: SOCIAL STATUS

[1] *Notebooks*, pp. 541–42.
[2] *Voyage*, pp. 183–84.
[3] See Freeman (1983), Brown (1991), esp. chapter 3, and Degler (1991).
[4] See Hill and Kaplan (1988), esp. pp. 282–83.
[5] Hewlett (1988). Differences in fertility between *kombeti* and other men (7.89 offspring vs. 6.34 offspring) weren't statistically significant, because there were only nine *kombeti* to study; but the chances are much better than even that the difference would have held up if a larger sample were available. Among other cultures in which a link between status and reproductive success has been found are the Efe of Zaire and the Mukogodo of Kenya. For these and other references, see Betzig (1993a). See also Betzig (1993b) and Betzig (1986). Napoleon Chagnon (1979) was among the first to point to unequal reproductive success in "egalitarian" societies.
[6] Murdock (1945), p. 89.
[7] Ardrey (1970), p. 121. Darwin did little speculating about the evolutionary roots of human social hierarchy, but when he did—as in discussing the possibly hereditary nature of "obedience to the leader of the community"—he sometimes seemed to lean toward "good of the group" logic. See *Descent*, vol. 1, p. 85.
[8] Ardrey (1970), p. 107. See also Wilson (1975), p. 281.
[9] Williams (1966), p. 218. For a nice step-by-step demonstration of how simply status hierarchies can evolve, see Stone (1989).
[10] See Maynard Smith (1982), chapter 2, or the summary of Maynard Smith's logic in Dawkins (1976), chapter 5.
[11] Submissive sparrows painted the darker, dominant color were mercilessly harassed. But if the birds were painted as dominants *and* given a raised testosterone level, they actually *became* successful dominants. Given this finding that a bit more melanin and testosterone could turn a submissive bird into a dominant, one would, says Maynard Smith, expect all birds to opt for the dominant strategy if it conferred a marked reproductive advantage. Yet they don't. See Maynard Smith (1982), pp. 82–84.
[12] See Betzig (1993a) for citations. For the link between status and reproductive success in other species, see Clutton-Brock (1988).
[13] See Lippitt et al. (1958). For more recent work see, e.g., Jones (1984).
[14] See Strayer and Trudel (1984) and Russon and Waite (1991).
[15] Atzwanger (1993).
[16] Mitchell and Maple (1985). For an experiment explicitly comparing human and nonhuman hierarchy-forming dynamics, see Barchas and Fisek (1984).
[17] *Expression*, pp. 261, 263.
[18] Weisfeld and Beresford (1982).
[19] Cited in Weisfeld and Beresford (1982), p. 117.
[20] On the emotions governing status competition, see Weisfeld (1980). In general, on hierarchy in humans and other primates, see Ellyson and Dovidio (1985). A subtle but perhaps universal expression of hierarchy has to do with who watches whom

the most, and under what circumstances, in conversation and other social interaction. A well-known paper that called attention to this index of status in primates is Chance (1967).

[21] On vervets: McGuire, Raleigh, and Brammer (1984). Raleigh and McGuire (1989) show what a crucial, if subtle, role female vervets sometimes play in the selection of the dominant male. On fraternity officers: unpublished data from McGuire (personal communication; also noted in *New York Times*, Sept. 27, 1983, p. C3).

[22] McGuire didn't check the serotonin levels of the fraternity officers before they were elected to office, and thus can't entirely exclude the possibility that the levels were just as high before this elevation in status. But various circumstantial evidence, including analogy with nonhuman primates (whose serotonin levels *were* checked before as well as after their rise in stature), leads him to believe that such elevations of status raise human serotonin levels considerably.

[23] Raleigh et al. (unpublished manuscript).

[24] On serotonin generally, see Kramer (1993) and Masters and McGuire (1994).

[25] On cheating: Aronson and Mettee (1968). On impulsive crimes: papers by Linnoila et al. in chapter 6 of Masters and McGuire (1994).

[26] As with reciprocal altruism, we shouldn't here overdramatize the case against group selectionism. If three primates are out hunting, and one of them contains a newly minted gene that facilitates a submissive role, and thus better team dynamics, their collective success could bring so much gain to that individual as to outweigh the costs of low relative status (e.g., a smaller share of the kill); one fourth of fifty pounds of meat is more than one third of twenty-five. This kind of dynamic (which some biologists would call group selection and others wouldn't) is certainly conceivable. Still, even in this scenario, the most valuable genes of all would be those that bring neither submission nor dominance, but a capacity for either, as circumstances warrant.

[27] De Waal (1982), p. 87.

[28] See de Waal (1984) and Goodall (1986) for detailed accounts of status-seeking, and its stakes, among male and female chimpanzees.

[29] Tannen (1990), p. 77.

[30] Daly and Wilson (1983), p. 79.

[31] The reader may sense a paradox. Earlier we said status is determined largely by environment. Now we stress the role of genes in variation in status among males. But we never said *all* variation in traits conducive to status is due to environmental differences. Indeed, for any genetically based trait to have ever been favored by natural selection there must have been *some* genetic variation within the population. Otherwise how could natural selection favor the trait *over* anything? Still, this favoring does have the effect of narrowing the range of variation. In this case, for example, natural selection would cull genes not conducive to successful status competition. The general pattern is that mutation and sexual recombination continually create variation, and natural selection continually compresses that variation, confining it to a fairly narrow range.

[32] CCD, vol. 2, p. 29.

[33] *Descent*, vol. 2, p. 326.

[34] *Descent*, vol. 2, pp. 368–69.

[35] Perusse (1993).

[36] Quoted in Symons (1979), p. 162.

[37] Low (1989).

[38] On chimps: de Waal (1982), pp. 56–58. On bonobos: de Waal (1989), p. 212; Kano (1990), p. 68. Whether human females are more or less "ambitious" than chimp females is an interesting question. Certainly there's a sense in which they're more competitive. As we've seen, high male parental investment has given them something to compete for (though the competition seems to be less intensely driven by natural selection than is male competition for females). On the other hand, female chimps, lacking a regular mate, must often assume primary responsibility for the protection and social advancement of their offspring, a duty that puts a premium both on their physical aggressiveness and on their pursuit of social status.

[39] Stone (1989).

[40] Goodall (1986), pp. 426–27.

[41] De Waal (1989), p. 69.

[42] De Waal (1982), p. 98.

[43] De Waal (1982), p. 196.

[44] De Waal (1982), p. 114. See de Waal (1989) for a broad discussion of reconciliation rituals among primates.

[45] De Waal (1982), p. 117.

[46] Goodall (1986), p. 431.

[47] De Waal (1982), p. 207.

[48] On the correlation between "attention structure" and dominance hierarchies, see Abramovitch (1980) and Chance (1967).

[49] See Weisfeld (1980), p. 277. See Stone (1989), pp. 22–23, on why this makes sense as a ladder-climbing technique.

[50] De Waal (1982), pp. 211–12.

[51] De Waal (1982), pp. 56, 136, 150–51.

[52] Freedman (1980), p. 336.

[53] Benedict (1934), p. 15.

[54] Benedict (1934), p. 99.

[55] Glenn Weisfeld (personal communication) has stressed this connection between status and values.

[56] See Chagnon (1968), chapters 1 and 5.

CHAPTER 13: DECEPTION AND SELF-DECEPTION

[1] CCD, vol. 4, p. 140.

[2] On fireflies, see Lloyd (1986). On orchids, snakes, and butterflies, see Trivers (1985), chapter 16.

[3] Goffman (1959), p. 17.

[4] Dawkins (1976), p. vi. Alexander (1974), p. 377, had noted that "sincerity represents a valuable social asset even when it derives from a real failure to recognize the reproductively selfish background and effects of one's own behavior. . . . [S]election may have consistently favored tendencies for humans not to be aware of what they are really doing or why they are doing it." See also Alexander (1975), p. 96, and Wallace (1973).

[5] Donald Symons and Leda Cosmides (personal communication) have outlined some of these problems with studying self-deception. See Greenwald (1988) for a clear

discussion of possible forms of self-deception, and of which forms are most likely to occur.

[6] CCD, vol. 2, pp. 438–39.

[7] *Papers,* vol. 2, p. 198.

[8] Glantz and Pearce (1989), Glantz and Pearce (1990).

[9] *Descent,* vol. 1, p. 99.

[10] See Lancaster (1986).

[11] *Descent,* vol. 1, p. 164.

[12] *New York Times,* May 17, 1988, pp. C1, C6.

[13] *Autobiography,* p. 139.

[14] *Bartlett's Book of Familiar Quotations,* 15th edition.

[15] Loftus (1992).

[16] See, e.g., Fitch (1970) and Streufert and Streufert (1969). The literature in this area has been criticized by some, such as Miller and M. Ross (1975) and Nisbett and L. Ross (1980), pp. 231–37. Miller and M. Ross note that the data can often be explained not just by a self-serving bias per se, but by the mechanics of human information processing. This is true but points only to the need for more subtle experiments—and, indeed, when Miller himself later conducted one such experiment, the self-serving bias explanation held up; see Miller (1976). Nisbett and L. Ross note that there clearly are cases in which people hold themselves more responsible for failure than for success. This too is true and well exemplifies the promise of evolutionary psychology; as it provides us with a surer grasp of the function of, for example, self-esteem, it should help us explain how some people come to give themselves too much credit and some too little—what sorts of situations and developmental environments are conducive to each tendency. Indeed, the remainder of this chapter is intended to provide some clues along these lines.

[17] LLCD, vol. 1, p. 137.

[18] See Krebs, Denton, and Higgins (1988), pp. 115–16.

[19] See, e.g., Buss and Dedden (1990).

[20] Desmond and Moore (1991), p. 491.

[21] See Stone (1989).

[22] Hartung (1988), p. 173.

[23] Trivers (1985), p. 417.

[24] Nesse (1990a), p. 273.

[25] CCD, vol. 6, p. 429.

[26] CCD, vol. 6, p. 430.

[27] See Dawkins and Krebs (1978).

[28] Alexander (1987), p. 128.

[29] For citations, see Aronson (1980), pp. 138–39. An alternative interpretation of these results is that the subjects irrationally feared that the person would find out about the derogation and retaliate by delivering especially severe shocks.

[30] These techniques were outlined by S. H. Schwartz and J. A. Howard, as cited in MacDonald (1988b).

[31] See Hilgard, Atkinson, and Atkinson (1975), p. 52.

[32] See Krebs, Denton, and Higgins (1988), p. 109; Gazzaniga (1992), chapter 6.

[33] Quoted in Timothy Ferris, *The Mind's Sky* (Bantam Books, 1992), p. 80.

[34] Barkow (1989), p. 104.

[35] CCD, vol. 1, p. 412.

[36] Bowlby (1991), p. 107.

[37] See Bowlby (1991), p. 363; *Origin*, pp. 54–55; *Autobiography*, p. 49. Erasmus is quoted in Gruber (1981), p. 51.

[38] Bowlby (1991), p. 363.

[39] Malinowski (1929), p. 91.

[40] See Cosmides and Tooby (1989), p. 77.

[41] *Autobiography*, p. 123.

[42] Trivers (1985), p. 420.

[43] See Aronson (1980), p. 109. See also Levine and Murphy (1943).

[44] See Greenwald (1980) and Trivers (1985), p. 418.

[45] Cited in Miller and Ross (1975), p. 217. Much the same effect is found in Ross and Sicoly (1979), experiment 2.

[46] This episode is recounted in Desmond and Moore (1991), pp. 495–99.

[47] *Expression*, p. 237.

[48] CCD, vol. 1, pp. 96, 98, 124, 126. It is exceedingly likely, though not quite certain, that the gift from Jenyns preceded the praise of Jenyns. Darwin asks Henslow to thank Jenyns for the "magnificent present" two days after he relayed the praise to Fox. Since this was the first letter to Henslow in months, it would have been his first opportunity to pass on his thanks via Henslow in a long time. The chances that the gift had arrived within the past forty-eight hours are slim. In any event, Darwin and Jenyns went on to work out a satisfactory relationship—"the more I see the more I like him" (CCD, vol. 1, p. 124)—and Darwin never again mentioned his "weak mind."

[49] *New York Times*, Oct. 14, p. A1.

[50] See, e.g., Chagnon (1968).

[51] Whether this sort of shaping would involve group selection depends in part on what the psychology of war is said to consist of. If you believe people are programmed to behave with true selflessness, so that jumping on hand grenades to save comrades is an act characteristic of our species, then the explanation may indeed call for group selection. But if you believe that a warrior tends to exploit his comrades—gets them to take the gravest risks, after which he will gladly do the raping and pillaging—then group selection looms less large. See Tooby and Cosmides (1988) for the view that interband conflict could help shape various kinds of adaptations for coalitional aggression without group selection. See also the Appendix of this book, item 6.

[52] Trivers (1971), p. 51.

[53] CCD, vol. 3, p. 366.

CHAPTER 14: DARWIN'S TRIUMPH

[1] CCD, vol. 6, p. 346.

[2] LLCD, vol. 3, p. 361.

[3] Brent (1983), pp. 517–18.

[4] Clark (1984), p. 214.

[5] Clark (1984), p. 3.

[6] CCD, vol. 1, pp. 85, 89; *Autobiography*, p. 63.

[7] Bowlby (1991), pp. 71–74.

[8] Brent (1983), p. 85.

[9] *Autobiography*, p. 55.

[10] *Autobiography*, p. 55.

[11] *New York Times*, May 17, 1988, p. C1.

[12] *Autobiography*, p. 64.

[13] CCD, vol. 1, p. 110.

[14] Desmond and Moore (1991), p. 81.

[15] Goodall (1986), p. 431.

[16] CCD, vol. 1, pp. 140, 142; "as perfect as nature can make him": Bowlby (1991), p. 124.

[17] CCD, vol. 1, pp. 143, 141; "perverted manner": CCD, vol. 2, p. 80.

[18] CCD, vol. 1, pp. 469, 503; "geological hammer": *Autobiography*, p. 82.

[19] CCD, vol. 1, pp. 57, 62; Brent (1983), p. 81; Desmond and Moore (1991), p. 76.

[20] CCD, vol. 1, pp. 416–17.

[21] Laura Betzig (personal communication).

[22] CCD, vol. 1, pp. 369, 508.

[23] CCD, vol. 1, p. 460; *Papers*, pp. 41–43.

[24] Bowlby (1991), p. 210.

[25] CCD, vol. 1, pp. 524, 532–33.

[26] Gruber (1981), p. 90.

[27] CCD, vol. 1, p. 517.

[28] See Thibaut and Riecken (1955). In those few cases in which the "low status" individual, rather than the "high status" individual, was seen by the subject as complying with the subject's influence for "internal reasons" (rather than under social pressure from the subject), the effect was reversed: the subject showed a particular increase in his liking for the "low status" individual. The authors suggest that this—perception of the locus of causality—could be the real independent variable, not the social-status variable. I would contend, though, that in those cases where the subject whom the authors designated "low-status" was perceived as having "internal reasons" for compliance, he was in fact being seen by subjects as giving off "high status" cues (even though he had described himself as having meager educational credentials, etc.). In fact, resistance to "social pressure," and a tendency to do things for "internal" reasons, is arguably part of the definition of "high status."

[29] CCD, vol. 2, p. 284; CCD, vol. 1, p. 512.

[30] *Autobiography*, p. 101. Darwin's overall assessment of Lyell in the autobiography is quite positive.

[31] Asch (1955).

[32] Verplanck (1955).

[33] Zimmerman and Bauer (1956).

[34] Himmelfarb (1959), p. 210.

[35] CCD, vol. 6, p. 445.

[36] Sulloway (1991), p. 32.

[37] Sulloway (1991), p. 32. Some might quibble with Sulloway's use of the term "self-esteem." Persistent but sporadic self-doubt isn't necessarily the same thing as low self-esteem. Indeed, a man with truly low self-esteem might never have mustered the bursts of audacity it took to challenge the received view of human creation. There are hints that evolutionary psychologists may disentangle "low self-esteem" (with its chronic self-doubt) and "insecurity" (with its periodic self-doubt), and show them to be calibrated by the early social environment for distinct reasons. But, in any event, Sulloway seems right in his larger point—that the painful self-doubt

instilled in Darwin by his early environment, including his father, helped his career.
[38] Bowlby (1991), pp. 70–73. Sulloway (1991) notes that if, as Bowlby says, Darwin's father fostered Darwin's self-doubt, then some of Darwin's scientific accomplishment can be traced to his father, though Sulloway doesn't suggest that the father was acting under the influence of a mental adaptation designed for that purpose.

[39] Bowlby does acknowledge some practical utility in Darwin's deference to authority. He suggests that Darwin, in being "respectful" of older men's opinions, "unlike some brash young men," may have been well equipped to win important sponsors. But he adds: "Welcome though such attitudes may be in youth, they can become excessive in later years. Then, not only is the status of others given undue respect and their opinions given undue weight but, far more important than that, the person's own worth and own opinions may be commensurately undervalued" (p. 72). Maybe, maybe not. Bowlby cites Darwin's extreme deference to authority figures, such as the great scientist Lord Kelvin, who made what seemed at the time a formidable criticism of Darwin's theory. Darwin amended the theory to accommodate such critics, and as a result the successive editions of the *Origin* became paler and paler versions of the original; the sixth and final edition is a poorer guide to Darwin's theory, as now understood, than the first. But even this social appeasement may not have been unfortunate in sheerly Darwinian terms; such flexibility probably helped the theory's stature during Darwin's life, thus preserving his social status and helping his immediate descendants capitalize on the Darwin name.

[40] See Aronson (1980), pp. 64–67.

[41] Brent (1983), p. 376.

[42] CCD, vol. 6, pp. 250, 256. Other authors have interpreted this conversation much as Brent did. See, e.g., Bowlby (1991), pp. 270–71, 279.

[43] *Autobiography*, p. 162.

[44] See LLCD, vol. 2, p. 156, for evidence of the effect of Lyell and Hooker being Darwin's sponsors.

[45] LLCD, vol. 2, pp. 238, 241.

[46] LLCD, vol. 2, pp. 165–66.

[47] LLCD, vol. 3, pp. 8–9.

[48] See Bowlby (1991), p. 254.

[49] *Autobiography*, p. 105.

[50] LLCD, vol. 2, p. 237.

[51] CCD, vol. 6, p. 432.

[52] CCD, vol. 6, pp. 100, 387, 514, 521.

[53] LLCD, vol. 2, p. 116.

[54] LLCD, vol. 2, pp. 116–17.

[55] LLCD, vol. 2, pp. 117–19. Others—e.g., Gould (1980), p. 48—have taken a similar view of this communication.

[56] Quoted in Rachels (1990), p. 34.

[57] *Papers*, vol. 2, p. 4.

[58] This view is especially defensible given ongoing changes in the rules of science. In the century before Darwin's, a scientist who was clearly the first to communicate a theory—as Darwin was in sending his theory to Gray—would have been accorded priority, even if he had never published the theory. This tradition had greatly declined by the mid-nineteenth century but hadn't wholly died out. (Sulloway, personal communication.)

[59] Rachels (1990) makes this point. Rachels, one of the few observers to have taken a harsh view of Darwin's treatment of Wallace, depicts the episode much as I do: the triumph of a powerful clique over a naïf.

[60] Quoted in Clark (1984), p. 119.

[61] Eiseley (1958), p. 292.

[62] Desmond and Moore (1991), p. 569.

[63] Clark (1984), p. 115.

[64] LLCD, vol. 2, p. 117.

[65] Brent (1983), p. 415.

[66] LLCD, vol. 2, p. 145.

[67] Bowlby (1991), pp. 88–89.

[68] See epigraph to this chapter.

[69] LLCD, vol. 2, p. 128.

[70] *Autobiography*, p. 124.

[71] But helping an ally greatly at *extremely* low cost can make sense sheerly in terms of putting the ally in better position to later serve you.

[72] See Alexander (1987).

[73] *Papers*, vol. 2, p. 4.

[74] Clark (1984), p. 119.

[75] Bowlby (1991), pp. 60, 73.

CHAPTER 15: DARWINIAN (AND FREUDIAN) CYNICISM

[1] *Notebooks*, p. 538.

[2] Richard Alexander has said that "within-group amity" often implies "between-group enmity."

[3] One milestone in this transition was the 1918 book *Eminent Victorians*, in which Lytton Strachey gleefully punctured Victorian pretense, finding, for example, rampant egotism in Florence Nightingale.

[4] For an authoritative treatment of Darwinian and other biological aspects of Freudianism, see Sulloway (1979a), esp. chapter 7.

[5] Daly and Wilson (1990b).

[6] Nesse (1991b).

[7] CCD, vol. 2, p. 439.

[8] Brent (1983), p. 24.

[9] Desmond and Moore (1991), p. 138.

[10] Quoted in Bowlby (1991), p. 350.

[11] See Buss (1991), pp. 473–77, and Tooby and Cosmides (1990a).

[12] *Autobiography*, p. 123.

[13] Freud (1922), pp. 79–80.

[14] Freud (1922), p. 80. Freud elsewhere formulated elaborate rules in an attempt to explain exceptions to the general tendency to ward off painful memories.

[15] MacLean (1983), p. 88. For a crisp tour through the evolution of the brain, see Jastrow (1981).

[16] Nesse and Lloyd (1992), p. 614.

[17] Slavin (1990).

[18] See Nesse and Lloyd (1992), p. 608.

[19] Nesse and Lloyd (1992), p. 611.

²⁰ Freud (1930), pp. 58, 62.

²¹ Freud (1922), p. 296.

²² See Connor (1989), chapters 1 and 6; Graham, Doherty, and Malek (1992); and Wyschogrod (1990), esp. pp. xiii–xxvii.

CHAPTER 16: EVOLUTIONARY ETHICS

¹ *Notebooks*, pp. 550, 629.

² *Descent*, p. 73.

³ Clark (1984), p. 197.

⁴ Hofstadter (1944), p. 45.

⁵ Rachels (1990), p. 62.

⁶ Rachels (1990), p. 65. See the first part of chapter 2 for a crisp summary of Spencer's ethics and other attempts to extract values from evolutionism.

⁷ See Rachels (1990), pp. 66–70.

⁸ The common claim that David Hume first identified the naturalistic fallacy is debatable. See Glossop (1967), esp. p. 533.

⁹ Mill (1874), pp. 385, 391, 398–99.

¹⁰ *Encyclopedia of Philosophy*, Macmillan, vol. 5, p. 319.

¹¹ LLCD, vol. 2, p. 312.

¹² *Descent*, vol. 2, p. 393.

¹³ Some authors—e.g., Richards (1987), pp. 234–41—have stressed the differences between Darwin's ethics and utilitarianism as classically formulated. There certainly were differences, as Darwin himself noted, but they have more to do with the derivation of the doctrine than with its practical application. (See note 22, below.) As Gruber (1981), p. 64, argues, Darwin "accepted the utilitarian ethic" in at least "a very general sense"; he evaluated actions "in terms of their actual consequences for living beings, not in terms of some supposedly timeless foreordained moral code." This approach to moral evaluation may seem unexceptional today, when so much ethical discourse is conducted explicitly in such terms, but in the nineteenth century, it distinguished Darwin's and Mill's ethics from large sectors of discourse. And there was another, equally crucial commonality between the two thinkers: even if Darwin used the term "welfare" where Mill used "happiness," both men counted every human being's welfare/happiness equally in the moral calculus. This egalitarianism, as we'll discuss later in this chapter, lies near the heart of utilitarianism. And it is a primary reason why utilitarianism was a left-wing doctrine in Victorian England. On Darwin's admiration for Mill's moral and political philosophy, see ED, vol. 2, p. 169.

¹⁴ See MacIntyre (1966), p. 251.

¹⁵ See Mill (1863), pp. 307–8, and Alan Ryan's introduction, p. 49.

¹⁶ Mill (1863), pp. 274–75.

¹⁷ That Mill was a "rule-based" utilitarian is not beyond debate. But see Mill (1863), pp. 291, 295, for evidence that he was. On act-based vs. rule-based utilitarianism generally, see Smart (1973).

¹⁸ Mill (1863), p. 288.

¹⁹ Actually, the very existence of pleasure and pain—of subjective experience generally—is a deeper mystery than many people, including many evolutionists, realize (though John Maynard Smith has noted it in passing). See Wright (1992).

[20] Gruber (1981), pp. 64, 66.

[21] Desmond and Moore (1991), p. 120.

[22] Actually, one might plausibly doubt whether what is commonly taken as Darwin's rationale for the primacy of the "greatest good or welfare of the community"—in particular a single paragraph on p. 98 of *Descent*, vol. 1—is in fact a rationale for it. Like so many discussions bearing on utilitarianism, this passage somewhat blurs the line between prescriptive and descriptive; it isn't perfectly clear (to me, at least) whether Darwin is saying that people *should* worry about the "good" or "welfare" rather than the "happiness" of the community, or that people in fact, by their nature, *do* worry about "good" or "welfare" rather than "happiness." Of course, this very sort of blurriness is frequently symptomatic of the naturalistic fallacy, the inference of *should* from *do*. Also vaguely suggestive of the naturalistic fallacy is Darwin's definition of the general good as "the means by which the greatest possible number of individuals can be reared in full vigour and health, with all their faculties perfect, under the conditions to which they are exposed." No doubt Darwin's beliefs about the origin of the moral sentiments—that they evolved for the "good of the group"—made the naturalistic fallacy tempting. That is: since evolution seemed to have designed the moral impulses to further the moral values that he himself had grown up believing in, there was no strong reason to distrust nature as a guide to right—at least, not in this particular context. Still, as we've seen in this chapter, Darwin was in other contexts emphatic in rejecting claims for nature's moral authority.

[23] Again (see previous note), Darwin, something of a group-selectionist, didn't see how thoroughly individual selfishness permeated nature's designs. Thus he could, on the one hand, be horrified by a cat's playing with mice but, on the other, view the human moral sentiments more sanguinely than a modern evolutionist might.

[24] Huxley (1984), pp. 80, 83.

[25] Singer (1981), p. 168.

[26] Williams (1989), p. 208.

[27] Alan Ryan's introduction to Mill (1863).

[28] See Betzig (1988).

[29] *Descent*, vol. 1, pp. 88–89.

CHAPTER 17: BLAMING THE VICTIM

[1] *Descent*, vol. 2, p. 393; *Notebooks*, p. 571.

[2] But see Daly and Wilson (1988), chapter 11, for one clear treatment of the issue of determinism and culpability.

[3] Ruse (1986), pp. 242–43, notes this contradiction.

[4] Mill (1863), p. 334.

[5] Matthew 5:44, 5:39; Exodus 21:24.

[6] Though Mill might have thus gotten off the hook by endorsing blame and punishment reluctantly, for their practical value, he emphatically did not take that route. Repaying good for good and evil for evil, he wrote, "is not only included within the idea of Justice as we have defined it, but is a proper object of that intensity of sentiment, which places the Just, in human estimation, above the simply expedient" (Mill [1863], p. 334).

[7] Dawkins (1982), p. 11.

[8] On Darwin's materialism and determinism, see Gruber (1981) and Richards (1987).

[9] *Notebooks*, pp. 526, 535. "Example of others" and "teaching of others" don't exhaust the list of environmental influences, of course. But his point, plainly, was that everything boils down to heredity and environment.

[10] *Notebooks*, p. 536.

[11] Archetypal environmentalist Skinner worked to expose the myth of free will in *Beyond Freedom and Dignity*, and argued that the notions of blame and credit exist solely for their practical value, not because they make philosophical sense. What he didn't know was that these notions were created by natural selection, which had implicitly recognized their practical value.

[12] *Notebooks*, p. 608.

[13] *Notebooks*, p. 608.

[14] *Notebooks*, p. 608.

[15] *Notebooks*, p. 614.

[16] Daly and Wilson (1988), p. 269.

[17] See Saletan and Watzman (1989).

[18] *Notebooks*, p. 608. The full passage, as transcribed by the editors, reads: "One must view a wrecked man, like a sickly one—We cannot help loathing a diseased offensive object, so we view wickedness.—it would however be more proper to pity than to hate & be disgusted." I suspect that "wrecked man" is a mistranscription of "wicked man." That's one reason I've used "wicked man" in my paraphrasal; but in any event, the subsequent use of "wickedness" loosely justifies this paraphrasal.

[19] Though Mill himself didn't much champion this view of punishment, it was favored by his father and by earlier thinkers in the utilitarian lineage, including the eighteenth-century Italian legal theorist Cesare Beccaria.

[20] Daly and Wilson (1988), p. 256.

[21] Bowlby (1991), p. 352.

[22] Axelrod (1987).

[23] Quoted in Franklin (1987), pp. 246–47.

[24] I'm using "pragmatic" in the (arguably corrupt) sense in which William James used the term, not in the stricter sense in which Charles S. Peirce, the founder of the pragmatic school of philosophy, used it.

[25] Mill (1859), p. 61.

[26] See Himmelfarb (1974), pp. 273–75. Himmelfarb views Mill as someone with morally conservative tendencies that were quite muted in some of his work (e.g., much of *On Liberty*) by the influence of his radical wife.

[27] Mill (1859), p. 104.

[28] Mill (1859), p. 61. See Himmelfarb (1974), chapter 6, and Himmelfarb (1968), p. 143, for the view that *On Liberty* really came at a quite liberated time in England's social history, and for evidence that Mill himself sometimes acknowledged as much.

[29] Mill (1859), p. 78.

[30] Mill (1874), p. 393. Again, see Himmelfarb—(1974), and (1968), chapter 4—on this sort of tension within Mill's writing.

[31] *Notebooks*, p. 608.

[32] *Notebooks*, p. 608.

CHAPTER 18: DARWIN GETS RELIGION

[1] *Autobiography*, p. 91.
[2] *Autobiography*, pp. 85–87.
[3] *Autobiography*, p. 93.
[4] Smiles (1859), pp. 16, 333; *Descent*, vol. 1, p. 101.
[5] Singer (1989), p. 631.
[6] "Buddha's Farewell Address," Burtt (1982), p. 47.
[7] Campbell (1975), p. 1103.
[8] See Dawkins (1976), p. 207, and, more generally, that entire chapter on "memes"; see also chapter 6 in Dawkins (1982).
[9] Symons (1979), p. 207.
[10] "The Way of Truth," Burtt (1982), p. 68.
[11] "The Way of Truth," Burtt (1982), p. 66.
[12] Matthew 6:19, 6:27.
[13] *Bhagavad Gita* II:55–58 (Edgerton [1944], p. 15).
[14] Ecclesiastes 6:7.
[15] Matthew 19:30.
[16] *Bhagavad Gita* II:44,52 (Edgerton [1944], pp. 13, 14).
[17] "The Way of Truth," Burtt (1982), p. 65.
[18] Ecclesiastes, 6:9.
[19] John 8:7, Matthew 7:5.
[20] "The Way of Truth," Burtt (1982), p. 65.
[21] "Truth Is Above Sectarian Dogmatism," Burtt (1982), p. 37.
[22] Singer (1981), pp. 112–14.
[23] *Descent*, vol. 1, pp. 100–101.
[24] Some modern interpretations treat this "war" as a metaphor for the struggle within the self: sensory desires must be attacked ferociously.
[25] Galatians 6:10 (emphasis added).
[26] Hartung (1993).
[27] See Johnson (1987).
[28] Campbell (1975), pp. 1103–4.
[29] *Bhagavad Gita* II:55 (Easwaran [1975], vol. 1, p. 105).
[30] *Bhagavad Gita* XIII:28 (Edgerton [1944], p. 68).
[31] Houghton (1957), p. 62.
[32] CCD, vol. 1, p. 496.
[33] Bowlby (1991), p. 352.
[34] Bowlby (1991), p. 450.
[35] LLCD, vol. 1, p. 124.
[36] ED, vol. 2, p. 253. Francis Darwin, in LLCD, has it as "I am not the least afraid to die."
[37] *Autobiography*, pp. 94–95.
[38] *Autobiography*, p. 95.

Bibliography

Abramovitch, Rona (1980) "Attention Structures in Hierarchically Organized Groups," in Omark, Strayer, and Freedman. (1980).

Alexander, Richard D. (1974) "The Evolution of Social Behavior," *Annual Review of Ecology and Systematics* 5:325–83.

——— (1975) "The Search for a General Theory of Behavior," *Behavioral Science* 10:77–100.

——— (1979) *Darwinism and Human Affairs,* Seattle: University of Washington Press.

——— (1987) *The Biology of Moral Systems,* Hawthorne, N.Y.: Aldine de Gruyter.

Alexander, Richard D., and Katherine M. Noonan (1979) "Concealment of Ovulation, Parental Care, and Human Social Evolution," in Chagnon and Irons (1979).

Alexander, Richard D., et al. (1979) "Sexual Dimorphisms and Breeding Systems in Pinnipeds, Ungulates, Primates and Humans," in Chagnon and Irons (1979).

Alland, Alexander, ed. (1985) *Human Nature: Darwin's View,* New York: Columbia University Press.

Ardrey, Robert (1970) *The Social Contract,* New York: Atheneum.

Aronson, Elliot, ed. (1973) *Readings About the Social Animal,* San Francisco: W. H. Freeman.

Aronson, Elliot (1980) *The Social Animal,* San Francisco: W. H. Freeman.

Aronson Elliot, and David R. Mettee (1968) "Dishonest Behavior as a Function of Differential Levels of Induced Self-Esteem," *Journal of Personality and Social Psychology* 9:121–27. Reprinted in Aronson (1973).

Asch, Solomon E. (1955) "Opinions and Social Pressure," *Scientific American*, November. Reprinted in Aronson (1973).

Atzwanger, Klaus (1993) "Social Reciprocity and Success," paper presented at meeting of the Human Behavior and Evolution Society, Binghamton, N.Y.

Axelrod, Robert (1984) *The Evolution of Cooperation*, New York: Basic Books.

—— (1987) "Laws of Life," *The Sciences* 27:44–51.

Badcock, Christopher (1990) *Oedipus in Evolution: A New Theory of Sex*, Oxford: Basil Blackwell.

Badrian, Alison, and Noel Badrian (1984) "Social Organization of Pan paniscus in the Lomako Forest, Zaire," in Randall L. Susman, ed., *The Pygmy Chimpanzee: Evolutionary Biology and Behavior*, New York: Plenum.

Bailey, Michael (1993) "Can Behavior Genetics Contribute to Evolutionary Explanations of Behavior?" paper presented at meeting of the Human Behavior and Evolution Society, Binghamton, N.Y.

Baker, R. Robin, and Mark A. Bellis (1989) "Number of Sperm in Human Ejaculates Varies in Accordance with Sperm Competition Theory," *Animal Behaviour* 37:867–69.

—— (1993) "Human Sperm Competition: Ejaculate Adjustment by Males and the Function of Masturbation"; and "Human Sperm Competition: Ejaculate Manipulation by Females and a Function for the Female Orgasm," *Animal Behaviour* 46:861–909.

Barchas, Patricia R., and M. Hamit Fisek (1984) "Hierarchical Differentiation in Newly Formed Groups of Rhesus and Humans," in Patricia R. Barchas, ed., *Social Hierarchies*, Westport, Conn.: Greenwood Press.

Barkow, Jerome (1973) "Darwinian Psychological Anthropology: A Biosocial Approach," *Current Anthropology* 14:373–88.

—— (1980) "Prestige and Self-Esteem: A Biosocial Interpretation," in Omark, Strayer, and Freedman (1980).

—— (1989) *Darwin, Sex, and Status*, Toronto: University of Toronto Press.

—— (1992) "Beneath New Culture Is Old Psychology: Gossip and Social Stratification," in Barkow, Cosmides, and Tooby (1992).

Barkow, Jerome H., Leda Cosmides, and John Tooby (1992) *The Adapted Mind: Evolutionary Psychology and the Generation of Culture*, New York: Oxford University Press.

Barlow, Nora, ed. (1959) *The Autobiography of Charles Darwin*, New York: Harcourt Brace.

Barrett, Paul H., ed. (1977) *The Collected Papers of Charles Darwin*, Chicago: University of Chicago Press.

Barrett, Paul H., et al., eds. (1987) *Charles Darwin's Notebooks, 1836–1844*, Ithaca, N.Y.: Cornell University Press.

Barret-Ducrocq, Françoise (1989) *Love in the Time of Victoria: Sexuality and Desire Among Working-Class Men and Women in Nineteenth-Century London*, translated by John Howe, New York: Penguin, 1992.

Bateman, A. J. (1948) "Intra-sexual Selection in *Drosophila*," *Heredity* 2:349–68.

Benedict, Ruth (1934) *Patterns of Culture*, Boston: Houghton-Mifflin Sentry edition, 1959.

Benshoof, Lee, and Randy Thornhill (1979) "The Evolution of Monogamy and Concealed Ovulation in Humans," *Journal of Social and Biological Structures* 2:95–106.

Betzig, Laura L. (1982) "Despotism and Differential Reproduction: A Cross-Cultural Correlation of Conflict Asymmetry, Hierarchy, and Degree of Polygyny," *Ethology and Sociobiology* 3:209–21.

——— (1986) *Despotism and Differential Reproduction: A Darwinian View of History*, New York: Aldine de Gruyter.

——— (1988) "Redistribution: Equity or Exploitation?" in Betzig, Borgerhoff Mulder, and Turke (1988).

——— (1989) "Causes of Conjugal Dissolution: A Cross-Cultural Study," *Current Anthropology* 30:654–76.

——— (1993a) "Where Are the Bastards' Daddies?" *Behavioral and Brain Sciences* 16:285–95.

——— (1993b) "Sex, Succession, and Stratification in the First Six Civilizations," in Lee Ellis, ed. *Social Stratification and Socioeconomic Inequality*, New York: Praeger.

Betzig, Laura, Monique Borgerhoff Mulder, and Paul Turke, eds. (1988) *Human Reproductive Behaviour: A Darwinian Perspective*, New York: Cambridge University Press.

Betzig, Laura, and Paul Turke (1986) "Parental Investment by Sex on Ifaluk," *Ethology and Sociobiology* 7:29–37.

Bonner, John Tyler (1980) *The Evolution of Culture in Animals*, Princeton, N.J.: Princeton University Press.

——— (1988) *The Evolution of Complexity by Means of Natural Selection*, Princeton, N.J.: Princeton University Press.

Bonner, John Tyler, and Robert M. May (1981) "Introduction," in Darwin (1871).

Boone, James L. III (1988) "Parental Investment, Social Subordination, and Population Processes Among the 15th and 16th Century Portuguese Nobility," in Betzig, Borgerhoff Mulder, and Turke (1988).

Bowlby, John (1991) *Charles Darwin: A New Life*, New York: Norton.

Bowler, Peter J. (1990) *Charles Darwin: The Man and His Influence*, Oxford: Basil Blackwell.

Brent, Peter (1981) *Charles Darwin: A Man of Enlarged Curiosity*, New York: Norton, 1983.

Briggs, Asa (1955) *Victorian People: A Reassessment of Persons and Themes, 1851–67*, Chicago: University of Chicago Press, 1972.

Brown, Donald E. (1991) *Human Universals*, New York: McGraw-Hill.

Browne, Janet, and Michael Neve, eds. (1989) Charles Darwin's *Voyage of the Beagle*, New York: Penguin Books.

Buehlman, Kim Therese, J. M. Gottman, and L. F. Katz (1992) "How a Couple View Their Past Predicts Their Future: Predicting Divorce from an Oral History Interview," *Journal of Family Psychology* 5:295–318.

Burkhardt, Frederick, and Sydney Smith, eds. (1985–91) *The Correspondence of Charles Darwin*, 8 vols., Cambridge: Cambridge University Press.

Burtt, E. A., ed. (1982) *The Teachings of the Compassionate Buddha*, New York: New American Library.

Buss, David (1989) "Sex Differences in Human Mate Preferences: Evolutionary Hypotheses Tested in 37 Cultures," *Behavioral and Brain Sciences* 12:1–49.

––––––– (1991) "Evolutionary Personality Psychology," *Annual Review of Psychology* 42:459–91.

––––––– (1994) *The Evolution of Desire: Strategies of Human Mating*, New York: Basic Books.

Buss, David, and Lisa A. Dedden (1990) "Derogation of Competitors," *Journal of Social and Personal Relationships* 7:395–422.

Buss, David, and D. P. Schmitt (1993) "Sexual Strategies Theory: An Evolutionary Perspective on Human Mating," *Psychological Review* 100:204–32.

Buss, David, et al. (1992) "Sex Differences in Jealousy: Evolution, Physiology, and Psychology," *Psychological Science* 3:251–55.

Campbell, Donald T. (1975) "On the Conflicts Between Biological and Social Evolution and Between Psychology and Moral Tradition," *American Psychologist* 30:1103–26.

Cashdan, Elizabeth (1993) "Attracting Mates: Effects of Paternal Investment on Mate Attraction," *Ethology and Sociobiology* 14:1–24.

Cavalli-Sforza, Luigi, et al. (1988) "Reconstruction of Human Evolution: Bringing Together Genetic, Archaeological, and Linguistic Data," *Proceedings of the National Academy of Science* 85:6002–6.

Chagnon, Napoleon (1968) *Yanomamö: The Fierce People*, New York: Holt, Rinehart and Winston.

—— (1979) "Is Reproductive Success Equal in Egalitarian Societies?" in Chagnon and Irons (1979).

—— (1988) "Life Histories, Blood Revenge, and Warfare in a Tribal Population," *Science* 239:985–92.

Chagnon, Napoleon, and William Irons, eds. (1979) *Evolutionary Biology and Human Social Behavior: An Anthropological Perspective*, North Scituate, Mass.: Duxbury Press.

Chance, Michael (1967) "Attention Structure as the Basis of Primate Rank Orders," *Man* 2:503–18.

Charmie, Joseph, and Samar Nsuly (1981) "Sex Differences in Remarriage and Spouse Selection," *Demography* 18:335–48.

Clark, Ronald W. (1984) *The Survival of Charles Darwin*, New York: Avon Books, 1986.

Clutton-Brock, Timothy, ed. (1988) *Reproductive Success: Studies of Individual Variation in Contrasting Breeding Systems*, Chicago: University of Chicago Press.

Clutton-Brock, T. H., and A.C.J. Vincent (1991) "Sexual Selection and the Potential Reproductive Rates of Males and Females," *Nature*, 351:58–60.

Colp, Ralph, Jr. (1981) "Charles Darwin, Dr. Edward Lane, and the 'Singular Trial' of *Robinson v. Robinson and Lane*," *Journal of the History of Medicine and Allied Sciences*, 36:205–13.

Connor, Steven (1989) *Postmodernist Culture: An Introduction to Theories of the Contemporary*, Oxford: Basil Blackwell.

Cosmides, Leda, and John Tooby (1987) "From Evolution to Behavior: Evolutionary Psychology as the Missing Link," in John Dupre, ed., *The Latest on the Best: Essays on Evolution and Optimality*, Cambridge, Mass.: MIT Press.

—— (1989) "Evolutionary Psychology and the Generation of Culture" (part 2), *Ethology and Sociobiology* 10:51–97.

—— (1992) "Cognitive Adaptations for Social Exchange," in Barkow et al. (1992).

Crawford, Charles B., B. E. Salter, and K. L. Lang (1989) "Human Grief: Is Its Intensity Related to the Reproductive Value of the Deceased?" *Ethology and Sociobiology* 10:297–307.

Crispell, Diane (1992) "The Brave New World of Men," *American Demographics*, January.

Cronin, Helena (1991) *The Ant and the Peacock: Altruism and Sexual Selection from Darwin to Today*, New York: Cambridge University Press.

Daly, Martin, Margo Wilson, and S. J. Weghorst (1982) "Male Sexual Jealousy," *Ethology and Sociobiology* 3:11–27.

Daly, Martin, and Margo Wilson (1980) "Discriminative Parental Solicitude: A Biological Perspective," *Journal of Marriage and the Family* 42:277–88.

——— (1983) *Sex, Evolution, and Behavior,* Boston: Willard Grant.

——— (1988) *Homicide,* Hawthorne, N.Y.: Aldine de Gruyter.

——— (1990a) "Killing the Competition: Female/Female and Male/Male Homicide," *Human Nature* 1:81–107.

——— (1990b) "Is Parent-Offspring Conflict Sex-Linked? Freudian and Darwinian Models," *Journal of Personality* 58:163–89.

Darwin, Charles (1859) *The Origin of Species,* New York: Penguin Books, 1968.

——— (1871) *The Descent of Man, and Selection in Relation to Sex,* Princeton, N.J.: Princeton University Press, 1981 (facsimile edition).

——— (1872) *The Expression of the Emotions in Man and Animals,* Chicago: University of Chicago Press edition, 1965.

Darwin, Francis, ed. (1888) *Life and Letters of Charles Darwin,* 3 vols., New York: Johnson Reprint Corp., 1969.

Dawkins, Richard (1976) *The Selfish Gene,* New York: Oxford University Press.

——— (1982) *The Extended Phenotype,* New York: Oxford University Press, 1989.

——— (1986) *The Blind Watchmaker,* New York: W. W. Norton and Co.

Dawkins, Richard, and John R. Krebs (1978) "Animal Signals: Information or Manipulation?" in J. R. Krebs and N. B. Davies, eds., *Behavioural Ecology,* Oxford: Basil Blackwell.

Degler, Carl N. (1991) *In Search of Human Nature: The Decline and Revival of Darwinism in American Social Thought,* New York: Oxford University Press.

Desmond, Adrian, and James Moore (1991) *Darwin: The Life of a Tormented Evolutionist,* New York: Warner Books.

Devore, Irven (1969) "The Evolution of Human Society," in J. F. Eisenberg and Wilton S. Dillon, eds., *Man and Beast: Comparative Social Behavior,* Washington, D.C.: Smithsonian Institution Press.

de Waal, Frans (1982) *Chimpanzee Politics,* Baltimore: Johns Hopkins University Press, 1989.

——— (1984) "Sex Differences in the Formation of Coalitions Among Chimpanzees," *Ethology and Sociobiology* 5:239–55.

——— (1989) *Peacemaking Among Primates,* Cambridge, Mass.: Harvard University Press.

de Waal, Frans, and Lesleigh Luttrell (1988) "Mechanisms of Social Reciprocity in Three Primate Species: Symmetrical Relationship Characteristics or Cognition?" *Ethology and Sociobiology* 9:101–18.

Dickemann, Mildred (1979) "Female Infanticide, Reproductive Strategies," and "Social Stratification: A Preliminary Model," in Chagnon and Irons (1979).

Dobzhansky, Theodosius (1962) *Mankind Evolving: The Evolution of the Human Species,* New Haven: Yale University Press.

Draper, Patricia, and Jay Belsky (1990) "Personality Development in Evolutionary Perspective," *Journal of Personality* 58:141–61.

Draper, Patricia, and Henry Harpending (1982) "Father Absence and Reproductive Strategy: An Evolutionary Perspective," *Journal of Anthropological Research* 38:255–73.

—— (1988) "A Sociobiological Perspective on the Development of Human Reproductive Strategies," in MacDonald (1988a).

Dugatkin, Lee Alan (1992) "The Evolution of the 'Con Artist,' " *Ethology and Sociobiology* 13:3–18.

Durant, John R. (1985) "The Ascent of Nature in Darwin's *Descent of Man,*" in David Kohn, ed., *The Darwinian Heritage,* Princeton, N.J.: Princeton University Press.

Easwaran, Eknath (1975) *The Bhagavad Gita for Daily Living,* 3 vols., Berkeley, Cal.: Blue Mountain Center of Meditation.

Edgerton, Franklin (1944), translation of *The Bhagavad Gita,* Cambridge, Mass.: Harvard University Press, 1972.

Eiseley, Loren (1958) *Darwin's Century,* New York: Anchor Books, 1961.

Ellis, Bruce, and Donald Symons (1990) "Sex Differences in Sexual Fantasy: an Evolutionary Psychological Approach," *Journal of Sex Research* 27:527–55.

Ellyson, S. L., and J. F. Dovidio, eds. (1985) *Power, Dominance, and Nonverbal Behavior,* New York: Springer-Verlag.

Essock Vitale, Susan M., and Michael T. McGuire (1988) "What 70 Million Years Hath Wrought: Sexual Histories and Reproductive Success of a Random Sample of American Women," in Betzig, Borgerhoff Mulder, and Turke (1988).

Fausto-Sterling, Anne (1985) *Myths of Gender,* New York: Basic Books.

Fisher, Ronald A. (1930) *The Genetical Theory of Natural Selection,* Oxford: Clarendon Press.

Fitch, Gordon (1970) "Effects of Self-Esteem, Perceived Performance, and Choice on Causal Attributions," *Journal of Personality and Social Psychology* 16:311–15.

Fletcher, David J. C., and Charles D. Michener, eds. (1987) *Kin Recognition in Animals,* New York: John Wiley & Sons.

Frank, Robert (1985) *Choosing the Right Pond: Human Behavior and the Quest for Status,* New York: Oxford University Press.

—— (1990) "A Theory of Moral Sentiments," paper presented at meeting of the Human Behavior and Evolution Society, Los Angeles.

Franklin, Jon (1987) *Molecules of the Mind,* New York: Atheneum.

Freeman, Derek (1983) *Margaret Mead and Samoa: The Making and Unmaking of an Anthropological Myth*, Cambridge, Mass.: Harvard University Press.

Freedman, Daniel G. (1980) "Cross-Cultural Notes on Status Hierarchies," in Omark, Strayer, and Freedman (1980).

Freeman, R. B. (1978) *Charles Darwin: A Companion*, Kent (England): Wm. Dawson and Sons.

Freud, Sigmund (1922) *A General Introduction to Psychoanalysis*, translated by Joan Riviere, New York: Washington Square Press, 1960.

—— (1930) *Civilization and Its Discontents*, translated by James Strachey, New York: Norton, 1961.

Gangestad, Steven W., and Jeffrey A. Simpson (1990) "Toward an Evolutionary History of Female Sociosexual Variation," *Journal of Personality* 58:69–95.

Gaulin, Steven J. C., and James S. Boster (1990) "Dowry as Female Competition," *American Anthropologist* 92:994–1005.

Gaulin, Steven J. C., and Randall W. FitzGerald (1986) "Sex Differences in Spatial Ability: An Evolutionary Hypothesis and Test," *American Naturalist* 127:74–88.

Gaulin, Steven J. C., and Carole J. Robbins (1991) "Trivers-Willard Effect in Contemporary North American Society," *American Journal of Physical Anthropology*, 85:61–69.

Gazzaniga, Michael (1992) *Nature's Mind: The Impact of Darwinian Selection on Thinking, Emotions, Sexuality, Language, and Intelligence*, New York: Basic Books.

Gergen, Kenneth J., M. S. Greenberg, and R. H. Willis, eds. (1980) *Social Exchange: Advances in Theory and Research*, New York: Plenum Press.

Ghiselin, Michael T. (1973) "Darwin and Evolutionary Psychology," *Science* 179:964–68.

Gilligan, Carol (1982) *In a Different Voice: Psychological Theory and Women's Development*, Cambridge, Mass.: Harvard University Press.

Glantz, Kalman, and John K. Pearce (1989) *Exiles from Eden: Psychotherapy from an Evolutionary Perspective*, New York: Norton.

—— (1990) "Towards an Evolution-Based Classification of Psychological Disorders," paper presented at meeting of the Human Behavior and Evolution Society, Los Angeles.

Glossop, Ronald J. (1967) "The Nature of Hume's Ethics," *Philosophy and Phenomenological Research* 27:527–36.

Goffman, Erving (1959) *The Presentation of Self in Everyday Life*, New York: Anchor/Doubleday.

Goodall, Jane (1986) *The Chimpanzees of Gombe: Patterns of Behavior*, Cambridge, Mass.: Harvard University Press.

Gould, Stephen Jay (1980) *The Panda's Thumb*, New York: Norton.

Graham, Elspeth, J. Doherty, and M. Malek (1992) "The Context and Language of Postmodernism," in Doherty, Graham, and Malek, eds., *Postmodernism and the Social Sciences,* London: MacMillan.

Grammer, Karl, J. Dittami, and B. Fischmann (1993) "Changes in Female Sexual Advertisement According to Menstrual Cycle," paper presented at meeting of the Human Behavior and Evolution Society, Syracuse, N.Y.

Greene, John C. (1961) *Darwin and the Modern World View,* New York: New American Library, 1963.

Greenwald, Anthony G. (1980) "The Totalitarian Ego: Fabrication and Revision of Personal History," *American Psychologist,* 357:603–18.

———— (1988) "Self-Knowledge and Self-Deception," in Lockard and Paulhus, eds. (1988).

Gronell, Ann M., (1984) "Courtship, Spawning and Social Organization of the Pipefish, *Corythoichthys intestinalis* (Pisces: Syngnathidae), with Notes on two Congeneric Species," *Zeitschrift für Tierpsychologie* 65:1–24.

Grote, John (1870) *An Examination of the Utilitarian Philosophy,* Cambridge: Deighton, Bell and Co.

Gruber, Howard E. (1981) *Darwin on Man: A Psychological Study of Scientific Creativity,* Chicago: University of Chicago Press.

Gruter, Margaret, and Roger D. Master, eds. (1986) *Ostracism: A Social and Biological Phenomenon,* New York: Elsevier.

Haldane, J.B.S. (1955) "Population Genetics," *New Biology* 18:34–51.

Hamilton, William D. (1963) "The Evolution of Altruistic Behavior," *American Naturalist* 97:354–56.

———— (1964) "The Genetical Evolution of Social Behaviour," parts 1 and 2, *Journal of Theoretical Biology* 7:1–52.

Harcourt, A. H., et al. (1981) "Testis Weight, Body Weight and Breeding System in Primates," *Nature,* 293:55–57.

Harpending, Henry C., and Jay Sobus (1987) "Sociopathy as an Adaptation," *Ethology and Sociobiology,* 8:63S–72S.

Hartung, John (1982) "Polygyny and the Inheritance of Wealth," *Current Anthropology,* 23:1–12.

———— (1988) "Deceiving Down: Conjectures on the Management of Subordinate Status," in Lockard and Paulhus (1988).

———— (1993) "Love Thy Neighbor: Prospects for Morality," unpublished manuscript.

Hewlett, Barry S. (1988) "Sexual Selection and Paternal Investment Among Aka Pygmies," in Betzig, Borgerhoff Mulder, and Turke (1988).

434

Hilgard, Ernest R., R. C. Atkinson, and Rita L. Atkinson (1975) *Introduction to Psychology*, New York: Harcourt Brace Jovanovich.

Hill, Elizabeth (1988) "The Menstrual Cycle and Components of Human Female Sexual Behaviour," *Journal of Social and Biological Structures* 11:443–55.

Hill, Elizabeth, and P. A. Wenzl (1981) "Variation in Ornamentation and Behavior in a Discotheque for Females Observed at Different Menstrual Phases," paper presented at meeting of the Animal Behavior Society, Knoxville, Tenn.

Hill, Kim, and Hillard Kaplan (1988) "Trade-offs in Male and Female Reproductive Strategies Among the Ache," parts 1 and 2, in Betzig, Borgerhoff Mulder, and Turke (1988).

Himmelfarb, Gertrude (1959) *Darwin and the Darwinian Revolution*, Garden City, N.Y.: Doubleday.

—— (1968) *Victorian Minds*, New York: Knopf.

—— (1974) *On Liberty & Liberalism: The Case of John Stuart Mill*, San Francisco: ICS Press, 1990.

—— (1987) *Marriage and Morals among the Victorians and Other Essays*, New York: Vintage.

Hofstadter, Richard (1944) *Social Darwinism in American Thought*, Boston: Beacon Press, 1955.

Houghton, Walter E. (1957) *The Victorian Frame of Mind, 1830–1870*, New Haven, Conn.: Yale University Press.

Howard, Jonathan (1982) *Darwin*, Oxford: Oxford University Press.

Hrdy, Sarah Blaffer (1981) *The Woman That Never Evolved*, Cambridge, Mass.: Harvard University Press.

—— (1987) "Sex-biased Parental Investment Among Primates and Other Mammals: A Critical Evaluation of the Trivers-Willard Hypothesis," in Richard J. Gelles and Jane B. Lancaster, eds., *Child Abuse and Neglect: Biosocial Dimensions*, Hawthorne, N.Y.: Aldine de Gruyter.

Hrdy, Sarah Blaffer, and Debra S. Judge (1993) "Darwin and the Puzzle of Primogeniture," *Human Nature* 4:1–45.

Humphrey, Nicholas K. (1976) "The Social Function of Intellect," in P.P.G. Bateson and R. A. Hinde, eds., *Growing Points in Ethology*, Cambridge: Cambridge University Press. Reprinted in Richard Byrne and Andrew Whiten, eds., *Machiavellian Intelligence*, Oxford: Oxford University Press, 1988.

Huxley, Thomas H. (1894) *Evolution and Ethics*, Princeton, N.J.: Princeton University Press, 1989.

Irons, William (1991) "How Did Morality Evolve?" *Zygon* 26:49–89.

Irvine, William (1955) *Apes, Angels, and Victorians: The Story of Darwin, Huxley, and Evolution*, New York: McGraw-Hill.

435

Jankowiak, William, and Ted Fisher (1992) "A Cross-Cultural Perspective on Romantic Love," *Ethnology* 31:149–55.

Jastrow, Robert (1981) *The Enchanted Loom: Mind in the Universe,* New York: Simon and Schuster.

Johnson, Gary R. (1987) "In the Name of the Fatherland: An Analysis of Kin Term Usage in Patriotic Speech and Literature," *International Politicial Science Review* 8:165–74.

Jones, Diane Carlson (1984) "Dominance and Affiliation as Factors in the Social Organization of Same-Sex Groups of Elementary School Children," *Ethology and Sociobiology* 5:193–202.

Kagan, Jerome, and Sharon Lamb, eds. (1987) *The Emergence of Morality in Young Children,* Chicago: University of Chicago Press.

Kahn, Joan R., and Kathryn A. London (1991) "Premarital Sex and the Risk of Divorce," *Journal of Marriage and the Family* 53:845–55.

Kano, Takayoshi (1990) "The Bonobos' Peaceable Kingdom," *Natural History,* November.

Kenrick, Douglas T., Sara E. Gutierres, and Laurie L. Goldberg (1989) "Influence of Popular Erotica on Judgments of Strangers and Mates," *Journal of Experimental Social Psychology* 25:159–67.

Kenrick, Douglas T., et al. (1990) "Evolution, Traits, and the Stages of Human Courtship: Qualifying the Parental Investment Model," *Journal of Personality* 58:97–115.

Kinzey, Warren G., ed. (1987) *The Evolution of Human Behavior: Primate Models,* Albany, N.Y.: State University of New York Press.

Kitcher, Philip (1985) *Vaulting Ambition: Sociobiology and the Quest for Human Nature,* Cambridge, Mass.: MIT Press.

Konner, Melvin (1982) *The Tangled Wing: Biological Constraints on the Human Spirit,* New York: Harper Colophon Books, 1983.

——— (1990) *Why the Reckless Survive . . . and Other Secrets of Human Nature,* New York: Viking.

Krebs, Dennis, K. Denton, and N. C. Higgins (1988) "On the Evolution of Self-Knowledge and Self-Deception," in MacDonald (1988a).

Krout, Maurice H. (1931) "The Psychology of Children's Lies," *Journal of Abnormal and Social Psychology* 26:1–27.

Lancaster, Jane G. (1986) "Primate Social Behavior and Ostracism," *Ethology and Sociobiology* 7:215–25. Reprinted in Gruter and Masters, eds. (1986).

Lehrman, Karen (1994) "Flirting with Courtship," in Eric Liu, ed., *Next: Young American Writers on the New Generation,* New York: Norton, 1994.

Leighton, Donna Robbins (1987) "Gibbons: Territoriality and Monogamy," in Smuts et al., eds. (1987).

Levine, Jerome M., and Gardner Murphy (1943) "The Learning and Forgetting of Controversial Material," *Journal of Abnormal and Social Psychology*, vol. 38. Reprinted in Maccoby, Newcomb, and Hartley (1958).

Levinsohn, Florence Hamlish (1990) "Breaking Up Is Still Hard to Do," *Chicago Tribune Sunday Magazine*, October 21.

Lippitt, Ronald, et al. (1958) "The Dynamics of Power: A Field Study of Social Influence in Groups of Children," in Maccoby, Newcomb, and Hartley (1958).

Litchfield, Henrietta, ed. (1915) *Emma Darwin: A Century of Family Letters, 1792–1896*, 2 vols., New York: Appleton and Co.

Lloyd, Elizabeth (1988) *The Structure and Confirmation of Evolutionary Theory*, Westport, Conn.: Greenwood Press.

Lloyd, James E. (1986) "Firefly Communication and Deception: 'Oh, What a Tangled Web,' " in Mitchell and Thompson (1986).

Lockard, Joan S. (1980) "Speculations on the Adaptive Significance of Self-Deception," in Lockard, ed., *The Evolution of Human Social Behavior*, New York: Elsevier, 1980.

Lockard, Joan S., and Delroy L. Paulhus, eds. (1988) *Self-Deception: An Adaptive Mechanism*, Englewood Cliffs, N.J.: Prentice Hall.

Loehlin, John C. (1992) *Genes and Environment in Personality Development*, Newbury Park, Cal.: Sage.

Loftus, Elizabeth (1992) "The Evolution of Memory," paper presented at Gruter Institute Conference on the Uses of Biology in the Study of Law, Squaw Valley, Cal.

Lomborg, Bjorn (1993) "The Structure of Solutions in the Iterated Prisoner's Dilemma," paper presented at Gruter Institute Conference on the Uses of Biology in the Study of Law, Squaw Valley, Cal.

Low, Bobbi S. (1989) "Cross-Cultural Patterns in the Training of Children: An Evolutionary Perspective," *Journal of Comparative Psychology* 103:311–19.

Maccoby, Eleanor E., T. M. Newcomb, and E. L. Hartley, eds. (1958) *Readings in Social Psychology*, New York: Holt, Rinehart and Winston.

MacDonald, Kevin, ed. (1988a) *Sociobiological Perspectives on Human Development*, New York: Springer-Verlag.

MacDonald, Kevin (1988b) "Sociobiology and the Cognitive-Developmental Tradition in Moral Development," in MacDonald (1988a).

—— (1990) "Mechanisms of Sexual Egalitarianism in Western Europe," *Ethology and Sociobiology* 11:195–238.

McGuire, M. T., M. J. Raleigh, and G. L. Brammer (1984) "Adaptation, Selection, and Benefit-Cost Balances: Implications of Behavioral-Physiological Studies of Social Dominance in Male Vervet Monkeys," *Ethology and Sociobiology* 5:269–77.

MacIntyre, Alasdair (1966) *A Short History of Ethics,* New York: Macmillan.

MacLean, Paul D. (1983) "A Triangular Brief on the Evolution of Brain and Law," in Margaret Gruter and Paul Bohannan, *Law, Biology, and Culture,* Santa Barbara, Cal.: Ross-Erikson, Inc.

Malinowski, Bronislaw (1929) *The Sexual Life of Savages in North-Western Melanesia: An Ethnographic Account of Courtship, Marriage and Family Life Among the Natives of the Trobriand Islands, British New Guinea,* New York: Harcourt, Brace.

Mann, Janet (1992) "Nurturance or Negligence: Maternal Psychology and Behavioral Preference Among Preterm Twins," in Barkow, Cosmides, and Tooby (1992).

Marcus, Steven (1974) *The Other Victorians: A Study of Sexuality and Pornography in Mid-Nineteenth-Century England,* New York: Basic Books.

Masters, Roger D., and Michael T. McGuire, eds. (1994) *The Neurotransmitter Revolution: Serotonin, Social Behavior, and the Law,* Carbondale, Ill.: Southern Illinois University Press.

Maynard Smith, John (1974) "The Theory of Games and the Evolution of Animal Conflict," *Journal of Theoretical Biology* 47:209–21.

——— (1982) *Evolution and the Theory of Games,* Cambridge: Cambridge University Press.

Mead, Margaret (1928) *Coming of Age in Samoa: A Psychological Study of Primitive Youth for Western Civilisation,* New York: Morrow, 1961.

Mealey, Linda, and Wade Mackey (1990) "Variation in Offspring Sex Ratio in Women of Differing Social Status," *Ethology and Sociobiology* 11:83–95.

Mill, John Stuart (1859) *On Liberty,* in Mill, *On Liberty and Other Writings,* New York: Cambridge University Press, 1989.

——— (1863) "Utilitarianism," in Mill and Jeremy Bentham, *Utilitarianism and Other Essays,* New York: Penguin, 1987.

——— (1874) "Nature," reprinted in vol. 10 of J. M. Robson, ed., *Collected Works of John Stuart Mill,* Toronto: University of Toronto Press, 1969.

Miller, Dale T. (1976) "Ego Involvement and Attributions for Success and Failure," *Journal of Personality and Social Psychology* 34:901–6.

Miller, Dale T., and Michael Ross (1975) "Self-Serving Biases in the Attribution of Causality: Fact or Fiction?" *Psychological Bulletin* 82:213–25.

Mitchell, G., and Terry L. Maple (1985) "Dominance in Nonhuman Primates," in Ellyson and Dovidio (1985).

Mitchell, Robert W., and Nicholas S. Thompson, eds. (1986) *Deception: Perspectives on Human and Nonhuman Deceit,* Albany, N.Y.: State University of New York Press.

Montgomerie, Robert (1991) "Mating Systems and the Fingerprinting Revolution," paper delivered at meeting of the Human Behavior and Evolution Society, Hamilton, Ontario.

Morris, Desmond (1967) *The Naked Ape,* New York: McGraw-Hill.

Murdock, George P. (1934) *Our Primitive Contemporaries,* Toronto: Macmillan.

———— (1945) "The Common Denominator of Cultures," in George P. Murdock, *Culture and Society,* Pittsburgh: Pittsburgh University Press, 1965.

———— (1949) *Social Structure,* New York: Macmillan.

Nesse, Randolph M. (1990a) "Evolutionary Explanations of Emotions," *Human Nature* 1:261–89.

———— (1990b) "The Evolutionary Functions of Repression and the Ego Defenses," *Journal of the American Academy of Psychoanalysis* 18:260–85.

———— (1991a) "Psychiatry," in Mary Maxwell, ed., *The Sociobiological Imagination,* Albany: State University of New York Press, 1991.

———— (1991b) "What Good Is Feeling Bad?" *The Sciences,* 31:30–37.

Nesse, Randolph, and Alan Lloyd (1992) "The Evolution of Psychodynamic Mechanisms," in Barkow, Cosmides, and Tooby (1992).

Nesse, Randolph, and George Williams (1995) *Why We Get Sick: The New Science of Darwinian Medicine,* New York: Times Books.

Nisbett, Richard, and Lee Ross (1980) *Human Inference: Strategies and Shortcomings of Social Judgment,* Englewood Cliffs, N.J.: Prentice Hall.

Nishida, Toshisada, and Mariko Hiraiwa-Hasegawa (1987) "Chimpanzees and Bonobos: Cooperative Relationships Among Males," in Smuts et al. (1987).

Omark, Donald R., F. F. Strayer, and D. G. Freedman, eds. (1980) *Dominance Relations: An Ethological View of Human Conflict and Social Interaction,* New York: Garland.

Orians, Gordon H. (1969) "On the Evolution of Mating Systems in Birds and Mammals," *American Naturalist* 103:589–603.

Palmer, Craig (1989) "Is Rape a Cultural Universal? A Reexamination of the Ethnographic Data," *Ethnology* 28:1–16.

Pedersen, F. A. (1991) "Secular Trends in Human Sex Ratios: Their Influence on Individual and Family Behavior," *Human Nature* 3:271–91.

Perusse, Daniel (1993) "Cultural and Reproductive Success in Industrial Societies: Testing the Relationship at the Proximate and Ultimate Levels," *Behavioral and Brain Sciences* 16:267–322.

Piaget, Jean (1932) *The Moral Judgment of the Child,* New York: Free Press, 1965.

Pinker, Steven (1994) *The Language Instinct,* New York: Morrow.

Plomin, R., and D. Daniels (1987) "Why Are Children in the Same Family So Different from Each Other?" *Behavioral and Brain Sciences* 10:1–6.

Price, J. S. (1967) "The Dominance Hierarchy and the Evolution of Mental Illness," *Lancet* 2:243.

Rachels, James (1990) *Created from Animals: The Moral Implications of Darwinism*, New York: Oxford University Press.

Raleigh, Michael J., and Michael T. McGuire (1989) "Female Influences on Male Dominance Acquisition in Captive Vervet Monkeys, *Cercopithecus aethiops sabaeus*," *Animal Behaviour* 38:59–67.

Raleigh, Michael J., M. T. McGuire, G. L. Brammer, D. B. Pollack, and Arthur Yuwiler "Serotonergic Mechanisms Promote Dominance Acquisition in Adult Male Vervet Monkeys" (unpublished paper).

Rapoport, Anatol (1960) *Fights, Games, and Debates*, Ann Arbor: University of Michigan Press.

Rasmussen, Dennis (1981) "Pair-bond Strength and Stability and Reproductive Success," *Psychological Review* 88:274–90.

Richards, Robert J. (1987) *Darwin and the Emergence of Evolutionary Theories of Mind and Behavior*, Chicago: University of Chicago Press.

Ridley, Matt (1994) *The Red Queen: Sex and the Evolution of Human Nature*, New York: Macmillan.

Riesman, David (1950) *The Lonely Crowd*, New Haven, Conn.: Yale University Press.

Rodman, Peter S., and John C. Mitani (1987) "Orangutans: Sexual Dimorphism in a Solitary Species," in Smuts et al. (1987).

Rose, Phyllis (1983) *Parallel Lives: Five Victorian Marriages*, New York: Vintage, 1984.

Ross, Michael, and Fiore Sicoly (1979) "Egocentric Biases in Availability and Attribution," *Journal of Personality and Social Psychology* 37:322–36.

Rothstein, Stephen I., and Raymond Pierotti (1988) "Distinctions Among Reciprocal Altruism, Kin Selection, and Cooperation and a Model for Initial Evolution of Beneficent Behavior," *Ethology and Sociobiology* 9:189–209.

Ruse, Michael (1986) *Taking Darwin Seriously: A Naturalistic Approach to Philosophy*, Oxford: Basil Blackwell.

Russon, A. E., and B. E. Waite (1991) "Patterns of Dominance and Imitation in an Infant Peer Group," *Ethology and Sociobiology* 13:55–73.

Saletan, William, and Nancy Watzman (1989) "Marcus Welby, J.D.," *The New Republic*, April 17.

Saluter, Arlene F. (1990) "Marital Status and Living Arrangements," Current Population Reports Series P-20, No. 450, Bureau of the Census, U.S. Dept. of Commerce.

Schelling, Thomas (1960) *The Strategy of Conflict,* Cambridge, Mass.: Harvard University Press.

Schweder, Richard A., M. Mahapatra, and J. G. Miller (1987) "Culture and Moral Development," in Kagan and Lamb (1987).

Short, R. V. (1976) "The Evolution of Human Reproduction," in *Proceedings of the Royal Society B* 195:3–24.

Shostak, Marjorie (1981) *Nisa: The Life and Words of a !Kung Woman,* New York: Vintage, 1983.

Simpson, George Gaylord (1947) "The Search for an Ethic," in Simpson, *The Meaning of Evolution,* New Haven, Conn.: Yale University Press.

Simpson, Jeffry A., S. W. Gangestad, and M. Bick (1993) "Personality and Nonverbal Social Behavior: An Ethological Perspective on Relationship Initiation," *Journal of Experimental Social Psychology* 29:434–61.

Singer, Peter (1981) *The Expanding Circle,* New York: Farrar, Straus and Giroux.

——— (1984) "Ethics and Sociobiology," *Zygon* 19:139–51.

——— (1989) "Ethics," *Encyclopedia Britannica* 18: 627–48.

——— (1993) *How Are We to Live? Ethics in an Age of Self-Interest,* Melbourne: Text Publishing Company.

Skinner, B. F. (1948) *Walden II,* New York: Macmillan.

——— (1972) *Beyond Freedom and Dignity,* New York: Knopf.

Slavin, Malcolm O. (1990) "The Dual Meaning of Repression and the Adaptive Design of the Human Psyche," *Journal of the American Academy of Psychoanalysis* 18:307–41.

Smart, J.J.C. (1973) "An Outline of a System of Utilitarian Ethics," in Smart and Bernard Williams, *Utilitarianism, For and Against,* Cambridge: Cambridge University Press.

Smiles, Samuel (1859) *Self-Help,* London: John Murray. Revised and enlarged edition, New York Publishing Company.

Smith, Martin S., Bradley J. Kish, and Charles B. Crawford (1987) "Inheritance of Wealth as Human Kin Investment," *Ethology and Sociobiology* 8:171–82.

Smuts, Barbara, et al., eds. (1987) *Primate Societies,* Chicago: University of Chicago Press.

Stewart, Kelly J., and Alexander H. Harcourt (1987) "Gorillas: Variation in Female Relationships," in Smuts et al. (1987).

Stone, Lawrence (1977) *The Family, Sex and Marriage in England 1500–1800,* New York: Harper Torchbook, 1979.

——— (1985) "Sex in the West," *The New Republic,* July 8.

——— (1990) *Road to Divorce: England, 1530–1987,* Oxford: Oxford University Press.

Stone, Valerie E. (1989) *Perception of Status: An Evolutionary Analysis of Nonverbal Status Cues,* Ph.D. dissertation, Department of Psychology, Stanford University.

Strachey, Lytton (1918) *Eminent Victorians,* New York: Harcourt Brace.

Strahlendorf, Peter W. (1991) *Evolutionary Jurisprudence: Darwinian Theory in Juridical Science,* S.J.D. thesis, Toronto, Ontario.

Strayer, F. F., and M. Trudel (1984) "Developmental Changes in the Nature and Function of Social Dominance Among Young Children," *Ethology and Sociobiology* 5:279–95.

Streufert, Siegfried, and Susan C. Streufert (1969) "Effects of Conceptual Structure, Failure, and Success on Attribution of Causality and Interpersonal Attitudes," *Journal of Personality and Social Psychology* 11:138–47.

Sulloway, Frank J. (1979a) *Freud, Biologist of the Mind: Behind the Psychoanalytic Legend,* New York: Basic Books.

——— (1979b) "Geographic Isolation in Darwin's Thinking: The Vicissitudes of a Crucial Idea," *Studies in History of Biology* 3:23–65.

——— (1982) "Darwin's Conversion: The *Beagle* Voyage and Its Aftermath," *Journal of the History of Biology* 15:325–96.

——— (1984) "Darwin and the Galapagos," *Biological Journal of the Linnean Society* 21:29–59.

——— (1991) "Darwinian Psychobiography," *New York Review of Books,* Oct. 10.

——— (in preparation) *Born to Rebel: Radical Thinking in Science and Social Thought,* Massachusetts Institute of Technology, Cambridge, Mass.

——— (in press) "Birth Order and Evolutionary Psychology: A Meta-Analytic Overview," *Psychological Inquiry.*

Susman, Randall L. (1987) "Pygmy Chimpanzees and Common Chimpanzees: Models for the Behavioral Ecology of the Earliest Hominids," in Kinzey (1987).

Symons, Donald (1979) *The Evolution of Human Sexuality,* New York: Oxford University Press.

——— (1982) "Another Woman That Never Existed," *Quarterly Review of Biology* 57:297–300.

——— (1985) "Darwinism and Contemporary Marriage," in Kingsley Davis, ed., *Contemporary Marriage,* New York: Russell Sage Foundation, 1985.

——— (1989) "A Critique of Darwinian Anthropology," *Ethology and Sociobiology* 10:131–44.

—— (1990) "Adaptiveness and Adaptation," *Ethology and Sociobiology* 11:427–44.

Tannen, Deborah (1990) *You Just Don't Understand: Women and Men in Conversation,* New York: Morrow.

Taylor, Charles E., and Michael T. McGuire (1988) "Reciprocal Altruism: Fifteen Years Later," *Ethology and Sociobiology* 9:67–72.

Teismann, Mark W., and Donald L. Mosher (1978) "Jealous Conflict in Dating Couples," *Psychological Reports* 42:1211–16.

Thibaut, John W., and Henry W. Riecken (1955) "Some Determinants and Consequences of the Perception of Social Causality," *Journal of Personality* 24:113–33. Reprinted in Maccoby, Newcomb, and Hartley (1958).

Thomson, Elizabeth, and Ugo Colella (1992) "Cohabitation and Marital Stability: Quality or Commitment?" *Journal of Marriage and the Family* 54:259–67.

Thornhill, Randy (1976) "Sexual Selection and Paternal Investment in Insects," *American Naturalist* 110:153–63.

Thornhill, Randy, and Nancy Thornhill (1983) "Human Rape: An Evolutionary Analysis," *Ethology and Sociobiology* 4:137–73.

Tiger, Lionel (1969) *Men in Groups,* New York: Random House.

Tooby, John (1987) "The Emergence of Evolutionary Psychology," in D. Pines, ed., *Emerging Syntheses in Science,* Santa Fe, N.M.: Santa Fe Institute.

Tooby, John, and Leda Cosmides (1988) "The Evolution of War and Its Cognitive Foundations," Institute for Evolutionary Studies Technical Report, 88–91.

—— (1989) "The Innate versus the Manifest: How Universal Does Universal Have to Be?" *Behavioral and Brain Sciences* 12:36–37.

—— (1990a) "On the Universality of Human Nature and the Uniqueness of the Individual: The Role of Genetics and Adaptation," *Journal of Personality* 58:1:17–67.

—— (1990b) "The Past Explains the Present: Emotional Adaptations and the Structure of Ancestral Environments," *Ethology and Sociobiology* 11:375–421.

—— (1992) "The Psychological Foundations of Culture," in Barkow, Cosmides, and Tooby (1992).

Tooby, John, and Irven DeVore (1987) "The Reconstruction of Hominid Behavioral Evolution," in Kinzey (1987).

Tooke, William, and Lori Camire (1990) "Patterns of Deception in Intersexual and Intrasexual Mating Strategies," *Ethology and Sociobiology* 12:345–64.

Trivers, Robert (1971) "The Evolution of Reciprocal Altruism," *Quarterly Review of Biology* 46:35–56.

—— (1972) "Parental Investment and Sexual Selection," in Bernard Campbell, ed., *Sexual Selection and the Descent of Man,* Chicago: Aldine de Gruyter.

—— (1974) "Parent-Offspring Conflict," *American Zoologist* 14:249–64.

—— (1985) *Social Evolution*, Menlo Park, Cal.: Benjamin/Cummings.

Trivers, Robert L., and Dan E. Willard (1973) "Natural Selection of Parental Ability to Vary the Sex Ratio of Offspring," *Science* 179:90–91.

Tucker, William (1993) "Monogamy and Its Discontents," *National Review*, October 4.

Vasek, Marie E. (1986) "Lying as a Skill: The Development of Deception in Children," in Mitchell and Thompson (1986).

Verplanck, William S. (1955) "The Control of the Content of Conversation: Reinforcement of Statements of Opinion," *Journal of Abnormal and Social Psychology* 51:668–76. Reprinted in Maccoby, Newcomb, and Hartley (1958).

Wallace, Bruce (1973) "Misinformation, Fitness, and Selection," *American Naturalist* 107:1–7.

Walsh, Anthony (1993) "Love Styles, Masculinity/Femininity, Physical Attractiveness and Sexual Behavior: A Test of Evolutionary Theory," *Ethology and Sociobiology* 14:25–38.

Wedgwood, Barbara, and Hensleigh Wedgwood (1980) *The Wedgwood Circle, 1730–1897: Four Generations of a Family and Their Friends*, Westfield, N.Y.: Eastview Editions.

Weisfeld, Glenn E. (1980) "Social Dominance and Human Motivation," in Omark, Strayer, and Freedman (1980).

Weisfeld, Glenn E., and Jody M. Beresford (1982) "Erectness of Posture as an Indicator of Dominance or Success in Humans," *Motivation and Emotion* 6:113–29.

Wells, P. A. (1987) "Kin Recognition in Humans," in Fletcher and Michener (1987).

West-Eberhard, Mary Jane (1991) "Sexual Selection and Social Behavior," in Michael H. Robinson and Lionel Tiger, eds., *Man and Beast Revisited*, Washington, D.C.: Smithsonian Institution Press.

Whitehead, Barbara Dafoe (1993) "Dan Quayle Was Right," *The Atlantic Monthly*, April.

Whyte, Lancelot Law (1967) "Unconscious," in *The Encyclopedia of Philosophy* (New York: Macmillan) 8:185–88.

Wiederman, Michael W., and Elizabeth Rice Allgeier (1992) "Gender Differences in Mate Selection Criteria: Sociobiological or Socioeconomic Explanation?" *Ethology and Sociobiology* 13:115–24.

Wilkinson, Gerald S. (1990) "Food Sharing in Vampire Bats," *Scientific American*, February.

Williams, George C. (1966) *Adaptation and Natural Selection: A Critique of Some Current Evolutionary Thought*, Princeton, N.J.: Princeton University Press, 1974.

—— (1975) *Sex and Evolution,* Princeton, N.J.: Princeton University Press.

—— (1989) "A Sociobiological Expansion of *Evolution and Ethics,*" a preface in Huxley (1894).

Williams, George C., and Randolph Nesse (1991) "The Dawn of Darwinian Medicine," *Quarterly Review of Biology* 66:1–22.

Wills, Christopher (1989) *The Wisdom of the Genes: New Pathways in Evolution,* New York: Basic Books.

Wilson, David S. (1989) "Levels of Selection: An Alternative to Individualism in Biology and the Social Sciences," *Social Networks* 11:257–72.

Wilson, David S., and Elliott Sober (1989) "Reviving the Superorganism," *Journal of Theoretical Biology* 136:337–56.

—— (in press) "Reintroducing Group Selection to the Human Behavioral Sciences," *Behavioral and Brain Sciences.*

Wilson, Edward O. (1975) *Sociobiology: The New Synthesis,* Cambridge, Mass.: Harvard University Press.

—— (1978) *On Human Nature,* Cambridge, Mass.: Harvard University Press.

—— (1987) "Kin Recognition: An Introductory Synopsis," in Fletcher and Michener (1987).

Wilson, James Q. (1993) *The Moral Sense,* New York: Free Press.

Wilson, Margo, and Martin Daly (1990) "The Age-Crime Relationship and the False Dichotomy of Biological versus Sociological Explanations," paper presented at meeting of Human Behavior and Evolution Society, Los Angeles.

—— (1992) "The Man Who Mistook His Wife for a Chattel," in Barkow, Cosmides, and Tooby (1992).

Wolfe, Linda D. (1991) "Human Evolution and the Sexual Behavior of Female Primates," in James D. Loy and Calvin B. Peters, eds., *Understanding Behavior,* New York: Oxford University Press, 1991.

Wrangham, Richard (1987) "The Significance of African Apes for Reconstructing Human Social Evolution," in Kinzey (1987).

Wright, Robert (1987) "Alcohol and Free Will," *The New Republic,* December 14.

—— (1990) "The Intelligence Test," *The New Republic,* January 29.

—— (1992) "Why Is It Like Something to Be Alive?" in William Shore, ed., *Mysteries of Life and the Universe,* New York: Harcourt Brace Jovanovich, 1992.

Wyschogrod, Edith (1990) *Saints and Postmodernism,* Chicago: University of Chicago Press.

Young, G. M. (1936) *Portrait of an Age: Victorian England,* Oxford: Oxford University Press, 1989.

Zimmerman, Claire, and Raymond A. Bauer (1956) "The Effect of an Audience upon What Is Remembered," *Public Opinion Quarterly* 20:238–48. Reprinted in Maccoby, Newcomb, and Hartley (1958).

INDEX

conditioning, behavioral, 8, 261
conscience, 182, 212–13, 214, 216, 222, 226, 308–9, 328, 377–78
  of Darwin, 15, 210–26, 308–10, 316, 317, 318, 377
  in high-status people, 343
  malleable, 215–16
  savviness and, 306
  Victorian, 218–23, 224–25, 328
  see also morality
conservatism, 40, 345, 348, 354
  family values and, 102, 105
  moral, 13, 14, 140, 141, 142, 362
  political, 13, 102, 105, 362
  poverty and, 105
contraception, 37, 44, 63, 67, 124, 125, 136, 142, 248, 286, 388
cooperation, 200, 216, 218, 221, 222, 241, 321, 330
  see also altruism
Copernicus, Nicolaus, 47
Cosmides, Leda, 26, 106, 107
courage, 186
cows, 156, 161, 162, 185
creationism, 24, 43
credit, assignment of, 267, 281, 349, 350
crickets, Mormon, 49, 54
crime, criminals, 100, 102, 222–23, 244, 348, 387
  biochemical defenses for, 352–53
  psychological maladies as defenses for, 352, 354–55
  serotonin and, 351
  see also punishment
crows, 185
cuckoldry, *see* infidelity, female
cultural (environmental) determinism, 244, 349
cultural evolution, 99
  conscience and, 222
cynicism, 313–26, 366, 373–74, 376–77
  postmodern, 325–26

Daly, Martin, 36, 66, 72, 100, 103, 120, 208, 222, 315, 352, 355

Darwin, Annie (daughter), 155, 166, 167, 177–79
Darwin, Caroline (sister), 110–11, 181, 213–14, 276
Darwin, Catherine (sister), 109–10, 112, 122
Darwin, Charles:
  allies of, 299–301, 316
  ambition of, 23, 288–89
  autobiography of, 14, 116, 128, 179, 218, 266–67, 276, 280, 289, 308, 364
  as barnacle expert, 231, 233–35, 296, 298–99
  on *Beagle*, 19, 22, 109, 110, 112, 113, 121, 180, 181, 182, 211, 225, 231, 247, 276, 291, 292, 293, 364, 376
  boasting of, 266–67, 269
  career ascent of, 112–13, 115–16, 118–19, 247, 288
  character of, 14–15, 23, 109, 112, 118, 210, 287, 301–2, 310, 317
  childhood lies of, 216–18
  children of, 109, 117, 129, 130, 155–56, 166, 167, 168, 176–79, 211, 218, 265, 267–68
  clerical career planned for, 21–22, 111
  conscience and moral feelings of, 15, 210–26, 308–10, 316, 317, 318, 377
  country home of, 129, 130, 230
  credentials of, 298–99
  criticism feared by, 232, 297–98
  death of, 287–88, 378, 379
  debts of, 276
  Descent of Man, 3, 33, 55, 93, 180, 181, 186, 189, 190, 210, 247, 327, 328, 345, 372
  Expression of the Emotions in Man and Animals, 189, 284
  family as moral influence on, 213–15
  Fuegian Indians and, 180–83, 184, 194, 225, 236–37, 238
  Geological Observations on South America, 230
  Geological Observations on the Volcanic Islands, 229–30

Illustration Credits

Illustrations between pages 276 and 277:

Sigmund Freud: Mary Evans Picture Gallery, London; Sigmund Freud Copyrights, Ltd.
Leaf-mimicking katydid: Animals Animals, New York
False coral snake: Animals Animals, New York

Yeroen, Luit, and Nikkie: Frans de Waal, Emory University, from *Chimpanzee Politics*
Luit jumping on female chimpanzee: Frans de Waal, Emory University, from *Chimpanzee Politics*

Charles Lyell: National Portrait Gallery, London
Joseph Hooker: National Portrait Gallery, London
Alfred Russel Wallace: National Portrait Gallery, London

Charles Darwin around 1855: Cambridge University Library

Illustrations between pages 340 and 341:

Anti-sociobiology poster: Library of Congress
Herbert Spencer: National Portrait Gallery, London

John Stuart Mill: National Portrait Gallery, London
Thomas Henry Huxley: From *Life and Letters of Thomas Henry Huxley,* edited by Leonard Huxley

George Williams in 1994: Barry Munger
Samuel Smiles: National Portrait Gallery, London

Charles Darwin around 1882: Cambridge University Library